VOLUME ONE HUNDRED AND TEN

INTERNATIONAL REVIEW OF NEUROBIOLOGY

Metal Related Neurodegenerative Disease

INTERNATIONAL REVIEW OF NEUROBIOLOGY

VOLUME 110

SERIES EDITORS

R. ADRON HARRIS

Waggoner Center for Alcohol and Drug Addiction Research
The University of Texas at Austin
Austin, Texas, USA

PETER JENNER

Division of Pharmacology and Therapeutics
GKT School of Biomedical Sciences
King's College, London, UK

VOLUME ONE HUNDRED AND TEN

INTERNATIONAL REVIEW OF NEUROBIOLOGY

Metal Related Neurodegenerative Disease

Edited by

KAILASH P. BHATIA

Institute of Neurology, University College of London
London, United Kingdom

SUSANNE A. SCHNEIDER

Department of Neurology
University Kiel, Kiel, Germany

ELSEVIER

AMSTERDAM • BOSTON • HEIDELBERG • LONDON
NEW YORK • OXFORD • PARIS • SAN DIEGO
SAN FRANCISCO • SINGAPORE • SYDNEY • TOKYO
Academic Press is an imprint of Elsevier

Academic Press is an imprint of Elsevier
32 Jamestown Road, London NW1 7BY, UK
Radarweg 29, PO Box 211, 1000 AE Amsterdam, The Netherlands
The Boulevard, Langford Lane, Kidlington, Oxford, OX5 1GB, UK
225 Wyman Street, Waltham, MA 02451, USA
525 B Street, Suite 1800, San Diego, CA 92101-4495, USA

First edition 2013

ISBN: 978-0-12-410502-7
ISSN: 0074-7742

For information on all Academic Press publications
visit our website at store.elsevier.com

Printed and bound by CPI Group (UK) Ltd, Croydon, CR0 4YY

CONTENTS

CONTRIBUTORS

Annu Aggarwal
Wilson Disease Clinic, Kokilaben Dhirubhai Ambani Hospital and Medical Research Institute, Mumbai, India

Kailash P. Bhatia
Institute of Neurology, University College of London, London, United Kingdom

Mohit Bhatt
Wilson Disease Clinic, Kokilaben Dhirubhai Ambani Hospital and Medical Research Institute, Mumbai, India

Jose Miguel Bras
Department of Molecular Neuroscience, Institute of Neurology, University College of London, London, United Kingdom

Patrick F. Chinnery
Wellcome Centre for Mitochondrial Research, Institute of Genetic Medicine, International Centre for Life, Newcastle University, Newcastle upon Tyne, United Kingdom

Peter T. Clayton
Clinical and Molecular Genetics Unit, UCL Institute of Child Health, London, United Kingdom

Monika Dezortova
MR Unit, Department of Diagnostic and Interventional Radiology, Institute for Clinical and Experimental Medicine, Prague, Czech Republic

Eve M. Dumas
Department of Neurology, Leiden University Medical Centre, Leiden, The Netherlands

Petr Dusek
Department of Neurology and Center of Clinical Neuroscience, Charles University in Prague, 1st Faculty of Medicine and General University Hospital, Prague, Czech Republic, and Institut für interventionelle und diagnostische Neuroradiologie, Universitätsmedizin Göttingen, Göttingen, Germany

J.B.M.M. Ferreira
Keizo Asami Laboratory, Federal University of Pernambuco, Recife, Pernambuco, Brazil

Allison Gregory
Department of Molecular and Medical Genetics, Oregon Health and Science University, Portland, OR, USA

Tobias B. Haack
Institute of Human Genetics, Technische Universität München, and Institute of Human Genetics, Helmholtz Zentrum München, Munich, Germany

Monika Hartig
Institute of Human Genetics, Technische Universität München, and German Network for Mitochondrial Disorders (mitoNET), Munich, Germany

Susan J. Hayflick
Department of Molecular & Medical Genetics; Department of Pediatrics, and Department of Neurology, Oregon Health & Science University, Portland, OR, USA

Penny Hogarth
Department of Molecular and Medical Genetics, and Department of Neurology, Oregon Health and Science University, Portland, OR, USA

Kurt A. Jellinger
Institute of Clinical Neurobiology, Vienna, Austria

Matthew P. Keasey
Keizo Asami Laboratory, Federal University of Pernambuco, Recife, Pernambuco, Brazil

Michael J. Keogh
Wellcome Centre for Mitochondrial Research, Institute of Genetic Medicine, International Centre for Life, Newcastle University, Newcastle upon Tyne, United Kingdom

Thomas Klopstock
German Network for Mitochondrial Disorders (mitoNET); DZNE – German Center for Neurodegenerative Diseases, and Department of Neurology, Friedrich-Baur-Institute, Ludwig-Maximilians-University, Munich, Germany

Satoshi Kono
First Department of Medicine, Hamamatsu University School of Medicine, Hamamatsu, Japan

Michael C. Kruer
Sanford Children's Health Research Center, Sanford Children's Hospital, Sioux Falls, South Dakota, USA

Manju A. Kurian
Neurosciences Unit, UCL-Institute of Child Health, and Department of Neurology, Great Ormond Street Hospital, London, United Kingdom

Roberta R. Lemos
Universidad de Santiago de Compostela, and Fundación Pública Galega de Medicina Xenómica-SERGAS-IDIS (Hospital Clínico Universitario), Santiago de Compostela, Spain

Thomas Meitinger
Institute of Human Genetics, Technische Universität München; German Network for Mitochondrial Disorders (mitoNET); Institute of Human Genetics, Helmholtz Zentrum München, and DZNE – German Center for Neurodegenerative Diseases, Munich, Germany

Philippa B. Mills
Clinical and Molecular Genetics Unit, UCL Institute of Child Health, London, United Kingdom

Christopher M. Morris
Medical Toxicology Centre, Wolfson Unit of Clinical Pharmacology, Newcastle University, Newcastle Upon Tyne, United Kingdom

Nardo Nardocci
Department of Child Neurology, Fondazione IRCCS Istituto Neurologico "Carlo Besta", Milan, Italy

Joao R.M. Oliveira
Keizo Asami Laboratory, and Neuropsychiatry Department, Federal University of Pernambuco, Recife, Pernambuco, Brazil

Holger Prokisch
Institute of Human Genetics, Technische Universität München; German Network for Mitochondrial Disorders (mitoNET), and Institute of Human Genetics, Helmholtz Zentrum München, Munich, Germany

Raymund A.C. Roos
Department of Neurology, Leiden University Medical Centre, Leiden, The Netherlands

Susanne A. Schneider
Department of Neurology, University Kiel, Kiel, Germany

Karin Tuschl
Clinical and Molecular Genetics Unit, UCL Institute of Child Health, London, United Kingdom

Simon J.A. van den Bogaard
Department of Neurology, Leiden University Medical Centre, Leiden, The Netherlands

Jens Wuerfel
Institut für interventionelle und diagnostische Neuroradiologie, Universitätsmedizin Göttingen, Göttingen, and NeuroCure Clinical Research Center, Charité University Medicine and Max Delbrück Center for Molecular Medicine, Berlin, Germany

Giovanna Zorzi
Department of Child Neurology, Fondazione IRCCS Istituto Neurologico "Carlo Besta", Milan, Italy

PREFACE

EDITORIAL: METAL-RELATED NEUROLOGICAL DISORDERS: SEVERAL NEW GENES AND BETTER UNDERSTANDING

It is a captivating era to be working in neurology as recent years have engendered fascinating developments. To illustrate it, this special edition is dedicated to metal-related neurological disease where new insights into the underlying genetic underpinnings have emerged at an ever-increasing pace. For example, in the field on iron-related disorders (i.e., syndromes of neurodegeneration with brain iron accumulation, NBIA) since the milestone paper on PKAN published 10 years ago (Hayflick, 2003), in the last 5 years, at least five other NBIA syndromes have been genetically delineated, the latest being the new X-linked dominant disorder, beta-propeller protein-associated neurodegeneration. This molecular genetic revolution is illustrated in the chapters 2, 3 and 4 as well as 5 and 6 on NBIA disorders, some dedicated to specific NBIA subtypes.

Neurogenetic workup has also provided valuable insights into conditions due to neural accumulation of other metals such as manganese, copper, or calcium, with yet more new genes on the threshold chapters 12, 13 and 14. The increasing number of genetic tests that have become commercially available (most recently as gene panels at more and more affordable costs) is of great clinical relevance and calls for an informed selection of the right test(s). The results should be presented with appropriate specific genetic counseling. Furthermore, setting our patients in touch with patient alliances may allow them and their carers to benefit from the experience of similarly affected patients (see Table 1).

New genetic approaches confirm that heritable factors underlie a larger proportion of patients than anticipated and genetic mutations should thus also be suspected in patients with a negative family history (e.g., due to *de novo* mutations). Thus, molecular confirmation enables a precise diagnosis. At the same time, there have been pathological advances, which can be correlated to molecular findings and (see the chapter 8 on pathology) which may challenge some of the previously held concepts for, for example, the presence of Lewy body pathology in some NBIA disorders like PLAN (but not PKAN), and this link makes it clear that the study of rare disorders

Table 1 Incomplete summary of patient alliance groups supporting patients with diseases covered in this issue

Country	Name	Contact person	Address	E-mail	Website
NBIA disorders					
United Kingdom	NBIA (under development)	N. Ali	86 Clare Road, Maidenhead, Berkshire SL6 4DQ, United Kingdom	naveali@gmail.com	
France	AIDNAI—Association Internationale DNAI	J. L. Vasseur		jeanloup.vasseur@free.fr	http://asso.orpha.net/DNAI/cgi-bin/DNAI/index.php?lng=fr
Germany	Hoffnungsbaum e.V.	A. Klucken	Velbert, Germany	Hoffnungsbaum@aol.com	www.hoffnungsbaum.de
Italy	AISNAF—Associazione Italiana Sindromi Neurodegenerative Da Accumulo Di Ferro	N. Scalise	Via Fellino 5, 87068 Rossano CS, Italy	nscalise@libero.it or fabry.sca@hotmail.it	www.aisnaf.org
Spain	ENACH Asociación	A. Lopez	C/Bruselas 5 1°A, 41012, Seville, Spain	sveanl@hotmail.com	www.enachasociacion.blogspot.com.es
USA	NBIA Disorders Association	P. Wood	2082 Monaco St; El Cajon, CA 92019-4235, USA	info@NBIAdisorders.org	www.NBIAdisorders.org

Neuroacanthocytosis					
United Kingdom	Advocacy for Neuroacanthocytosis	G. Irvine	32 Launceston Place London W8 5RN, United Kingdom	glenn@naadvocacy.org	www.naadvocacy.org
Wilson's disease					
Germany	Morbus Wilson e.V.	Ms Rasp	Leiblstraße 2, 83024 Rosenheim, Germany	morbus.wilson@t-online.de	www.morbus-wilson.de
Switzerland	Morbus Wilson Schweiz		Eich 25, CH-5742 Kölliken, Switzerland	morbus-wilson@hispeed.ch	www.morbus-wilson.ch
United Kingdom	Wilson's Disease Support Group	L. Hart	36, Audley Drive, Lenton Abbey, Nottingham NG9 2SF, United Kingdom	linda@wilsonsdisease.org.uk	www.wilsonsdisease.org.uk
USA	Wilson Disease Association	M Graper	5572 North Diversey Blvd. Milwaukee, WI 53217, USA	info@wilsonsdisease.org	www.wilsonsdisease.org
Restless legs syndrome					
Europe	European Alliance for Restless Legs Syndrome			www.earls.eu with links to national groups in Europe	

may have ramifications and help in the understanding of more common conditions like Parkinsons's disease (PD).

Improved imaging techniques with higher resolution (with 7 Tesla MRI available in a growing number of centers) and new sequences with increased sensitivity to certain tissues, for example, T2* or susceptibility-weighted imaging to detect iron, lead to radiological–phenotypic–genotypic correlations (outlines in the chapter 9 on neuroimaging).

However, the new insights have at the same time revealed the complexity of metal-related diseases and also that the broader pathophysiology of these disease entities remains ill-understood, and we have only just begun to unravel the complex puzzle of genes, the environment, and the aging brain. Mitochondrial dysfunction, abnormal lysosomal metabolism (discussed in the chapter 11 on lysosomal storage disorders), and altered lipid pathways are all suspected to play a role in these Mendelian disorders. In addition, the boundaries toward sporadic nondegenerative disease i.e., restless legs syndrome, RLS and neurodegenerative disease of the elderly (i.e., Parkinson's disease) blur and may be less well defined than previously thought.

Currently, treatment of metal-related neurological disorders remains symptomatic and knowledge of a molecular diagnosis has no direct consequence on treatment strategies. However, improved scientific understanding will eventually hopefully provide the basis for the development of specific and even curative therapies for our patients.

Susanne A. Schneider
Department of Neurology, University Kiel, Kiel, Germany
Kailash P. Bhatia
Institute of Neurology, University College of London,
London, United Kingdom

REFERENCE

Hayflick, S. J., Westaway, S. K., Levinson, B., Zhou, B., Johnson, M. A., Ching, K. H., et al. (2003 Jan 2). Genetic, clinical, and radiographic delineation of Hallervorden-Spatz syndrome. *The New England Journal of Medicine, 348*(1), 33–40.

CHAPTER ONE

The Relevance of Metals in the Pathophysiology of Neurodegeneration, Pathological Considerations

Kurt A. Jellinger[1]
Institute of Clinical Neurobiology, Vienna, Austria
[1]Corresponding author: e-mail address: kurt.jellinger@univie.ac.at

Contents

Abstract

Neurodegenerative disorders are featured by a variety of pathological conditions that share similar critical processes, such as oxidative stress, free radical activity, proteinaceous aggregations, mitochondrial dysfunctions, and energy failure. They are mediated or triggered by an imbalance of metal ions leading to changes of critical biological systems and initiating a cascade of events finally leading to neurodegeneration and cell death. Their causes are multifactorial, and although the source of the shift in oxidative homeostasis is still unclear, current evidence points to changes in the balance of redox transition metals, especially iron, copper, and other trace metals. They are present at elevated levels in Alzheimer disease, Parkinson disease, multisystem atrophy, etc., while in other neurodegenerative disorders, copper, zinc, aluminum, and manganese are involved. This chapter will review the recent advances of the role of metals in the pathogenesis and pathophysiology of major neurodegenerative diseases and discuss the use of chelating agents as potential therapies for metal-related disorders.

International Review of Neurobiology, Volume 110
ISSN 0074-7742
http://dx.doi.org/10.1016/B978-0-12-410502-7.00002-8

1. INTRODUCTION

The classification of neurodegenerative diseases, previously based on the anatomical systems involved by neuronal loss, has been progressively replaced by molecular diagnoses, based on the presence of abnormal protein aggregates that accumulate in the nervous system (for review, see Jellinger (2013)). There is extensive evidence for the association of protein aggregation and neurodegeneration in many such disorders (Jellinger, 2010; Kovacs, Botond, & Budka, 2010), providing a link between the major pathogenic processes of protein aggregation and oxidative damage (Halliwell, 2006; Ramalingam & Kim, 2012). Essential metals are crucial for the maintenance of cell homeostasis. The neurobiology of metals is now growing in interest, since it has been linked to many neurodegenerative diseases. Metals such as iron (Fe), copper (Cu), and manganese (Mn) appear to play an important role in oxidative stress (OS), particularly in mitochondria, protein misfolding, and aggregation leading eventually to neurodegeneration (Farina, Avila, da Rocha, & Aschner, 2013; Jomova & Valko, 2011; Jomova, Vondrakova, Lawson, & Valko, 2010; Kozlowski et al., 2009; Molina-Holgado, Hider, Gaeta, Williams, & Francis, 2007; Muhoberac & Vidal, 2013; Tabner, El-Agnaf, German, Fullwood, & Allsop, 2005; Ward, Crichton & Dexter, 2013; Zatta, Lucchini, van Rensburg, & Taylor, 2003). This is probably the result of a progressive deterioration of the metal regulatory systems and dysfunctional metal transport by brain barriers (Yokel, 2006; Zheng & Monnot, 2012), which appear highly enhanced under various pathological conditions, causing increased OS (Ansari & Scheff, 2010) and favoring abnormal metal–protein interactions. There is accumulating evidence that altered metal homeostasis may be involved in aggregation of intrinsically disordered proteins (Breydo & Uversky, 2011), for example, interacting on αSyn structure and aggregation propensity (Santner & Uversky, 2010). This supports a crucial involvement of metals in neurodegeneration (Bolognin, Messori, & Zatta, 2009; Farina, Avila, da Rocha, & Aschner, 2013; Jomova & Valko, 2011; Mot et al., 2011; Savelieff, Lee, Liu, & Lim, 2013) (see Table 1.1). The causes of abnormal protein (mis)folding and consequent protein accumulation in the central nervous system (CNS) need to be elucidated. Nevertheless, genetic and environmental factors are involved (Gomes, 2013). It is still unclear how their formation leads

Table 1.1 Neurodegenerative disorders with possible metal-associated pathology

Disorder implicated	Metal implicated	Metalloproteins or enzymes
Alzheimer disease	Copper, iron, zinc, aluminum ↑	Aβ, APP, tau protein ↑
Parkinson disease	Iron ↑, zinc ↑	α-Synuclein ↑, neuromelanin ↓, ferritin ↓
Lewy body dementia	Iron ↑	α-Synuclein ↑
Multisystem atrophy	Iron ↑, copper ↑	α-Synuclein ↑
Progressive supranuclear palsy	Iron ↑	Tau protein ↑
Creutzfeldt–Jakob disease	Copper ↑, iron ↑	Prion protein
Amyotrophic lateral sclerosis	Copper ↑, iron ↑, zinc ↑	Superoxide dismutase 1 ↑, TDP-43 ↑
Friedreich ataxia	Iron ↑	Frataxin ↑
Multiple sclerosis	Iron ↑	
Wilson disease	Copper ↑	Ceruloplasmin deficiency
Menkes disease	Copper ↓	Cytochrome oxidase ↑
PKAN/NBIA type 1	Iron ↑	Pantothenate kinase deficiency
Huntington disease	Iron ↑, calcium ↑, manganese ↑	Huntingtin ↑
Infantile neuroaxon dystrophy (PLA2G6-associated NBIA)	Iron ↑	
Neuroferritinopathy	Iron ↑	Ferritin light-chain gene (FTL1)
Down syndrome	Copper ↑, iron ↑, zinc ↑	Aβ ↑, tau protein ↑

to neuronal dysfunction and death, but recent studies suggest that small soluble oligomeric aggregates are more likely to be toxic than insoluble fibrils (Gavrin, Denny, & Saiah, 2012). Metal ions are involved in the formation of potentially toxic oligomer species (Nubling et al., 2012).

Redox-active metals like Fe and Cu are critical for correct functioning of the CNS. They have essential roles in many important processes including but not limited to protein stabilization, enzyme catalysis, and energy production. Aluminum (Al), Zn, Cu, and Fe cause conformational changes of β-amyloid protein (Aβ), while Al causes accumulation of Aβ and tau protein (Kawahara, 2005; Walton, 2013). Al and Cu are involved in initiating or propagating neuroinflammation (Campbell, 2006).

An increased content of Fe, Al, and Cu in aged human brain tissues may reflect aspects of aging and neurodegeneration in individual cases (House, Esiri, Forster, Ince, & Exley, 2012).

Biometals are important not only in neurodegenerative processes in advanced age but also in rare ones of childhood (Parker, Koistinaho, White, & Kanninen, 2013). They undergo redox cycling reactions and have the ability to produce reactive radicals such as superoxide and nitric oxide in biological systems (Allsop, Mayes, Moore, Masad, & Tabner, 2008; Gaeta & Hider, 2005; Rivera-Mancia et al., 2010). Disruption of metal ion homeostasis producing great amounts of redox-active material may lead to OS, a state where increased formation of reactive oxygen species (ROS) overwhelms antioxidant protection and subsequently induces DNA damage, lipid peroxidation, formation of reactive aldehydes, and protein modification/misfolding. They are also overwhelming ubiquitin–proteasome degradation systems, which can result in accumulation of pathological proteins within intracellular inclusions and irreversible cellular damage over time, all symptomatic for neurodegenerative disorders (Crichton, Dexter, & Ward, 2011; Oteiza, Mackenzie, & Verstraeten, 2004; Ramalingam & Kim, 2012). ROS have an essential role in the regulation of synaptic plasticity (Massaad & Klann, 2011). The underlying mechanism of action for all these metals involves formation of the superoxide radical and hydroxyl radical, mainly via the Haber–Weiss reaction, wherein superoxide (O2•$^-$) participates in the reduction of ferric Fe^{3+} to ferrous Fe^{2+}. It is followed by the Fenton reaction, in which ferrous Fe^{2+} and hydrogen peroxide (H_2O_2) react to generate ferric Fe^{3+}, hydrogen peroxide (OH^-), and the reactive hydroxyl radical (•OH). Cu(1+) and Cu^{3+} ions may substitute for Fe^{2+} and Fe^{3+} ions in the Haber–Weiss reaction; peroxynitrite can be formed from the reaction of NO• with O2•$^-$. (Greenough, Camakaris, & Bush, 2012; Halliwell, 2006; Jomova & Valko, 2011; Kell, 2010) (Fig. 1.1). These reactions are finally producing malondialdehyde, 4-hydroxynonenal, and other exocyclic DNA adducts (Barnham, Masters, & Bush, 2004). Chemical alterations of deoxyribose, purine, and pyrimidine bases and destruction of cellular elements, including protein and DNA through lipid peroxidation, and mitochondrial dysfunction ultimately lead to cellular death (Bonda

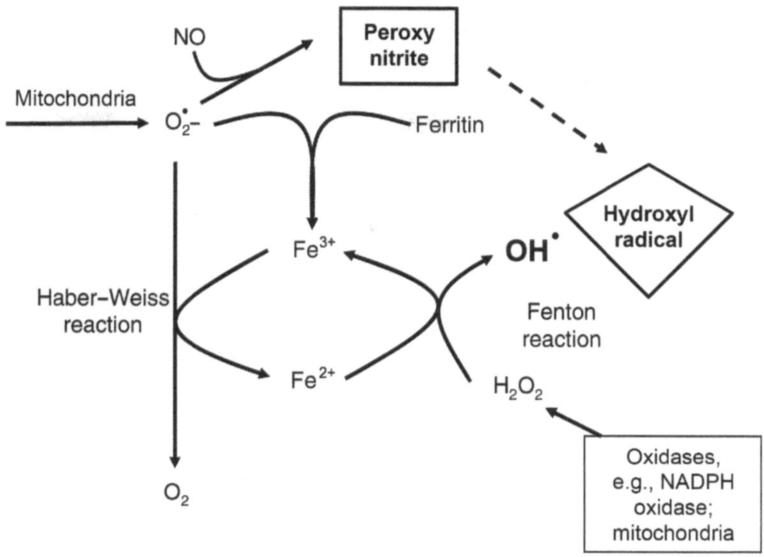

Figure 1.1 The Haber–Weiss and Fenton reactions combine using poorly liganded iron in a catalytic cycle to produce the damaging hydroxyl radical. Poorly liganded iron can also be liberated via the destruction of heme and other iron-containing substances. Peroxynitrite anion ($ONOO^-$) is produced by the reaction of superoxide and nitric oxide (NO•). *Modified from Kell, 2010.*

et al., 2011; Lovell & Markesbery, 2006; Palmieri & Sblendorio, 2007). Catalysis of Fenton reactions by ions like Cu(1+) and Fe^{2+} is so damaging to the biological milieu that right from their entry, they need to be strictly controlled until they arrive at their storage site. This chaperoning occurs usually by proteins involved in transport, delivery, and distribution processes (Kozlowski et al., 2009).

Aluminum, Zn, Cu, and Fe cause conformational changes in Aβ protein (Bush, 2003; Drago, Bolognin, & Zatta, 2008; Greenough et al., 2012; Shcherbatykh & Carpenter, 2007), and a relationship between Al and Fe homeostases has been suggested (Kawahara & Kato-Negishi, 2011; Walton, 2013). Endogenous antioxidant mechanisms, such as superoxide dismutase (SOD), glutathione (GSH)/GSH peroxidase, and catalase, may serve to defend against damage caused by ROS (Jomova & Valko, 2011; Jomova et al., 2010). On the other hand, redox-negative metals, such as cadmium (Cd) and lead (Pb), show their toxic effects via bonding to sulfhydryl groups of proteins and depletion of GSH. A special position is occupied by the redox inert metal zinc (Zn), an essential component of numerous proteins involved in the defense against OS. Zn may enhance DNA damage via impairment of DNA repair mechanisms and has an impact on the immune system.

There is evidence that proteins frequently observed in age-related neurodegenerative disorders, such as Aβ and neurofibrillary tangles (NFT) in Alzheimer disease (AD), are in fact defense mechanisms attempting to diminish ROS but become pathological as the disease progresses and proteins aggregate (Smith, Casadesus, Joseph, & Perry, 2002). There is now strong evidence that each of the major protein participants in AD pathology has important interactions with transitional metals: amyloid precursor protein (APP) is the neuronal Fe export ferroxidase with a major interaction with ferroportin; presenilins are needed for the import of around 50% of cellular Cu and Zn, and tau promotes the export of neuronal Fe by facilitating the trafficking of APP to the surface. The high-affinity features reported for the binding of Zn^{2+} to Aβ protein contrast with the low affinity of the interaction of this metal with αSyn, the molecular basis of which has recently been elucidated (Valiente-Gabioud et al., 2012). Therefore, amyloid, tau, and αSyn pathologies arise in a milieu of high metal flux, and the major components of AD pathology may contribute to the disease by failing in their metal transport roles (Bush, 2013). The role of metals in AD has been reviewed recently (Hajji, Calvert, Ritchie, & Sastre, 2013).

Mitochondrial dysfunction is a basic problem associated with many neurodegenerative processes as well as normal aging (Filosto et al., 2011; Lin & Beal, 2006; Oettinghaus, Licci, Scorrano, & Frank, 2012; Schon & Przedborski, 2011). A proposed mechanism of neuronal death is the response to OS and abnormal calcium (Ca^{2+}) signaling causing mitochondrial dysfunction (Barsukova, Bourdette, & Forte, 2011; Gleichmann & Mattson, 2011; Jeyaraju, Cisbani, & Pellegrini, 2009; Pivovarova & Andrews, 2010). Defects in mitochondrial respiration are responsible for the degeneration and death of neurons during the course of most, if not all, neurodegenerative disorders. This is now proven by failed mitochondrial maintenance including the scavenging of toxic ROS, degradation of damaged mitochondrial proteins, and protein turnover in mitochondria (Federico et al., 2012; Karbowski & Neutzner, 2012).

In the following, the role of major specific metals in neurodegeneration will be discussed. Mercury that like other toxic metals may enter motor neurons as a possible pathogenic factor in amyotrophic lateral sclerosis and other neurodegenerative disorders (Pamphlett & Kum Jew, 2013) will not be considered here.

2. THE ROLE OF SPECIFIC METALS IN NEURODEGENERATION

2.1. The role of iron

Iron is essential for normal neuronal functioning and is a crucial cofactor of a plethora of metabolic processes like oxygen transport, DNA synthesis, and mitochondrial respiration, while it can cause dysfunction by the generation of ROS, resulting in lipid peroxidation and neurotoxicity (Hare, Ayton, Bush, & Lei, 2013; Jellinger, 1999; Nunez et al., 2012; Salvador, 2010). Iron accumulates in the aging human brain (Hallgren & Sourander, 1958), where it is predominantly stored in oligodendrocytes and detoxified by its binding to ferritin (Hulet, Powers, & Connor, 1999). Quantitative imaging of brain Fe using postmortem magnetic resonance imaging (MRI) showed the highest concentrations in the globus pallidus, followed by putamen, caudate nucleus, thalamus, and white matter regions (Langkammer et al., 2010). Homeostasis of brain Fe involves a complex mechanism of interactions between metabolic, genetic, and developmental processes. Iron homeostasis, transport, and Fe-induced neurotoxicity have been reviewed recently (Ayton, Lei, & Bush, 2013; Crichton et al., 2011; Hagemeier, Geurts, & Zivadinov, 2012; Hare, Ayton, Bush, & Lei, 2013; Horowitz & Greenamyre, 2010; Ke & Qian, 2007; Rouault, 2013). Increased Fe concentrations due to misregulation of Fe transport and storage regulated by a number of proteins (transferrin receptor/Tfr, ferritin, and cytoplasmic Fe-regulating proteins) trigger a cascade of deleterious events that lead to neuronal dysfunction and death (Batista-Nascimento, Pimentel, Menezes, & Rodrigues-Pousada, 2012; Gille & Reichmann, 2011; Hagemeier et al., 2012; Mills, Dong, Wang, & Xu, 2010; Thomas & Jankovic, 2004). Fe is an intrinsic producer of ROS, and increased Fe promotes neurotoxicity because of hydroxyl radical formation, which results in GSH consumption, protein aggregation, lipid peroxidation, and nucleic acid modification (Kienzl, Jellinger, Stachelberger, & Linert, 1999; Nunez et al., 2012; Thompson, Shoham, & Connor, 2001). Free Fe can induce OS via the Fenton reaction (Halliwell, 2006). Dysregulation in Fe metabolism, Fe-mediated OS, and free radical formation may contribute to the pathogenesis of neurodegenerative diseases, and recent studies have suggested a connection between endoplasmic reticulum stress, Fe metabolism, and neurodegeneration (Liu & Connor, 2012). Hence, it is of utmost importance to maintain Fe homeostasis and to remove excessive Fe in order to indirectly ameliorate the detrimental effects of free radicals on cellular structures. Increase in heme oxygenase-1 (HO-1), a crucial enzyme in cellular metabolism, degrading heme and generating biliverdin (a powerful antioxidant), carbon monoxide, and Fe, has been observed in normal aging brain. This might lead to its increased susceptibility to OS, even though it promotes certain neuroprotective effects (Hirose, Ikematsu, & Tsuda, 2003; Schipper, Song, Zukor, Hascalovici, & Zeligman, 2009).

Iron accumulation in the brain colocalizing with sites of neuronal death (Barnham & Bush, 2008) is associated with numerous neurodegenerative disorders, including AD (Barnham & Bush, 2008; Duce et al., 2010; Smith et al., 2010; Squitti, 2012b), Parkinson disease (PD) (Berg, Riederer, & Gerlach, 2008; Boelmans et al., 2012; Sian-Hulsmann, Mandel, Youdim, & Riederer, 2011; Zecca, Youdim, Riederer, Connor, & Crichton, 2004), multisystem atrophy (Jellinger, 2003; Kurisaki, Yomono, Murayama, & Hebisawa, 2002), progressive supranuclear palsy (PSP) (Galazka-Friedman, Bauminger, Szlachta, & Friedman, 2012), Huntington disease (HD) (Rosas et al., 2012), sporadic Creutzfeldt-Jakob disease (Rowinska-Zyrek, Valensin, Luczkowski, & Kozlowski, 2013; Singh et al., 2013), and Friedreich ataxia (Gille & Reichmann, 2011; Koeppen et al., 2013; Pastore, 2013), most of them having an onset in adult or elder age. Various genetic disorders of Fe metabolism are related with neurodegeneration with brain iron accumulation (NBIA) such as pantothenate kinase-associated neurodegeneration (PKAN) (Duda & Jellinger, 2011; Dusek & Schneider, 2012; Gregory & Hayflick, 2011; Hogarth et al., 2013; Kruer et al., 2011; Li et al., 2012; Schipper, 2012; Schneider, Hardy, & Bhatia, 2012; Schneider, Zorzi, & Nardocci, 2013), PLA2G6-associated infantile neuroaxonal dystrophy (Jellinger, 2004), or neuroferritinopathy (Keogh, Jonas, Coulthard, Chinnery, & Burn, 2012; Lehn, Boyle, Brown, Airey, & Mellick, 2012; Vidal, Delisle, Rascol, & Ghetti, 2011). Many movement disorders are associated with dysregulation of brain Fe (Dusek, Jankovic, & Le, 2012; Taba, 2013). However, it is not clear whether Fe contributed to the progression of these disorders.

Intracytoplasmic accumulation of Fe^{2+} in oligodendrocytes may explain their high susceptibility to degeneration under conditions of OS induced by inflammation and mitochondrial dysfunction (Zhang et al., 2005). Oligodendrocyte destruction releases this accumulated Fe^{2+} into the extracellular space, where it might further amplify oxidative damage in axons and other cells.

Continuing and autocatalytic production based on positive feedback mechanisms of hydroxyl radicals via Fenton chemistry involving poorly liganded Fe has been considered a unifying mechanism of cell death in many neurodegenerative and other diseases (Kell, 2010). However, recent studies on the possible role of Fe suggested that the mechanisms leading to neuronal death in PD, AD, and PSP might be different, although all may be related to Fe-mediated OS (Galazka-Friedman et al., 2012).

2.1.1 Alzheimer disease

AD morphologically characterized by cerebral atrophy, extracellular Aβ (plaques and amyloid angiopathy), tau-positive NFT, and loss of synapses and neurons shares many similarities with PD and HD, not least the extensive evidence for OS, mitochondrial failure, and increased levels of iron (Hagemeier et al., 2012; Jellinger, 2010; Kell, 2010), while others found no significant difference (Schrag, Mueller, Oyoyo, Smith, & Kirsch, 2011). Oxidative damage is one of the earliest, if not the earliest, event in AD (Sultana & Butterfield, 2010) already demonstrable in mild cognitive impairment (MCI), the clinical precursor of AD. In AD, evidence from animal models (Belkacemi & Ramassamy, 2012) and studies from human patients (Smith et al., 2010) indicate that oxidative damage precedes pathological changes. Imbalance of Fe homeostasis and free radical generation are present in preclinical AD, but Fe imbalance is not necessarily unique in affected brain regions (Smith et al., 2010). OS is mainly localized to the synapses, with levels increasing in a disease-dependent fashion, implicating an involvement of OS in AD-related synaptic loss (Ansari & Scheff, 2010). An important pathological finding in AD was that Fe accumulation occurs in the same brain regions characterized by Aβ deposition (Zecca et al., 2004). Dysregulation of Fe metabolism in the most vulnerable regions of AD brains (hippocampus and amygdala) is the result of Fe-regulating protein 1 (IRP1), which maintains transferrin receptor mRNA in a stabilized state causing cells to behave as though they were in a permanent state of Fe deficiency (Walton, 2013).

Iron binds with high affinity to Aβ and tau protein, the major component of NFTs, possibly as an antioxidant mechanism attempting to diminish ROS (Honda, Casadesus, Petersen, Perry, & Smith, 2004), but they become pathological as the disease progresses and proteins aggregate. Fe, Al, Zn, and Cu influence the *in vitro* formation of amyloid fibrils (House et al., 2004) but differentially alter Aβ-42 aggregation and toxicity (Bolognin et al., 2011). The view that Aβ is itself toxic has been reassessed, showing that Aβ is not toxic in the absence of redox metal ions (Huang et al., 1999).

Aβ may act as an antioxidant and have a neuroprotective role by limiting Fe and other metals through sequestration and redox silencing (Atwood, Robinson, & Smith, 2002; Smith et al., 2002). However, these processes lead to a vicious cycle of increasing both Fe and Aβ and mediating Aβ-induced stress-activated protein kinase (JNK/SAPK) activation (Zhu et al., 2004). They may exacerbate the toxicity of Aβ and of other important proteins implicated in AD such as APP and β- and γ-secretases, as a result of increased OS (Castellani, Lee, Perry, & Smith, 2006; Tamagno et al., 2008). Increased APP formation in parallel with inhibition of α-secretase activity (cleaving APP within the Aβ domain leading to release of extracellular soluble sAPP-α fragment and precluding Aβ generation) induced by Fe overload would favor Aβ deposition (Crichton et al., 2011). Moreover, there is evidence that insoluble polymers of proteins such as Aβ and tau bind Fe (Castellani et al., 2007; Dickens & Franz, 2010; Jiang et al., 2009) and thereby act in an autocatalytic manner to promote further radical production and OS. Sublethal levels of Aβ oligomers decrease nonferritin-bound Fe uptake and do not potentiate Fe toxicity in primary hippocampal neurons (SanMartin, Paula-Lima, Hidalgo, & Nunez, 2012). There are many data that OS and perturbed intracellular calcium levels leading to mitochondrial dysfunction are central in the pathogenesis of AD, although the exact mechanisms are still debated (Hauptmann et al., 2009; Horowitz & Greenamyre, 2010; Karbowski & Neutzner, 2012; Lin & Beal, 2006; Ramalingam & Kim, 2012).

Modern imaging methods have shown elevated Fe levels mainly in early onset AD, signifying that excessive Fe may be a risk factor for the age of onset (Bartzokis, Tishler, Shin, Lu, & Cummings, 2004). Early-stage AD subjects show increased Fe content in the temporal cortex, while both Fe levels and associated redox activity were already seen in MCI (Smith et al., 2010). In APP/PS1 mice, an animal model of AD, elevated cortical Fe levels were observed in the early stage of disease development (Leskovjan et al., 2011). These and other studies indicated that Fe may be a contributing factor in the early AD development (Hagemeier et al., 2012).

2.1.2 Parkinson disease

PD is a prevalent neurodegenerative disorder neuropathologically featured by considerable neuronal loss in the dopaminergic substantia nigra compacta (SNc), with denervation of the nigrostriatal dopaminergic system, associated with Lewy pathology and α-synuclein (αSyn) deposition in the central, peripheral, and autonomic nervous systems (see Jellinger (2012)). PD is associated with extensive involvement of Fe (Sian-Hulsmann et al., 2011), with most extensive Fe deposition in substantia nigra (SN) and lateral globus pallidus (Gotz, Double, Gerlach, Youdim, & Riederer, 2004; Zecca et al., 2004), as well as in dopaminergic neurons of PD patients (Jellinger, 1999; Oakley et al., 2007). Increased levels of Fe were seen in early PD in the SN contralateral to the more affected side of the body (Martin, Wieler, & Gee, 2008), while the extent of Fe deposition has been related to severity of motor symptoms in PD (Wallis et al., 2008).

Free Fe in the SN can cause increases in OS, mitochondrial dysfunction, and neuroinflammation, eventually resulting in neuronal death (Dexter, 2013; Kienzl, Jellinger, Stachelberger, & Linert, 1999; Shamoto-Nagai et al., 2006). An increase in labile Fe in PD may cause production of damaging free radicals, as opposed to increases in total ferritin-bound Fe (Friedman, Galazka-Friedman, & Koziorowski, 2009). The overproduction of ROS may impair the cell's ability to sequester Fe, while some of the bound Fe retains redox-active properties that allow catalyzation of ROS formation (Faucheux et al., 2002).

The sources of increased Fe could be (1) dysregulation of Fe homeostatic control; (2) malfunction in molecules involved in sequestrating excess Fe, caused by pathological variation in Fe tissue distribution; (3) aging associated with accumulation of bound Fe in the brain; and (4) interlinkage of these putative pathologies (Sian-Hulsmann et al., 2011).

Iron uptake systems in PD brain are upregulated, whereas Fe storage in ferritin is downregulated. The ferritin deficiency has been attributed to the sustained activity of Fe regulating the protein-binding activity that suppresses ferritin synthesis (Faucheux et al., 2002). However, the mechanisms involved in these processes remain to be established. A breakdown of protective ferritin complex can lead to an overload of free Fe radicals, which in turn can be highly neurotoxic via the production of free radicals and lipid peroxidation of cell membranes. Increased levels of lipid peroxidation products, for example, malondialdehyde, and a corresponding reduction of polyunsaturated fatty acids are found in PD SN (Montine et al., 2004). Overexpression of αSyn by cells significantly increases Fe^{2+} levels (Zhou, Lan, Tan, & Lim, 2010). It is translocated to the mitochondria, where it impairs mitochondrial function and increases oxidative stress (Parihar, Parihar, Fujita, Hashimoto, & Ghafourifar, 2008; Shavali, Brown-Borg, Ebadi, & Porter, 2008). Under oxidative conditions, this may trigger a cascade of events leading to large intracellular αSyn aggregates and the dispersal of protective pigment to precipitate nigral cell death (Halliday et al., 2005; Lv, Jiang, Xu, Song, & Xie, 2011; Ruiperez,

Darios, & Davletov, 2010). Aggregation of αSyn may lead to accumulation of redox-active Fe in the cytosol with consequent release of Fe-generated free radicals. This concords with the view that Fe-mediated OS may be mediated via αSyn oligomerization that occurs during development of PD pathology (Olivares, Huang, Branden, Greig, & Rogers, 2009) and causes dopaminergic cell destruction (Stankiewicz et al., 2007). Thus, delivery of toxic amounts of Fe to nigral dopamine neuron mitochondria may contribute to the demise of these neurons in PD. There is an interaction between αSyn and redox-active metal ions, including Cu, which may induce chemical modification of the protein *in vitro*, and these modifications may be relevant for the aggregation properties of αSyn (Bisaglia, Tessari, Mammi, & Bubacco, 2009).

Independent of the cause of Fe overload, its interaction with other disease-promoting factors, in particular with mitochondrial dysfunction in PD, has been verified (Berg et al., 2008). Why dopaminergic neurons are especially susceptible to neurodegeneration is not entirely clear, although dopamine together with Fe probably catalyzes a toxic complex (Arreguin, Nelson, Padway, Shirazi, & Pierpont, 2009). The inhibition of dopamine synthesis results in a decreased storage of Fe in dopamine neurovesicles, which would result in enhanced cell death of dopaminergic neurons (Ortega, Cloetens, Deves, Carmona, & Bohic, 2007). Differential distribution of key proteins in individual cells within one brain region may influence their eventual vulnerability (Double, Reyes, Werry, & Halliday, 2010). The selective vulnerability of A9 nigral neurons rich in neuromelanin and caspase-3 but poor in glycolytic enzymes and calbindin may be related to increased Fe content, making them susceptible for OS (Fasano, Bergamasco, & Lopiano, 2006; Faucheux et al., 2003; Jellinger et al., 1992; Kell, 2010; Lv et al., 2011; Sian-Hulsmann et al., 2011; Wypijewska et al., 2010). Conversely, most dopaminergic neurons that resist degeneration are in the scantly populated dorsal tier of the SNc and contain calbindin and glycolytic enzymes, thus being involved only in the terminal disease stages (Double et al., 2010; Fearnley & Lees, 1991).

Misfolded αSyn is believed to disrupt mitochondrial function by inhibiting complex I of the electron transport chain resulting in the loss of mitochondrial membrane potential (Loeb, Yakunin, Saada, & Sharon, 2010). A decline in the activity of complex I has been found in the SN of PD patients (Bender et al., 2013), which is matched by a decrease in protein content apparently reflecting a mitochondrial DNA defect and reduced production of complex I subunits, which result in impaired proton pumping and electron transport. *In vitro* incubation of αSyn with mitochondria results in a dose dependent transport of αSyn to mitochondria and an inhibiting effect on complex I activity (Liu et al., 2009). Consequences are increase in ROS, proteasome inhibition, aberrant protein accumulation, and eventually cell death (Beach et al., 2004; Liu & Connor, 2012), that is caused by a multifactorial cascade of pathogenic events (Hirsch, Jenner, & Przedborski, 2013).

Lewy bodies, the pathological hallmark of PD, mostly consist of ubiquitinated αSyn, a lipoprotein important in various presynaptic processes, interacting with SNARE complexes to facilitate exocytosis (Chandra, Gallardo, Fernandez-Chacon, Schluter, & Sudhof, 2005; Spillantini et al., 1997). Iron promotes conformational changes within αSyn causing its aggregation (Febbraro, Giorgi, Caldarola, Loreni, & Romero-Ramos, 2012; Uversky, 2007), which can potentially increase OS and protein aggregation into Lewy bodies (Golts et al., 2002; Munch et al., 2000). Fe also accumulates within Lewy bodies (Castellani, Siedlak, Perry, & Smith, 2000; Takanashi et al., 2001), while chelation of Fe can partially inhibit this αSyn aggregation (Hashimoto, Takeda, Hsu, Takenouchi, & Masliah, 1999). Iron deposition may be secondary to other processes, because no increase in Fe levels in SN of preclinical and mild PD has been found (Dexter, Sian, Jenner, & Marsden, 1993; Riederer et al., 1989). There is ongoing debate as to whether Fe initially accumulates. Postmortem examination of PD brains revealed that much of the Fe is present in nigral dopamine neurons (Oakley et al., 2007; Salazar et al., 2008) but much more in nigral microglia (Gotz et al., 2004). Since most of these brains come from patients who died with end-stage PD, it cannot be determined whether the initial site of Fe accumulation is indeed microglia or whether it accumulates after phagocytosis of neurons that have accumulated Fe (Horowitz & Greenamyre, 2010). However, recent studies support the hypothesis that the increased Fe content of microglia comes from phagocytosis of Fe-laden dopaminergic neurons both in human PD and in the rotenone rat model. Under conditions of cellular Fe overload, Fe import machinery would be posttranscriptionally downregulated (Mastroberardino et al., 2009). Microglia activation and corresponding loss of dopaminergic terminals in the affected nigrostriatal system in early PD suggest that neuroinflammatory reactions contribute to the progressive neurodegenerative process (Hirsch & Hunot, 2009; Long-Smith, Sullivan, & Nolan, 2009; Przedborski, 2010).

Loss of neuromelanin-containing neurons may subsequently result in an increase of free redox-active Fe, which in turn causes OS through Fenton chemistry. Iron is normally stored in ferritin, but in PD, ferritin levels are decreased (Dexter et al., 1990; Faucheux et al., 2002; Mann et al., 1994). If the capacity of neurons to store Fe is exceeded, potentially toxic free Fe will accumulate. Neuromelanin may act in a protective manner by the sequestration of redox-active Fe as well as chelation of Fe in the neuronal cytosol (Sian-Hulsmann et al., 2011; Zucca et al., 2006). Neuromelanin in dopaminergic neurons binding free and bound Fe is decreased in PD, resulting in increased availability of redox-active (free) Fe in the SN (Linert et al., 1996). Iron is an important cofactor in tyrosine hydroxylase that is involved in dopamine biosynthesis and of monoamine oxidase (MAO), an enzyme that is involved in dopamine metabolism. Hydrogen peroxide, generated in this reaction, can be converted to ROS by "free" iron. Thus, neuromelanin-derived Fe is implicated in the mechanisms of ROS production (Fig. 1.2).

Mismanagement of cellular Fe resulting in the enhanced availability of free Fe may represent a key factor of OS in later stage PD. Although at present, it remains unresolved as to whether Fe represents the cause or consequence of dopaminergic neuronal cell death, Fe executes its cytotoxic effect via OS involving intracellular Ca^{2+} homeostasis, mitochondrial dysfunction, and endoplasmic reticulum stress, together with abnormal protein degradation by the ubiquitin–proteasome system in mitochondria resulting in the accumulation of αSyn and other proteins (Cali, Ottolini, & Brini, 2011). Altered proteolysis is associated with dopaminergic cell death and Lewy body formation in both sporadic and familial PD (Fornai et al., 2003), and a reduction in activity of proteasomal enzymes in the SNc was found in sporadic PD (McNaught, Belizaire, Isacson, Jenner, & Olanow, 2003). *In vitro* studies with recombinant proteins indicate that both wild type and mutant αSyn form amyloid-like fibrils similar to those found in LBs. The rate of aggregation is higher in mutant αSyn forms perhaps due to destabilization of the protein structures by the mutation. Intrastriatal inoculation of fibrillary αSyn in wild-type nontransgenic mice is sufficient to initiate LB formation, dopaminergic cell death, and cinical symptoms of PD, reinforcing the central role of αSyn misfolding and aggregation in PD pathogenesis (Luk et al., 2012). Current pharmacotherapies are aimed to restore normal Fe and new chelation strategies are under research (Heli, Mirtorabi, & Karimian, 2011; Mounsey & Teismann, 2012).

2.1.3 Multiple sclerosis

MS, a chronic inflammatory autoimmune disease of the CNS of hitherto uncertain causes, associated with demyelination and neurodegeneration

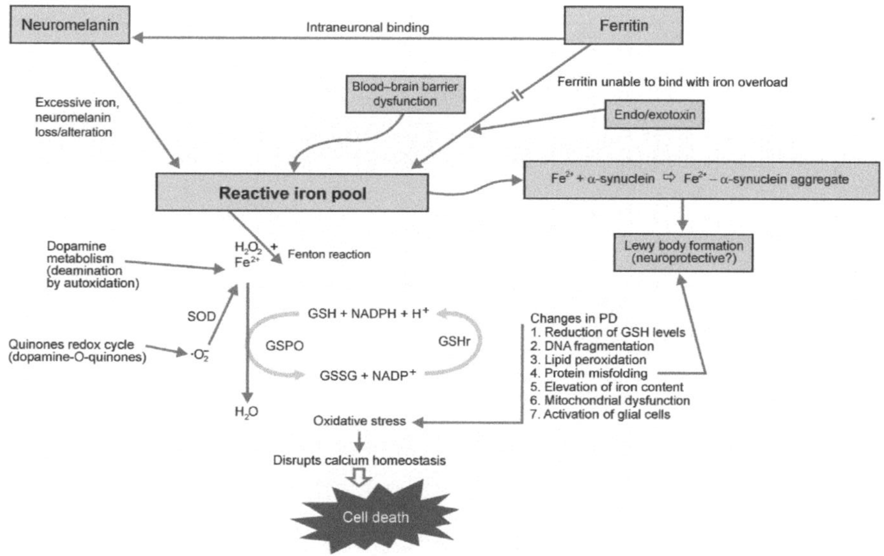

Figure 1.2 Synopsis of events leading to iron-mediated neuronal cell death in PD. Putative sources for iron release are ferritin, neuromelanin, toxins, and leaky blood–brain barrier. Reaction-free Fe may act via the Fenton reaction with hydrogen peroxide (H_2O_2) produced, for example, from monoamine oxidase (MAO)-metabolized dopamine, to yield toxic and hydroxyl (OH•) free radical species. In addition, reactive Fe may cause α-synuclein to aggregate, which may also lead to generation of highly reactive OH• and/or Lewy body formation (the latter probably neuroprotective). The highly reactive OH• radicals may induce a cascade of deleterious events such as oxidative stress (OS). Antioxidants such as superoxide dismutase (SOD) or glutathione (GSH) try to protect cells from the consequences of OS such as DNA damage, mitochondrial dysfunction, disturbances of cellular calcium homeostasis, and protein misfolding, resulting in excitoxicity and finally introducing cell death. The elevated Fe levels in SN may thus play an essential role in the degeneration of this area in PD. NADP, nicotinamide adenine dinucleotide phosphate; NADPH, NADP reduced; GSPO, glutathione peroxidase; GSHr, glutathione reductase; GSSG, glutathione oxidized; GSH, glutathione reduced. *(Modified from Sian-Hulsmann et al., 2011).*

(Lassmann, Bruck, & Lucchinetti, 2007), is suggested to be related to Fe dysregulation and deposition with implication of ROS, free radical formation, and, in particular, hydroxyl radicals (Bakshi et al., 2008; Haacke et al., 2009; LeVine & Chakrabarty, 2004). Phospholipids are major constituents of myelin and Fe accumulation has been related with myelin derangements. Major mechanisms underlying MS pathology include microglial activation, mitochondrial injury, OS, excitotoxicity, proteolytic enzyme production, calcium overload, and Fe accumulation (Fischer et al., 2012; Gonsette, 2008; Haider et al., 2011; Lassmann, van Horssen, & Mahad, 2012). MS patients have abnormal Fe deposition in and around plaques and in various parts of the brain, as shown by MRI methods (Craelius, Migdal, Luessenhop, Sugar, & Mihalakis, 1982; Habib et al., 2012; Ropele et al., 2011; Walsh et al., 2013). Fe accumulation in MS is an early pathogenic event in the demyelinating disease (Al-Radaideh et al., 2012; Levine, Bilgen, & Lynch, 2013). Evidence of abnormal homeostasis comes from the analyses of Fe and Fe-related proteins in CSF, blood, and postmortem MS brain sections (Khalil et al., 2011). Intracytoplasmic accumulation of Fe^{2+} in oligodendrocytes causing degeneration of these cells under OS releases this accumulated Fe into the extracellular space, which might further amplify oxidative damage in axons and other cells. In MS lesions, Fe^{2+} is taken up by activated macrophages and microglia (Bagnato, Hametner, & Welch, 2013). However, Fe^{2+}-containing microglia become, in part, dystrophic and may undergo fragmentation and degeneration (Lopes, Sparks, & Streit, 2008), leading to a second wave of Fe^{2+} liberation (Hametner et al., 2013; Lassmann et al., 2012; Reynolds & Crichton, 2013). Fe^{2+} accumulation in the extracellular space and its uptake into cells within the lesions might increase the susceptibility of the surrounding tissue to free radical-driven demyelination and neurodegeneration. Since this accumulation is an age-depending process, it is likely to be more pronounced in patients with progressive MS than in those in the relapsing–remitting stage (Lassmann et al., 2012). Polymorphisms in the genes coding for Fe binding and transport proteins in the presence of local Fe overload might be responsible for suboptimal Fe handling. This might account for the significant variability peculiar to MS phenotypes (Gemmati et al., 2012). However, Fe^{2+} accumulation is minimal or even absent in most currently used experimental animal models of MS, which may explain the lack of disease phenotypes that truly resemble progressive MS in these models. Furthermore, Fe deposition is independent of cellular inflammation in an experimental autoimmune encephalitis mouse model, which presents perivascular Fe deposits, providing evidence against the suggestion that Fe deposits account for inflammatory infiltrates observed in human MS (Williams et al., 2011). On the other hand, Fe is a sensitive biomarker for inflammation in MS lesions, and imaging of proinflammatory (M1) markers in Fe-containing macrophages in established MS lesions may provide an important information on disease status and treatment efficacy (Mehta et al., 2013). Further studies should establish the association between inflammation, Fe deposition, microglia activation, and neurodegeneration (Zivadinov, Weinstock-Guttman, & Pirko, 2011). New neuropathologic evidence argues against the autoimmune hypothesis and rather suggests that MS is a primary oligodendrogliopathy in which the inflammatory response may be a mere epiphenomenon (Nakahara, Maeda, Aiso, & Suzuki, 2012). Identifying the mechanisms of Fe accumulation and determining how Fe promotes pathogenesis in MS are important areas of future research (Levine et al., 2013).

2.1.4 Other neurodegenerative disorders

A variety of neurodegenerative diseases are the result of genetic defects leading to the accumulation of Fe. *Friedreich ataxia* is caused by the insertion of a trinucleotide repeat in the gene encoding a protein called frataxin (Santos et al., 2010; Seguin et al., 2010), which is a mitochondrial Fe chaperone protein involved in the insertion of Fe^{2+} in the mitochondrial respiratory chain (Huang et al., 2009; Pandolfo & Pastore, 2009). This defect leads to the liberation of mitochondrial Fe, inducing OS. Mitochondrial Fe detoxification was revealed as a primary function of frataxin limiting OS and preserving cell longevity in yeast cells (Gakh et al., 2006). This underlines strongly the problems of Fe release leading to cell death. *Amyotrophic lateral sclerosis* (ALS) that appears in both a familial and a sporadic form is another neurodegenerative disease in which Fe has been strongly implicated in some mechanisms as described earlier, in particular enhanced production of oxygen free radicals in the spinal cord (Kell, 2010; Ramalingam & Kim, 2012). An increase of ROS has been reported in both postmortem species and models of ALS, which may be the consequence of altered metabolism of Fe and Cu ions that share the property to undergo redox cycling and generate ROS (Carri, Ferri, Cozzolino, Calabrese, & Rotilio, 2003; Roos, Vesterberg, & Nordberg, 2006).

HD, an autosomal dominantly inherited progressive neurodegenerative disorder, occurs via the addition of trinucleotide CAG repeats within exon 1 of the gene encoding polyglutamine repeats in the huntingtin protein

(Bauer & Nukina, 2009). Huntingtin inclusion bodies act as centers of OS due to deranged Fe metabolism (Firdaus et al., 2006), as shown in gene knockout models of HD (Henshall et al., 2009; Lumsden, Henshall, Dayan, Lardelli, & Richards, 2007). Mutant huntingtin has been found to cause generalized transport defects in neurons, including mitochondrial transport and dynamics (Chang, Rintoul, Pandipati, & Reynolds, 2006; Orr et al., 2008; Trushina et al., 2004), suggesting that it, either directly or indirectly, can disrupt mitochondrial fusion and fission (Oettinghaus et al., 2012). Several less prevalent neurodegenerative disorders resulting from genetic defects leading to accumulation of Fe deposits include PKAN and neuroferritinopathy. Investigations of genetic pathways have elucidated the metabolic pathways in which they cause increase in Fe levels (Gregory & Hayflick, 2011; Ke & Qian, 2007; Schneider et al., 2012). For instance, in PKAN, deficient pantothenate kinase, a key regulatory enzyme in the biosynthesis of coenzyme A, critical to mitochondrial energy metabolism, resulting in cysteine synthesis, which itself is toxic, has a strong affinity to bind to Fe. It undergoes autoxidation in the presence of Fe and enhances Fe-induced lipid peroxidation (Gregory & Hayflick, 2011). There is evidence in neuroferritinopathy that excessive Fe levels cause oxidative damage (Miyajima et al., 1998). However, a causal link for these neurodegenerative disorders between Fe accumulation and the initiation or augmentation of pathology remains controversial, and increased Fe level may simply be caused by tissue loss.

2.2. The role of aluminum

Aluminum, an environmentally abundant nonredox trivalent cation, has essential neurotoxic effects. It has long been implicated in the pathogenesis of AD and other neurodegenerative diseases (Campbell, 2002; Gupta et al., 2005) since it is associated with the abnormal aggregation of Aβ (Domingo, 2006; Exley, 2005, 2006), tau aggregation (Mizoroki et al., 2007; Shcherbatykh & Carpenter, 2007), OS, and cellular dysfunction (Bharathi et al., 2008; Kawahara, 2005). Al^{3+} produces OS on its own and synergistically with Cu and Fe (see Walton (2013)). Al stabilizes intracellular Fe^{2+} by preventing its oxidation to Fe^{3+} but enhances Fe-initiated oxidative damage (Exley, 2006). A number of metals including Fe^{2+}, Fe^{3+}, Al^{3+}, Zn^{2+}, and Cu^{2+} are found colocalized with β-sheets of Aβ-42 in senile plaque cores in AD brain. This could act as a source of ROS and explain some of the oxidative damage found in the immediate vicinity of senile plaques (Exley, 2006). Al stabilizes Aβ oligomerization that is significant for neurotoxicity, and it induces human Aβ fibrils to aggregate and may represent a link between Al and amyloid cascade hypothesis of AD (Kawahara & Kato-Negishi, 2011). The interaction of Al with altered forms of tau in paired helical filaments (PHF) is likely to contribute to the formation of NFT because (1) Al and abnormally phosphorylated tau are colocalized in NFTs and (2) Al is known to preferentially interact with such phosphorylated proteins. Al has been demonstrated to bind selectively to PHF tau, induces it to aggregate, and retards its *in vivo* proteolysis (Shin, 1997). However, in a Drosophila model with chronic dietary Al overloading, Al-induced neurodegeneration is independent of Aβ and tau-associated toxicity, while the flies accumulated large amounts of Fe and ROS and exhibited elevated SOD2 activity, suggesting that Al toxicity is mediated through ROS and Fe accumulation (Wu, Du, Xue, Wu, & Zhou, 2012) (Fig. 1.3). Al metabolism is closely intertwined with Fe

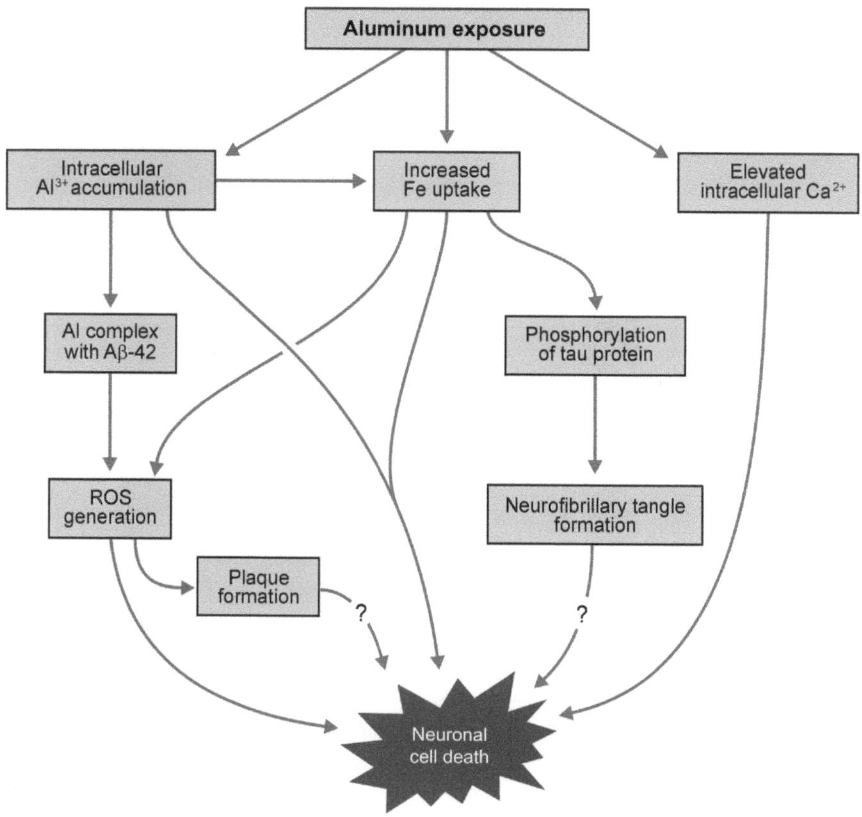

Figure 1.3 Possible mechanisms whereby aluminum causes cell death in AD.

metabolism, and a relationship between Al, Fe, and calcium homeostases has been suggested (Kawahara, 2005). Furthermore, Al may account for disorders of astroglial functions involved in neurodegeneration (Aremu & Meshitsuka, 2006). Intracellular Al is suggested to interfere with Ca^{2+} metabolism (homeostasis and signaling) in the aged brain and AD (Walton, 2012). Recent studies in AD discussing a possible relationship between Al and ferritin, the major Fe storage protein (De Sole et al., 2013), suggested new strategies to prevent potential effects of Al toxicity in preventing AD (Percy, Kruck, Pogue, & Lukiw, 2011).

There is evidence of Al accumulation in the tangle-bearing neurons and in amyloid fibers in the cores of senile plaques in AD brains (Bharathi et al., 2008; Miu & Benga, 2006; Yumoto, Kakimi, Ohsaki, & Ishikawa, 2009) and in tangle-bearing neurons of cases of ALS/Parkinson dementia complex of Guam. Abnormally high Al within the NFT is accompanied by excess concentration of Fe (Perl & Moalem, 2006). Intraneuronal Al accumulation enhances Fe uptake and expression of hyperphosphorylated tau in neuroblastoma cells.

Hippocampal CA1 cells from brains of aged humans with and without AD showed cytoplasmic pools of aggregates of an Al/hyperphosphorylated tau complex similar to that found in mature NFTs (Walton, 2010). Al plays a causative role in dialysis encephalopathy, but comparative studies have shown that there is no significant increase of AD morphology in dialysis-associated encephalopathy and long-term hemodialysis (Reusche, Koch, Lindner, Harrison, & Friedrich, 2001; Walton, 2013).

Several Al-based models of AD exhibit OS, hyperphosphorylated tau, and granulovacuolar degeneration (Walton, 2013). Rabbit brain-induced NFTs share some similarity with NFTs found in human AD (Savory, Herman, & Ghribi, 2006), while APP expression, distribution, and accumulation are altered by Al in a rodent model of AD (Walton & Wang, 2009). Al exposure induces AD-like histopathologic alterations in mouse brain, suggesting that chronic Al administration is responsible for oxidative cell damage that interferes with endoplasmic reticulum functions inducing Aβ accumulation and neurodegeneration (Rodella et al., 2008). Al-induced oxidative DNA damage and cell-cycle disruption have been demonstrated in different regions of rat brain (Kumar, Bal, & Gill, 2009). Al induces axonopathy in hippocampal regions of rat and rabbit brains (Walton & Wang, 2009). APP/PS1 transgenic mice treated with Al show a close relationship between overexpression of APP and PS1 genes and Al overload, suggesting that these mice exposed to Al have potential value for improving

AD models (Zhang et al., 2012). The contributions of Al to AD pathogenesis and neuropathology, involving APP and Aβ metabolism, formation and growth of tau pathology (NFTs), and neuron-to-neuron spreading of Al inducting the progression of AD, have been reviewed recently (Walton, 2013).

2.3. The role of copper

Copper is an essential metal present in high levels in the CNS and plays a pivotal role in maintaining homeostasis of other trace elements such as Fe and as a cofactor in mitochondrial ATP production, whereas an excess of Cu is injurious. Cu deficiency elicits glial and neuronal response typical of neurodegenerative disorders (Zucconi et al., 2007).

Several proteins are involved in the maintenance of Cu homeostasis, including Cu transporters and metal chaperones (Cerpa, Varela-Nallar, Reyes, Minniti, & Inestrosa, 2005). Redox reactions are the basis of Cu toxicity: it catalyzes the production of ROS in Fenton and Haber–Weiss reactions. The abnormalities in Cu homeostasis result in rare and severe neurodegenerative disorders such as Menkes and Wilson's disease and may play a role in AD and others (Davies et al., 2013) (Table 1.1). The localization of Cu and Cu transporters in the human brain has recently be determined, supporting a role for these proteins in Cu transport and providing novel data regarding Cu regulation (Davies et al., 2013). However, little is known about Cu distribution among various cell types or cell-specific regulation of Cu homeostasis (Lutsenko, Bhattacharjee, & Hubbard, 2010). In the brain, intracellular levels of Cu, which are almost entirely protein bound, heavily exceed extracellular levels. Cu also binds to APP and prion protein, both binding it through Cu-binding domains that also have been shown to reduce Cu *in vitro* (Cerpa et al., 2005). Furthermore, direct effects of Cu on synaptic transmission and plasticity have been found (Gaier, Eipper, & Mains, 2013). Astrocytes accumulate Cu and protect neurons from Cu toxicity by buffering Cu intracellularly (Tiffany-Castiglioni, Hong, & Qian, 2011). Sequestration of intracellular Cu involves GSH, which can mediate transfer of Cu to metallothionein (MT). Cu is a cofactor imperative for SOD1 activity, the levels of which are decreased in Menkes disease contributing to reduced activity of Cu-dependent enzymes that may be responsible for neurological damage in this disorder (Parker et al., 2013).

αSyn also has been shown to bind to Cu^{2+} (Brown, 2009). The determinants of aggregation enhancement of αSyn mediated by Cu^{2+} are the direct consequence of the formation of a αSyn–Cu^{2+} complex (Binolfi et al., 2010). Recent studies indicated that αSyn possesses at least two binding sites for Cu^{2+}, one of which is located in its N-terminal region (Ahmad, Burns, Fink, & Uversky, 2012).

Enhanced production of Aβ, under the action of increased OS and neurotoxicity of several metals, is suggested to represent a major event in the development of AD (Hardy & Selkoe, 2002). A central tenet of Aβ toxicity is linked with the presence of redox metals, mainly Cu and nonredox Zn (Bush, 2003; Bush & Curtain, 2008; Valko, Morris, & Cronin, 2005). Cu promotes the trafficking and redistribution of APP by increasing its exocytosis and reducing its endocytosis, respectively, thus regulating presumably the function of APP (Acevedo et al., 2011). Cu-mediated growth of Aβ fibrils occurs in the presence of oxidized and negatively charged liposomes (Shimanouchi, Onishi, Kitaura, Umakoshi, & Kuboi, 2011). The mechanism of ROS production by Cu–Aβ in relation to its aggregation state as well as the metal-transfer reaction from and to Aβ is crucial to understand the high toxicity of Aβ oligomers (Faller, 2009). Cu is known to bind Aβ with a high affinity via histidine and tyrosine residues (Hung, Bush, & Cherny, 2010). Cu in abnormally high concentrations and markers indicating OS have been found in amyloid plaques. Under *in vitro* conditions, the presence of Cu and Fe ions in the vicinity of Aβ catalyzes the formation of hydrogen peroxide, further catalytically converted to the reactive hydroxyl radical (Cuajungco et al., 2000), whereas Fe inhibits neurotoxicity induced by trace copper and biological reductants (White et al., 2004). In addition to Cu^{2+}, Aβ also binds Zn^{2+} and Fe^{3+} *in vitro* and the amounts of these metals are also markedly elevated in amyloid plaques of AD brains. Cu^{2+} interaction with Aβ promotes its neurotoxicity that correlates with the metal reduction ($Cu^{2+} \rightarrow Cu^{1+}$) and the generation of hydrogen peroxide, further catalytically converted to the reactive hydroxyl radical (Jomova et al., 2010). Catalytic oxidation by the Cu^{2+}–Aβ complex includes lipid peroxidation products such as 4-hydroxynonenal that accumulate in AD, and APP transgenic mice brains and oxidatively modified lipids can contribute to AD pathogenesis by promoting Aβ oligomerization (Hung et al., 2010). The role of Aβ adducts in the (catalytic) production of ROS and the subsequent degradation of Aβ by these reactive species is important (Chassaing et al., 2012).

Elevated labile Cu is associated with oxidative pathology, suggesting that dysregulation of Cu is intimately involved in the pathogenesis of AD (Eskici & Axelsen, 2012; James et al., 2012). Even though Cu dysfunction cannot be assumed as a determinant of the disease, its causative rather than associated role in AD pathology has been claimed (Linder, 2012; Squitti, 2012a). A Cu phenotype has been proposed for AD, based on defects found in genes involved in Cu metabolism, causing Cu dyshomeostasis (Squitti & Polimanti, 2013). *In vitro* studies have shown that the chelation of Cu and Zn by a suitable ligand inhibited the formation of ROS (Bush & Curtain, 2008). Liberation of Cu from amyloid plaques by chelating elements may be a useful tool for AD treatment (Geng, Li, Wu, Ren, & Qu, 2012).

2.4. The role of zinc

Zinc (Zn) is an important trace element, its major pool in the brain being present in specific synaptic vesicles in a subgroup of glutamatergic neurons. It has been shown to play a role in the transmodulation of synaptic transmission as well as in cortical plasticity (Barnham & Bush, 2008; Bitanihirwe & Cunningham, 2009; Gower-Winter & Levenson, 2012). Ionotropic glutamate receptors, such as NMDA, AMPA, and kainate receptors, are ligand-gated ion channels that mediate much of the excitatory neurotransmission in the brain. They are regulated by and facilitate the postsynaptic uptake of Zn (Morris & Levenson, 2012).

The pool of Zn constitutes about 8% of the total Zn content in the brain, while the remainder is bound to proteins. The intracellular concentration may be mediated both by efflux from the cell by Zn transporters and by complexing with apothionein to form MT, which may serve as the source of Zn for incorporation into proteins. However, Zn is released from MT by disulfides, increasing concentrations of which are formed under OS (Cuajungco & Lees, 1998). Zn can also be a potent inhibitor of nitric oxide synthase. Zn^{2+} is a potent toxic cation involved in neuronal injury. Toxic accumulation may result from either transsynaptic Zn^{2+} movement or cation mobilization from intracellular sites. To enter the cytosol, it can flux through glutamate receptor-associated channels or Zn^{2+}-sensitive membrane transporters, while MTs and mitochondria provide sites of intracellular Zn release. Intracellular Zn homeostasis is sensitive to acidosis, OS, and neuroinflammation. MT has control function in neuroinflammation (Manso, Adlard, Carrasco, Vasak, & Hidalgo, 2011). Interestingly, Zn^{2+} itself is a strong inducer of OS by promoting mitochondrial and extramitochondrial production of ROS and disruption of metabolic enzymatic activity, ultimately leading to apoptotic/necrotic processes (Capasso, Jeng, Malavolta, Mocchegiani, & Sensi, 2005; Frazzini, Rockabrand,

Mocchegiani, & Sensi, 2006). Zn dysmetabolism occurring during OS is important in a number of age-related neurodegenerative diseases where OS is implicated, such as AD, PD, and ALS.

Zn-induced augmentation of total SOD, Zn-SOD (SOD1), mitochondrial Mn-SOD (SOD2), and HO-1 is associated with increased OS and involvement of both cytosolic and mitochondrial machinery in Zn-induced αSyn, leading to degeneration of the dopaminergic system (Singh et al., 2011). On the other hand, Zn-MT is a potential mediator of antioxidant defense mechanisms in response to dopamine-induced stress, and MT-mediated Zn mobilization is an effective signal that is used by the cell to manage oxidized dopamine as an intrinsic neurotoxin (Eibl, Abdallah, & Ross, 2010). Zn7-metallothionein-3 (Zn7-MT-3) through Cu^{2+} removal from the αSyn–Cu complex efficiently prevents its deleterious redox activity and abolishes αSyn oligomerization and ROS production (Meloni & Vasak, 2011).

Zn inhibits enzymes degrading APP to nonamyloidogenic peptides that degrade the soluble form of Aβ (Capasso et al., 2005; Mocchegiani, Bertoni-Freddari, Marcellini, & Malavolta, 2005). Sequestration of Zn by Aβ deposits (Aβ oligomers and plaques) not only drives Aβ aggregation but also disrupts Zn homeostasis in regions vulnerable to AD pathology, resulting in intraneuronal Zn levels, which are either too low or excessively high. Zn^{2+} overload enhances APP cleavage and Aβ deposition in both mouse models of AD and human AD (Wang et al., 2010) by promoting Aβ aggregation via population shift of polymorphic states (Miller, Ma, & Nussinov, 2010). Disturbed metal homeostasis in the vicinity of Zn-enriched neurons may predispose formation of metal-induced Aβ aggregates, subsequent fibrillation of which can lead to Aβ formation (Tougu et al., 2009). These findings and alterations of the cerebral Zn pool, particularly synaptic Zn, in a mouse model of AD (Lee, Cho, Seo, Hwang, & Koh, 2012) confirmed the potential role for Zn in the pathogenesis of AD (Lyubartseva & Lovell, 2012). Zn transporter mRNA levels in postmortem AD brain increased significantly as the disease progressed and neuritic Braak stage increased. These changes could either reflect or cause the altered Zn distribution in AD, potentially increasing the likelihood of interaction between Zn and Aβ or tau protein (Beyer et al., 2012).

Excess Zn has been found in brain regions associated with NFT pathology and cognitive impairment that may be closely associated with anatomical damage and Zn dysmetabolism in the hippocampus due to Fe deficiency (Youdim, 2008). These data support the hypothesis linking extraneuronal

Aβ to intraneuronal NFTs, based on Aβ-induced alterations in Zn concentrations in neurons affecting stability of polymerized microtubules and their binding to MAP-tau (Craddock et al., 2012). Selective measurement of releasable synaptic Zn in human autopsy hippocampal tissues from AD brains has shown increase in extracellular Zn concentrations (Bjorklund, Sadagoparamanujam, & Taglialatela, 2012).

On the other hand, a growing number of reports indicate that Zn in micromolecular concentrations inhibits Aβ-induced toxicity (Cuajungco et al., 2000). The exact mechanisms of this protective effect of Zn are unclear; however, one of the reasons might be cytoprotection through blockage of the membrane Ca channel pore formed by Aβ-40 (Jomova et al., 2010). An additional protective role of Zn in AD is related to its competition with Cu (and Fe) to bind to Aβ. Preventing Cu from interacting with Aβ precludes the Cu–Aβ-induced formation of hydrogen peroxide and free radicals (Cuajungco et al., 2000). However, factors resulting in OS and nitrosative stress lead to abnormal metabolism of Aβ by uncontrolled flooding of the vesicular Zn pool (Cuajungco & Lees, 1998). Thus, while low levels of Zn protect against Aβ toxicity, excess of Zn released by oxidants could trigger neuronal death that is independent or even synergistic with the toxic affect of Aβ. At higher concentrations of Zn, its binding to Aβ forces the precipitation of Aβ to produce fibrillar Aβ aggregates (Cuajungco & Faget, 2003). Consequently, inflammatory responses to nonsoluble Aβ plaques are disruption of Zn homeostasis followed by uncontrolled Zn release, typical for OS. While under normal conditions, there is a sensitive balance between Zn, Cu, and Aβ metabolisms, OS and nitrative stress may perturb the balance that leads to uncontrolled Zn elevation and amyloid deposition finally inducing Aβ-mediated OS and cytotoxicity. Prion protein facilitates uptake of Zn into neuronal cells (Watt et al., 2012).

Mitochondrial inhibitor models of HD and PD (e.g., the MPTP and 6-hydroxydopamine (6-OHDA) models of PD or the 3-nitroproprionic acid model of HD) induce Zn accumulation and are attenuated by inhibition of Zn toxicity *in vitro* or *in vivo*, thus demonstrating preventable mechanisms of the resultant neurotoxicity (Sheline, Zhu, Zhang, Shi, & Cai, 2013).

ALS, a small proportion, about 2%, related to a mutation in the SOD1 gene, is associated with multiple pathological processes, including OS, glutamate excitoxicity, neuroinflammation, mitochondrial dysfunction, aggregation of proteins, and abnormalities in growth factors. The transitional metals Zn and Cu regulate SOD1 protein stability and activity, and disbalance of the homeostasis of these metals has been implicated in the pathogenesis of ALS (Vonk & Klomp, 2008). Equimolar Cu–Zn-SOD rapidly forms heterodimers with Zn-deficient SOD as a heterodimer with Cu–Zn-SOD and prevents aggregation. SOD1 binds Zn, and many of the mutant forms of this enzyme associated with ALS show altered Zn binding. It induces motor neuron death via a selective inhibition of brain-derived neurotrophic factor activity (Post, Eibl, & Ross, 2008). Aggregations of Cu–Zn-SODs have been observed in familial and sporadic ALS (Chattopadhyay & Valentine, 2009). The stabilization of Zn-deficient SOD with Cu–Zn-SOD may contribute to the dominant inheritance of ALS mutations (Roberts et al., 2007). Nonamyloid aggregates arising from mature Cu–Zn-SODs resemble those observed in human ALS (Hwang et al., 2010). Alterations in the expression of MTs, which regulate cellular levels of Zn, have been reported in mutant SOD1 mice, and deletion of MTs in these animals accelerates disease progression (Smith & Lee, 2007).

Emerging research has suggested that mutations and pathology associated with the TDP-43 gene and protein may be more common that SOD1 mutations in familial and sporadic ALS (Rothstein, 2009). TDP-43 accumulation and inclusion formation is characteristic not only in most sporadic ALS cases but also in those arising from mutations in several genes (Ince et al., 2011). TDP-43 redistribution in neurons is an early event in sporadic ALS (Giordana et al., 2010). Zn induces depletion and aggregation of endogenous TDP-43 and suggests that specific Zn-associated processes could affect TDP-43 metabolism in neurodegenerative diseases (Caragounis et al., 2010).

2.5. The role of manganese

Manganese (Mn) is an essential ubiquitous trace element and is required for many enzymatic reactions. Exposure to excessive Mn levels results in neurotoxicity to the extrapyramidal system; it appears to be regulated by a number of factors, including oxidative injury, production of cytotoxic levels of free radicals, depletion of peroxidase and catalase, and neuroinflammation (Dobson, Erikson, & Aschner, 2004; HaMai & Bondy, 2004; Milatovic, Zaja-Milatovic, Gupta, Yu, & Aschner, 2009). Mn can oxidize dopamine, generating reactive species, and also affect mitochondrial function, correlating with OS (Farina et al., 2013).

The highest concentrations of Mn in the brain are achieved in the basal ganglia, particularly in the striatum, which may precipitate a form of parkinsonism (Aschner, Erikson, Herrero Hernandez, & Tjalkens, 2009; Guilarte, 2010; Roth, 2009). While manganism shares many common features with PD, there are distinct differences between the two disorders suggesting that manganism might indirectly affect nigrostriatal dopaminergic function (Erikson, Syversen, Aschner, & Aschner, 2005; Olanow, 2004). Changes in the striatal proteome may exhibit gene–environment interactions between mutant huntingtin and Mn by suppressing phenotypes of YAC128Q mice (Wegrzynowicz, Holt, Friedman, & Bowman, 2012).

Mn is a putative modulator of dopamine biology (the primary target of Mn neurotoxicity) and has effects on mitochondrial dysfunction and disruption of cellular energy metabolism, glial activation, and disruption of synaptic transmission (Aschner et al., 2009). Synaptic mechanisms during the early phase of Mn intoxication may have importance for the progression of this pathological condition, involving the basal ganglia (Calabresi et al., 2001), with selective vulnerability of pallidal neurons (Spadoni et al., 2000). Although dopamine transporter scans show a slight increase in the putamen of manganism patients compared with controls, the data indicate that the presymptomatic dopaminergic terminals are not the main target of chronic Mn intoxication (Huang, Weng, Lu, Chu, & Yen, 2003). Other factors of Mn neurotoxicity are NF-κB induction, activation of nitric oxide by astrocytes as an early event in manganism (Liu, Sullivan, Madl, Legare, & Tjalkens, 2006), and dopamine interaction, which is not controlled by NF-κB as its inhibitor (SN50) does not prevent ROS (Prabhakaran, Ghosh, Chapman, & Gunasekar, 2008). αSyn overexpression enhances Mn-induced neurotoxicity through the NF-κB-mediated pathway and may play a role in dopaminergic cell death (Prabhakaran, Chapman, & Gunasekar, 2011). Glutamate regulation is affected by increased brain Mn levels, which early induce disturbances in GABA regulation as well as glutamate-related excitotoxicity (Fitsanakis, Au, Erikson, & Aschner, 2006). Neuropathology in nonhuman primates indicated decreased glutamine synthetase expression in the globus pallidus with otherwise normal markers of glutaminergic and GABAergic neurotransmission. Increased Aβ (A4) precursor-like protein 1 gene expression with multiple markers of neurodegeneration and glial cell activation was observed in the frontal cortex (Burton & Guilarte, 2009).

Chronic exposure to Mn decreases striatal dopamine turnover in human αSyn transgenic mice (Peneder et al., 2011) and induces microglia activation and dystrophy in the SN in nonhuman primates (Verina, Kiihl, Schneider, &

Guilarte, 2011). Although chronic exposure to Mn is not the initial causative agent provoking parkinsonism, there is evidence that this can predispose to the development of dystonic movements associated with PD (Roth, 2009). Repeated Mn exposure further has long-term effects on the regulation of excytotoxic dopamine release in the striatum, which may be involved in the mechanism underlying its toxicity (Khalid, Aoun, & Mathews, 2011). Mn dyshomeostasis can result from enhanced OS through synergistic actions of dopamine oxidation, while antioxidant factors may signify a potential therapeutic approach against Mn toxicity (Chakraborty & Aschner, 2012). In addition to the production of lesions similar (and different) to PD, there is a link between Mn and HD, with emphasis on huntingtin function and its potential for altering Mn homeostasis and toxicity (Bowman, Kwakye, Hernandez, & Aschner, 2011). High Mn and low Cu levels were found in prion protein plaques (Johnson et al., 2013). However, the distinct role of Mn needs further elucidation.

3. CONCLUSIONS AND FUTURE PERSPECTIVES

The basic pathogenic factors in a number of neurodegenerative diseases, such as OS, protein misfolding and aggregation, mitochondrial dysfunction, and energy failure, are mediated or triggered by an imbalance of metal ions initiating a cascade of events finally leading to dysfunction of neuronal networks. Many of the basic mechanisms involved in these processes remain to be established, and how ion homeostasis is maintained and disturbed in the brain is an area of increasing interest. The role of poorly liganded metals has been rather underappreciated in the past. A better understanding of the multifold factors involved in these processes is crucial for elucidating the pathophysiological mechanisms responsible for the excess of essential metal ions as well as for the development of treatment interventions that can disturb the chain of pathological events occurring in many neurodegenerative diseases, the etiology of most of them currently remains to be elucidated. It is widely unknown whether various metals, like iron, zinc, copper, and aluminum, have common or separate models of action. Understanding these processes requires an integrative multisystem-level approach that may lead to further progress in neurodegeneration research. Considering recent advances in neurobiology of metals, the complex relationship between biometal metabolism, genetic and environmental exposures, and the pathophysiology of many neurodegenerative diseases warrants further investigation. Evidence available from experimental and transgenic animal studies is convincing and together with human brain studies indicates important roles of metal ions in neurodegeneration, yet it does not provide conclusive answers about the causality of the metal-related processes or effective preventive and therapeutic strategies in human beings. Recent advances in metalloproteomics will contribute to a richer understanding of the mechanics and precise role of metalloenzymes and proteins in the brain (Lothian et al., 2013). Currently, very little can be done to slow the progress of neurodegeneration. Its multifaceted profile necessitates a change in paradigm toward the design of compounds that address several drugs simultaneously. The involvement of metal ions in a majority of these deleterious disorders has made them an emerging target for therapeutic future interventions. One approach has been to chelate and sequester the ions and thus limit their potential interfere with protein folding or render them unable to undergo oxidative processes (Li, Jankovic, & Le, 2011; Telpoukhovskaia & Orvig, 2013). Newer approaches suggest that redistributing metal ions has therapeutic benefits. Recent studies indicate that alleviating cellular Cu deficiency may be a possible way to limit neurodegeneration (Badrick & Jones, 2011), while redistribution of Fe (mobilization from transferrin) by iron chelating agents, for example, hydroxy pyridones, may have relevance for the treatment of many related disease types (Evans, Kong, & Hider, 2012). Recently, several compounds were developed to reduce through chelation metal ions that interact with both metal-induced Aβ aggregation and ROS generated by this and other aggregates. Designing drugs with multitargeted activity may be the next step in the treatment of neurodegenerative disorders (Geldenhuys & Van der Schyf, 2013a, 2013b), but further validation and evaluation of the efficacy of multitargeted design drugs will be necessary.

REFERENCES

Acevedo, K. M., Hung, Y. H., Dalziel, A. H., Li, Q. X., Laughton, K., Wikhe, K., et al. (2011). Copper promotes the trafficking of the amyloid precursor protein. *Journal of Biological Chemistry*, *286*(10), 8252–8262.

Ahmad, A., Burns, C. S., Fink, A. L., & Uversky, V. N. (2012). Peculiarities of copper binding to alpha-synuclein. *Journal of Biomolecular Structure and Dynamics*, *29*(4), 825–842.

Allsop, D., Mayes, J., Moore, S., Masad, A., & Tabner, B. J. (2008). Metal-dependent generation of reactive oxygen species from amyloid proteins implicated in neurodegenerative disease. *Biochemical Society Transactions*, *36*(Pt 6), 1293–1298.

Al-Radaideh, A. M., Wharton, S. J., Lim, S. Y., Tench, C. R., Morgan, P. S., Bowtell, R. W., et al. (2012). Increased iron accumulation occurs in the earliest stages of demyelinating disease: An ultra-high field susceptibility mapping study in Clinically Isolated Syndrome. *Multiple Sclerosis*, *19*, 896–903.

Ansari, M. A., & Scheff, S. W. (2010). Oxidative stress in the progression of Alzheimer disease in the frontal cortex. *Journal of Neuropathology and Experimental Neurology*, *69*(2), 155–167.

Aremu, D. A., & Meshitsuka, S. (2006). Some aspects of astroglial functions and aluminum implications for neurodegeneration. *Brain Research Reviews*, *52*(1), 193–200.

Arreguin, S., Nelson, P., Padway, S., Shirazi, M., & Pierpont, C. (2009). Dopamine complexes of iron in the etiology and pathogenesis of Parkinson's disease. *Journal of Inorganic Biochemistry*, *103*(1), 87–93.

Aschner, M., Erikson, K. M., Herrero Hernandez, E., & Tjalkens, R. (2009). Manganese and its role in Parkinson's disease: From transport to neuropathology. *Neuromolecular Medicine*, *11*(4), 252–266.

Atwood, C. S., Robinson, S. R., & Smith, M. A. (2002). Amyloid-beta: Redox-metal chelator and antioxidant. *Journal of Alzheimer's Disease*, *4*(3), 203–214.

Ayton, S., Lei, P., & Bush, A. I. (2013). Metallostasis in Alzheimer's disease. *Free Radical Biology & Medicine*, *62*, 76–89.

Badrick, A. C., & Jones, C. E. (2011). Reorganizing metals: The use of chelating compounds as potential therapies for metal-related neurodegenerative disease. *Current Topics in Medicinal Chemistry*, *11*(5), 543–552.

Bagnato, F., Hametner, S., & Welch, E. B. (2013). Visualizing iron in multiple sclerosis. *Magnetic Resonance Imaging*, *31*(3), 376–384.

Bakshi, R., Thompson, A. J., Rocca, M. A., Pelletier, D., Dousset, V., Barkhof, F., et al. (2008). MRI in multiple sclerosis: Current status and future prospects. *Lancet Neurology*, *7*(7), 615–625.

Barnham, K. J., & Bush, A. I. (2008). Metals in Alzheimer's and Parkinson's diseases. *Current Opinion in Chemical Biology*, *12*(2), 222–228.

Barnham, K. J., Masters, C. L., & Bush, A. I. (2004). Neurodegenerative diseases and oxidative stress. *Nature Reviews Drug Discovery*, *3*(3), 205–214.

Barsukova, A. G., Bourdette, D., & Forte, M. (2011). Mitochondrial calcium and its regulation in neurodegeneration induced by oxidative stress. *European Journal of Neuroscience*, *34*(3), 437–447.

Bartzokis, G., Tishler, T. A., Shin, I. S., Lu, P. H., & Cummings, J. L. (2004). Brain ferritin iron as a risk factor for age at onset in neurodegenerative diseases. *Annals of the New York Academy of Sciences*, *1012*, 224–236.

Batista-Nascimento, L., Pimentel, C., Menezes, R. A., & Rodrigues-Pousada, C. (2012). Iron and neurodegeneration: From cellular homeostasis to disease. *Oxidative Medicine and Cellular Longevity*, *2012*, 128647.

Bauer, P. O., & Nukina, N. (2009). The pathogenic mechanisms of polyglutamine diseases and current therapeutic strategies. *Journal of Neurochemistry*, *110*(6), 1737–1765.

Beach, T. G., Walker, D. G., Sue, L. I., Newell, A., Adler, C. C., & Joyce, J. N. (2004). Substantia nigra Marinesco bodies are associated with decreased striatal expression of dopaminergic markers. *Journal of Neuropathology and Experimental Neurology*, *63*(4), 329–337.

Belkacemi, A., & Ramassamy, C. (2012). Time sequence of oxidative stress in the brain from transgenic mouse models of Alzheimer's disease related to the amyloid-beta cascade. *Free Radical Biology & Medicine*, *52*(3), 593–600.

Bender, A., Desplats, P., Spencer, B., Rockenstein, E., Adame, A., Elstner, M., et al. (2013). TOM40 mediates mitochondrial dysfunction induced by alpha-synuclein accumulation in Parkinson's disease. *PLoS One*, *8*(4), e62277.

Berg, D., Riederer, P., & Gerlach, M. (2008). Contribution of disturbed iron metabolism to the pathogenesis of Parkinson's disease. *Future Neurology*, *3*, 447–461.

Beyer, N., Coulson, D. T., Heggarty, S., Ravid, R., Hellemans, J., Irvine, G. B., et al. (2012). Zinc transporter mRNA levels in Alzheimer's disease postmortem brain. *Journal of Alzheimer's Disease*, *29*(4), 863–873.

Bharathi, P., Vasudevaraju, P., Govindaraju, M., Palanisamy, A. P., Sambamurti, K., & Rao, K. S. (2008). Molecular toxicity of aluminium in relation to neurodegeneration. *The Indian Journal of Medical Research*, *128*(4), 545–556.

Binolfi, A., Rodriguez, E. E., Valensin, D., D'Amelio, N., Ippoliti, E., Obal, G., et al. (2010). Bioinorganic chemistry of Parkinson's disease: Structural determinants for the copper-mediated amyloid formation of alpha-synuclein. *Inorganic Chemistry*, *49*(22), 10668–10679.

Bisaglia, M., Tessari, I., Mammi, S., & Bubacco, L. (2009). Interaction between alpha-synuclein and metal ions, still looking for a role in the pathogenesis of Parkinson's disease. *Neuromolecular Medicine*, *11*(4), 239–251.

Bitanihirwe, B. K., & Cunningham, M. G. (2009). Zinc: The brain's dark horse. *Synapse*, *63*(11), 1029–1049.

Bjorklund, N. L., Sadagoparamanujam, V. M., & Taglialatela, G. (2012). Selective, quantitative measurement of releasable synaptic zinc in human autopsy hippocampal brain tissue from Alzheimer's disease patients. *Journal of Neuroscience Methods*, *203*(1), 146–151.

Boelmans, K., Holst, B., Hackius, M., Finsterbusch, J., Gerloff, C., Fiehler, J., et al. (2012). Brain iron deposition fingerprints in Parkinson's disease and progressive supranuclear palsy. *Movement Disorders*, *27*(3), 421–427.

Bolognin, S., Messori, L., Drago, D., Gabbiani, C., Cendron, L., & Zatta, P. (2011). Aluminum, copper, iron and zinc differentially alter amyloid-Abeta(1–42) aggregation and toxicity. *The International Journal of Biochemistry & Cell Biology*, *43*(6), 877–885.

Bolognin, S., Messori, L., & Zatta, P. (2009). Metal ion physiopathology in neurodegenerative disorders. *Neuromolecular Medicine*, *11*(4), 223–238.

Bonda, D. J., Lee, H. G., Blair, J. A., Zhu, X., Perry, G., & Smith, M. A. (2011). Role of metal dyshomeostasis in Alzheimer's disease. *Metallomics*, *3*(3), 267–270.

Bowman, A. B., Kwakye, G. F., Hernandez, E. H., & Aschner, M. (2011). Role of manganese in neurodegenerative diseases. *Journal of Trace Elements in Medicine and Biology*, *25*(4), 191–203.

Breydo, L., & Uversky, V. N. (2011). Role of metal ions in aggregation of intrinsically disordered proteins in neurodegenerative diseases. *Metallomics*, *3*(11), 1163–1180.

Brown, D. R. (2009). Brain proteins that mind metals: A neurodegenerative perspective. *Dalton Transactions*, 7(21), 4069–4076.

Burton, N. C., & Guilarte, T. R. (2009). Manganese neurotoxicity: Lessons learned from longitudinal studies in nonhuman primates. *Environmental Health Perspectives*, *117*(3), 325–332.

Bush, A. I. (2003). The metallobiology of Alzheimer's disease. *Trends in Neurosciences*, *26*(4), 207–214.

Bush, A. I. (2013). The metal theory of Alzheimer's disease. *Journal of Alzheimer's Disease*, *33*(Suppl 1), S277–S281.

Bush, A. I., & Curtain, C. C. (2008). Twenty years of metallo-neurobiology: Where to now? *European Biophysics Journal*, *37*(3), 241–245.

Calabresi, P., Ammassari-Teule, M., Gubellini, P., Sancesario, G., Morello, M., Centonze, D., et al. (2001). A synaptic mechanism underlying the behavioral abnormalities induced by manganese intoxication. *Neurobiology of Disease*, *8*(3), 419–432.

Cali, T., Ottolini, D., & Brini, M. (2011). Mitochondria, calcium, and endoplasmic reticulum stress in Parkinson's disease. *Biofactors*, *37*(3), 228–240.

Campbell, A. (2002). The potential role of aluminium in Alzheimer's disease. *Nephrology, Dialysis, Transplantation*, *17*(Suppl 2), 17–20.

Campbell, A. (2006). The role of aluminum and copper on neuroinflammation and Alzheimer's disease. *Journal of Alzheimer's Disease*, *10*(2–3), 165–172.

Capasso, M., Jeng, J. M., Malavolta, M., Mocchegiani, E., & Sensi, S. L. (2005). Zinc dyshomeostasis: A key modulator of neuronal injury. *Journal of Alzheimer's Disease*, *8*(2), 93–108, discussion 209–115.

Caragounis, A., Price, K. A., Soon, C. P., Filiz, G., Masters, C. L., Li, Q. X., et al. (2010). Zinc induces depletion and aggregation of endogenous TDP-43. *Free Radical Biology & Medicine, 48*(9), 1152–1161.

Carri, M. T., Ferri, A., Cozzolino, M., Calabrese, L., & Rotilio, G. (2003). Neurodegeneration in amyotrophic lateral sclerosis: The role of oxidative stress and altered homeostasis of metals. *Brain Research Bulletin, 61*(4), 365–374.

Castellani, R. J., Lee, H. G., Perry, G., & Smith, M. A. (2006). Antioxidant protection and neurodegenerative disease: The role of amyloid-beta and tau. *American Journal of Alzheimer's Disease and Other Dementias, 21*(2), 126–130.

Castellani, R. J., Moreira, P. I., Liu, G., Dobson, J., Perry, G., Smith, M. A., et al. (2007). Iron: The Redox-active center of oxidative stress in Alzheimer disease. *Neurochemical Research, 32*(10), 1640–1645.

Castellani, R. J., Siedlak, S. L., Perry, G., & Smith, M. A. (2000). Sequestration of iron by Lewy bodies in Parkinson's disease. *Acta Neuropathologica, 100*(2), 111–114.

Cerpa, W., Varela-Nallar, L., Reyes, A. E., Minniti, A. N., & Inestrosa, N. C. (2005). Is there a role for copper in neurodegenerative diseases? *Molecular Aspects of Medicine, 26*(4–5), 405–420.

Chakraborty, S., & Aschner, M. (2012). Altered manganese homeostasis: Implications for BLI-3-dependent dopaminergic neurodegeneration and SKN-1 protection in C. elegans. *Journal of Trace Elements in Medicine and Biology, 26*(2–3), 183–187.

Chandra, S., Gallardo, G., Fernandez-Chacon, R., Schluter, O. M., & Sudhof, T. C. (2005). Alpha-synuclein cooperates with CSPalpha in preventing neurodegeneration. *Cell, 123*(3), 383–396.

Chang, D. T., Rintoul, G. L., Pandipati, S., & Reynolds, I. J. (2006). Mutant huntingtin aggregates impair mitochondrial movement and trafficking in cortical neurons. *Neurobiology of Disease, 22*(2), 388–400.

Chassaing, S., Collin, F., Dorlet, P., Gout, J., Hureau, C., & Faller, P. (2012). Copper and heme-mediated a-beta toxicity: Redox chemistry, a-beta oxidations and anti-ROS compounds. *Current Topics in Medicinal Chemistry, 12*(22), 2573–2595.

Chattopadhyay, M., & Valentine, J. S. (2009). Aggregation of copper-zinc superoxide dismutase in familial and sporadic ALS. *Antioxidants & Redox Signaling, 11*(7), 1603–1614.

Craddock, T. J., Tuszynski, J. A., Chopra, D., Casey, N., Goldstein, L. E., Hameroff, S. R., et al. (2012). The zinc dyshomeostasis hypothesis of Alzheimer's disease. *PLoS One*, 7(3), e33552.

Craelius, W., Migdal, M. W., Luessenhop, C. P., Sugar, A., & Mihalakis, I. (1982). Iron deposits surrounding multiple sclerosis plaques. *Archives of Pathology & Laboratory Medicine, 106*(8), 397–399.

Crichton, R. R., Dexter, D. T., & Ward, R. J. (2011). Brain iron metabolism and its perturbation in neurological diseases. *Journal of Neural Transmission, 118*(3), 301–314.

Cuajungco, M. P., & Faget, K. Y. (2003). Zinc takes the center stage: Its paradoxical role in Alzheimer's disease. *Brain Research Brain Research Reviews, 41*(1), 44–56.

Cuajungco, M. P., Goldstein, L. E., Nunomura, A., Smith, M. A., Lim, J. T., Atwood, C. S., et al. (2000). Evidence that the beta-amyloid plaques of Alzheimer's disease represent the redox-silencing and entombment of abeta by zinc. *Journal of Biological Chemistry, 275*(26), 19439–19442.

Cuajungco, M. P., & Lees, G. J. (1998). Nitric oxide generators produce accumulation of chelatable zinc in hippocampal neuronal perikarya. *Brain Research, 799*(1), 118–129.

Davies, K. M., Hare, D. J., Cottam, V., Chen, N., Hilgers, L., Halliday, G., et al. (2013). Localization of copper and copper transporters in the human brain. *Metallomics, 5*(1), 43–51.

De Sole, P., Rossi, C., Chiarpotto, M., Ciasca, G., Bocca, B., Alimonti, A., et al. (2013). Possible relationship between Al/ferritin complex and Alzheimer's disease. *Clinical Biochemistry, 46*(1–2), 89–93.

Dexter, D. T., Carayon, A., Vidailhet, M., Ruberg, M., Agid, F., Agid, Y., et al. (1990). Decreased ferritin levels in brain in Parkinson's disease. *Journal of Neurochemistry, 55*(1), 16–20.

Dexter, D. T., Sian, J., Jenner, P., & Marsden, C. D. (1993). Implications of alterations in trace element levels in brain in Parkinson's disease and other neurological disorders affecting the basal ganglia. *Advances in Neurology, 60*, 273–281.

Dexter, D. (2013). Parkinson's disease: Involvement of iron and oxidative stress. In R. Ward, R. Crichton, & D. Dexter (Eds.), *Mechanisms and metal involvement in neurodegenerative diseases* (pp. 58–79). London, UK: The Royal Society of Chemistry.

Dickens, M. G., & Franz, K. J. (2010). A prochelator activated by hydrogen peroxide prevents metal-induced amyloid Beta aggregation. *Chembiochem, 11*(1), 59–62.

Dobson, A. W., Erikson, K. M., & Aschner, M. (2004). Manganese neurotoxicity. *Annals of the New York Academy of Sciences, 1012*, 115–128.

Domingo, J. L. (2006). Aluminum and other metals in Alzheimer's disease: A review of potential therapy with chelating agents. *Journal of Alzheimer's Disease, 10*(2–3), 331–341.

Double, K. L., Reyes, S., Werry, E. L., & Halliday, G. M. (2010). Selective cell death in neurodegeneration: Why are some neurons spared in vulnerable regions? *Progress in Neurobiology, 92*(3), 316–329.

Drago, D., Bolognin, S., & Zatta, P. (2008). Role of metal ions in the abeta oligomerization in Alzheimer's disease and in other neurological disorders. *Current Alzheimer Research, 5*(6), 500–507.

Duce, J. A., Tsatsanis, A., Cater, M. A., James, S. A., Robb, E., Wikhe, K., et al. (2010). Iron-export ferroxidase activity of beta-amyloid precursor protein is inhibited by zinc in Alzheimer's disease. *Cell, 142*(6), 857–867.

Duda, J. E., & Jellinger, K. A. (2011). Neurodegeneration with brain iron accumulation. In D. W. Dickson & R. O. Weller (Eds.), *Neurodegeneration: The molecular pathology of dementia and movement disorders* (2nd ed., pp. 446–455). Oxford: Blackwell.

Dusek, P., Jankovic, J., & Le, W. (2012). Iron dysregulation in movement disorders. *Neurobiology of Disease, 46*(1), 1–18.

Dusek, P., & Schneider, S. A. (2012). Neurodegeneration with brain iron accumulation. *Current Opinion in Neurology, 25*(4), 499–506.

Eibl, J. K., Abdallah, Z., & Ross, G. M. (2010). Zinc-metallothionein: A potential mediator of antioxidant defence mechanisms in response to dopamine-induced stress. *Canadian Journal of Physiology and Pharmacology, 88*(3), 305–312.

Erikson, K. M., Syversen, T., Aschner, J. L., & Aschner, M. (2005). Interactions between excessive manganese exposures and dietary iron-deficiency in neurodegeneration. *Environmental Toxicology and Pharmacology, 19*(3), 415–421.

Eskici, G., & Axelsen, P. H. (2012). Copper and oxidative stress in the pathogenesis of Alzheimer's disease. *Biochemistry, 51*(32), 6289–6311.

Evans, R. W., Kong, X., & Hider, R. C. (2012). Iron mobilization from transferrin by therapeutic iron chelating agents. *Biochimica et Biophysica Acta, 1820*(3), 282–290.

Exley, C. (2005). The aluminium-amyloid cascade hypothesis and Alzheimer's disease. *Subcellular Biochemistry, 38*, 225–234.

Exley, C. (2006). Aluminium and iron, but neither copper nor zinc, are key to the precipitation of beta-sheets of Abeta_{42} in senile plaque cores in Alzheimer's disease. *Journal of Alzheimer's Disease, 10*(2–3), 173–177.

Faller, P. (2009). Copper and zinc binding to amyloid-beta: Coordination, dynamics, aggregation, reactivity and metal-ion transfer. *Chembiochem, 10*(18), 2837–2845.

Farina, M., Avila, D. S., da Rocha, J. B., & Aschner, M. (2013). Metals, oxidative stress and neurodegeneration: A focus on iron, manganese and mercury. *Neurochemistry International, 62*(5), 575–594.

Fasano, M., Bergamasco, B., & Lopiano, L. (2006). Modifications of the iron-neuromelanin system in Parkinson's disease. *Journal of Neurochemistry*, *96*(4), 909–916.

Faucheux, B. A., Martin, M. E., Beaumont, C., Hauw, J. J., Agid, Y., & Hirsch, E. C. (2003). Neuromelanin associated redox-active iron is increased in the substantia nigra of patients with Parkinson's disease. *Journal of Neurochemistry*, *86*(5), 1142–1148.

Faucheux, B. A., Martin, M. E., Beaumont, C., Hunot, S., Hauw, J. J., Agid, Y., et al. (2002). Lack of up-regulation of ferritin is associated with sustained iron regulatory protein-1 binding activity in the substantia nigra of patients with Parkinson's disease. *Journal of Neurochemistry*, *83*(2), 320–330.

Fearnley, J. M., & Lees, A. J. (1991). Ageing and Parkinson's disease: Substantia nigra regional selectivity. *Brain*, *114*(Pt 5), 2283–2301.

Febbraro, F., Giorgi, M., Caldarola, S., Loreni, F., & Romero-Ramos, M. (2012). alpha-Synuclein expression is modulated at the translational level by iron. *Neuroreport*, *23*(9), 576–580.

Federico, A., Cardaioli, E., Da Pozzo, P., Formichi, P., Gallus, G. N., & Radi, E. (2012). Mitochondria, oxidative stress and neurodegeneration. *Journal of Neurological Sciences*, *322*(1–2), 254–262.

Filosto, M., Scarpelli, M., Cotelli, M. S., Vielmi, V., Todeschini, A., Gregorelli, V., et al. (2011). The role of mitochondria in neurodegenerative diseases. *Journal of Neurology*, *258*(10), 1763–1774.

Firdaus, W. J., Wyttenbach, A., Giuliano, P., Kretz-Remy, C., Currie, R. W., & Arrigo, A. P. (2006). Huntingtin inclusion bodies are iron-dependent centers of oxidative events. *The FEBS Journal*, *273*(23), 5428–5441.

Fischer, M. T., Sharma, R., Lim, J. L., Haider, L., Frischer, J. M., Drexhage, J., et al. (2012). NADPH oxidase expression in active multiple sclerosis lesions in relation to oxidative tissue damage and mitochondrial injury. *Brain*, *135*(Pt 3), 886–899.

Fitsanakis, V. A., Au, C., Erikson, K. M., & Aschner, M. (2006). The effects of manganese on glutamate, dopamine and gamma-aminobutyric acid regulation. *Neurochemistry International*, *48*(6–7), 426–433.

Fornai, F., Lenzi, P., Gesi, M., Ferrucci, M., Lazzeri, G., Natale, G., et al. (2003). Recent knowledge on molecular components of Lewy bodies discloses future therapeutic strategies in Parkinson's disease. *Current Drug Targets CNS and Neurological Disorders*, *2*(3), 149–152.

Frazzini, V., Rockabrand, E., Mocchegiani, E., & Sensi, S. L. (2006). Oxidative stress and brain aging: Is zinc the link? *Biogerontology*, 7(5–6), 307–314.

Friedman, A., Galazka-Friedman, J., & Koziorowski, D. (2009). Iron as a cause of Parkinson disease—A myth or a well established hypothesis? *Parkinsonism & Related Disorders*, *15*(Suppl 3), S212–S214.

Gaeta, A., & Hider, R. C. (2005). The crucial role of metal ions in neurodegeneration: The basis for a promising therapeutic strategy. *British Journal of Pharmacology*, *146*(8), 1041–1059.

Gaier, E. D., Eipper, B. A., & Mains, R. E. (2013). Copper signaling in the mammalian nervous system: Synaptic effects. *Journal of Neuroscience Research*, *91*(1), 2–19.

Gakh, O., Park, S., Liu, G., Macomber, L., Imlay, J. A., Ferreira, G. C., et al. (2006). Mitochondrial iron detoxification is a primary function of frataxin that limits oxidative damage and preserves cell longevity. *Human Molecular Genetics*, *15*(3), 467–479.

Galazka-Friedman, J., Bauminger, E. R., Szlachta, K., & Friedman, A. (2012). The role of iron in neurodegeneration–Mossbauer spectroscopy, electron microscopy, enzyme-linked immunosorbent assay and neuroimaging studies. *Journal of Physics: Condensed Matter*, *24*(24), 244106.

Gavrin, L. K., Denny, R. A., & Saiah, E. (2012). Small molecules that target protein misfolding. *Journal of Medicinal Chemistry*, *55*(24), 10823–10843.

Geldenhuys, W. J., & Van der Schyf, C. J. (2013a). Designing drugs with multi-target activity: The next step in the treatment of neurodegenerative disorders. *Expert Opinion on Drug Discovery*, *8*(2), 115–129.
Geldenhuys, W. J., & Van der Schyf, C. J. (2013b). Rationally designed multi-targeted agents against neurodegenerative diseases. *Current Medicinal Chemistry*, *20*(13), 1662–1672.
Gemmati, D., Zeri, G., Orioli, E., De Gaetano, F. E., Salvi, F., Bartolomei, I., et al. (2012). Polymorphisms in the genes coding for iron binding and transporting proteins are associated with disability, severity, and early progression in multiple sclerosis. *BMC Medical Genetics*, *13*, 70.
Geng, J., Li, M., Wu, L., Ren, J., & Qu, X. (2012). Liberation of copper from amyloid plaques: Making a risk factor useful for Alzheimer's disease treatment. *Journal of Medicinal Chemistry*, *55*(21), 9146–9155.
Gille, G., & Reichmann, H. (2011). Iron-dependent functions of mitochondria—Relation to neurodegeneration. *Journal of Neural Transmission*, *118*(3), 349–359.
Giordana, M. T., Piccinini, M., Grifoni, S., De Marco, G., Vercellino, M., Magistrello, M., et al. (2010). TDP-43 redistribution is an early event in sporadic amyotrophic lateral sclerosis. *Brain Pathology*, *20*(2), 351–360.
Gleichmann, M., & Mattson, M. P. (2011). Neuronal calcium homeostasis and dysregulation. *Antioxidants & Redox Signaling*, *14*(7), 1261–1273.
Golts, N., Snyder, H., Frasier, M., Theisler, C., Choi, P., & Wolozin, B. (2002). Magnesium inhibits spontaneous and iron-induced aggregation of alpha-synuclein. *Journal of Biological Chemistry*, *277*(18), 16116–16123.
Gomes, C. M. (2013). Protein misfolding in disease and small molecule therapies. *Current Topics in Medicinal Chemistry*, *12*(22), 2460–2469.
Gonsette, R. E. (2008). Neurodegeneration in multiple sclerosis: The role of oxidative stress and excitotoxicity. *Journal of Neurological Sciences*, *274*(1–2), 48–53.
Gotz, M. E., Double, K., Gerlach, M., Youdim, M. B., & Riederer, P. (2004). The relevance of iron in the pathogenesis of Parkinson's disease. *Annals of the New York Academy of Sciences*, *1012*, 193–208.
Gower-Winter, S. D., & Levenson, C. W. (2012). Zinc in the central nervous system: From molecules to behavior. *Biofactors*, *38*(3), 186–193.
Greenough, M. A., Camakaris, J., & Bush, A. I. (2012). Metal dyshomeostasis and oxidative stress in Alzheimer's disease. *Neurochemistry International*, *62*, 540–555.
Gregory, A., & Hayflick, S. J. (2011). Genetics of neurodegeneration with brain iron accumulation. *Current Neurology and Neuroscience Reports*, *11*(3), 254–261.
Guilarte, T. R. (2010). Manganese and Parkinson's disease: A critical review and new findings. *Environmental Health Perspectives*, *118*(8), 1071–1080.
Gupta, V. B., Anitha, S., Hegde, M. L., Zecca, L., Garruto, R. M., Ravid, R., et al. (2005). Aluminium in Alzheimer's disease: Are we still at a crossroad? *Cellular and Molecular Life Sciences*, *62*(2), 143–158.
Haacke, E. M., Makki, M., Ge, Y., Maheshwari, M., Sehgal, V., Hu, J., et al. (2009). Characterizing iron deposition in multiple sclerosis lesions using susceptibility weighted imaging. *Journal of Magnetic Resonance Imaging*, *29*(3), 537–544.
Habib, C. A., Liu, M., Bawany, N., Garbern, J., Krumbein, I., Mentzel, H. J., et al. (2012). Assessing abnormal iron content in the deep gray matter of patients with multiple sclerosis versus healthy controls. *AJNR American Journal of Neuroradiology*, *33*(2), 252–258.
Hagemeier, J., Geurts, J. J., & Zivadinov, R. (2012). Brain iron accumulation in aging and neurodegenerative disorders. *Expert Review of Neurotherapeutics*, *12*(12), 1467–1480.
Haider, L., Fischer, M. T., Frischer, J. M., Bauer, J., Hoftberger, R., Botond, G., et al. (2011). Oxidative damage in multiple sclerosis lesions. *Brain*, *134*(Pt 7), 1914–1924.

Hajji, N., Calvert, C., Ritchie, C. W., & Sastre, M. (2013). The role of metals in Alzheimer's disease. In R. Ward, R. Crichton, & D. Dexter (Eds.), *Mechanisms and metal involvement in neurodegenerative diseases* (pp. 80–97). London, UK: The Royal Society of Chemistry.

Hallgren, B., & Sourander, P. (1958). The effect of age on the non-haemin iron in the human brain. *Journal of Neurochemistry*, *3*(1), 41–51.

Halliday, G. M., Ophof, A., Broe, M., Jensen, P. H., Kettle, E., Fedorow, H., et al. (2005). Alpha-synuclein redistributes to neuromelanin lipid in the substantia nigra early in Parkinson's disease. *Brain*, *128*(Pt 11), 2654–2664.

Halliwell, B. (2006). Oxidative stress and neurodegeneration: Where are we now? *Journal of Neurochemistry*, *97*(6), 1634–1658.

HaMai, D., & Bondy, S. C. (2004). Oxidative basis of manganese neurotoxicity. *Annals of the New York Academy of Sciences*, *1012*, 129–141.

Hametner, S., Wimmer, I., Haider, L., Pfeifenbring, S., Bruck, W., & Lassmann, H. (2013). Iron and neurodegeneration in the multiple sclerosis brain. *Annals of Neurology*.

Hardy, J., & Selkoe, D. J. (2002). The amyloid hypothesis of Alzheimer's disease: Progress and problems on the road to therapeutics. *Science*, *297*(5580), 353–356.

Hare, D., Ayton, S., Bush, A., & Lei, P. (2013). A delicate balance: Iron metabolism and diseases of the brain. *Frontiers in Aging Neuroscience*, *5*, 34.

Hashimoto, M., Takeda, A., Hsu, L. J., Takenouchi, T., & Masliah, E. (1999). Role of cytochrome c as a stimulator of alpha-synuclein aggregation in Lewy body disease. *Journal of Biological Chemistry*, *274*(41), 28849–28852.

Hauptmann, S., Scherping, I., Drose, S., Brandt, U., Schulz, K. L., Jendrach, M., et al. (2009). Mitochondrial dysfunction: An early event in Alzheimer pathology accumulates with age in AD transgenic mice. *Neurobiology of Aging*, *30*(10), 1574–1586.

Heli, H., Mirtorabi, S., & Karimian, K. (2011). Advances in iron chelation: An update. *Expert Opinion on Therapeutic Patents*, *21*(6), 819–856.

Henshall, T. L., Tucker, B., Lumsden, A. L., Nornes, S., Lardelli, M. T., & Richards, R. I. (2009). Selective neuronal requirement for huntingtin in the developing zebrafish. *Human Molecular Genetics*, *18*(24), 4830–4842.

Hirose, W., Ikematsu, K., & Tsuda, R. (2003). Age-associated increases in heme oxygenase-1 and ferritin immunoreactivity in the autopsied brain. *Legal Medicine (Tokyo, Japan)*, *5*(Suppl.1), S360–S366.

Hirsch, E. C., & Hunot, S. (2009). Neuroinflammation in Parkinson's disease: A target for neuroprotection? *Lancet Neurology*, *8*(4), 382–397.

Hirsch, E. C., Jenner, P., & Przedborski, S. (2013). Pathogenesis of Parkinson's disease. *Movement Disorders: Official Journal of the Movement Disorder Society*, *28*(1), 24–30.

Hogarth, P., Gregory, A., Kruer, M. C., Sanford, L., Wagoner, W., Natowicz, M. R., et al. (2013). New NBIA subtype: Genetic, clinical, pathologic, and radiographic features of MPAN. *Neurology*, *80*(3), 268–275.

Honda, K., Casadesus, G., Petersen, R. B., Perry, G., & Smith, M. A. (2004). Oxidative stress and redox-active iron in Alzheimer's disease. *Annals of the New York Academy of Sciences*, *1012*, 179–182.

Horowitz, M. P., & Greenamyre, J. T. (2010). Mitochondrial iron metabolism and its role in neurodegeneration. *Journal of Alzheimer's Disease*, *20*(Suppl 2), S551–S568.

House, E., Collingwood, J., Khan, A., Korchazkina, O., Berthon, G., & Exley, C. (2004). Aluminium, iron, zinc and copper influence the in vitro formation of amyloid fibrils of Abeta42 in a manner which may have consequences for metal chelation therapy in Alzheimer's disease. *Journal of Alzheimer's Disease*, *6*(3), 291–301.

House, E., Esiri, M., Forster, G., Ince, P. G., & Exley, C. (2012). Aluminium, iron and copper in human brain tissues donated to the Medical Research Council's Cognitive Function and Ageing Study. *Metallomics*, *4*(1), 56–65.

Huang, M. L., Becker, E. M., Whitnall, M., Suryo Rahmanto, Y., Ponka, P., & Richardson, D. R. (2009). Elucidation of the mechanism of mitochondrial iron loading in Friedreich's ataxia by analysis of a mouse mutant. *Proceedings of the National Academy of Sciences of the United States of America, 106*(38), 16381–16386.

Huang, X., Cuajungco, M. P., Atwood, C. S., Hartshorn, M. A., Tyndall, J. D., Hanson, G. R., et al. (1999). Cu(II) potentiation of alzheimer abeta neurotoxicity. Correlation with cell-free hydrogen peroxide production and metal reduction. *Journal of Biological Chemistry, 274*(52), 37111–37116.

Huang, C. C., Weng, Y. H., Lu, C. S., Chu, N. S., & Yen, T. C. (2003). Dopamine transporter binding in chronic manganese intoxication. *Journal of Neurology, 250*(11), 1335–1339.

Hulet, S. W., Powers, S., & Connor, J. R. (1999). Distribution of transferrin and ferritin binding in normal and multiple sclerotic human brains. *Journal of Neurological Sciences, 165*(1), 48–55.

Hung, Y. H., Bush, A. I., & Cherny, R. A. (2010). Copper in the brain and Alzheimer's disease. *Journal of Biological Inorganic Chemistry, 15*(1), 61–76.

Hwang, Y. M., Stathopulos, P. B., Dimmick, K., Yang, H., Badiei, H. R., Tong, M. S., et al. (2010). Nonamyloid aggregates arising from mature copper/zinc superoxide dismutases resemble those observed in amyotrophic lateral sclerosis. *Journal of Biological Chemistry, 285*(53), 41701–41711.

Ince, P. G., Highley, J. R., Kirby, J., Wharton, S. B., Takahashi, H., Strong, M. J., et al. (2011). Molecular pathology and genetic advances in amyotrophic lateral sclerosis: An emerging molecular pathway and the significance of glial pathology. *Acta Neuropathologica, 122*(6), 657–671.

James, S. A., Volitakis, I., Adlard, P. A., Duce, J. A., Masters, C. L., Cherny, R. A., et al. (2012). Elevated labile Cu is associated with oxidative pathology in Alzheimer disease. *Free Radical Biology & Medicine, 52*(2), 298–302.

Jellinger, K. A. (1999). The role of iron in neurodegeneration: Prospects for pharmacotherapy of Parkinson's disease. *Drugs & Aging, 14*(2), 115–140.

Jellinger, K. A. (2003). Apoptosis vs. nonapoptotic mechanisms in neurodegeneration. In P. L. Wood (Ed.), *Neuroinflammation. Mechanisms and management* (pp. 29–88). Totowa, NJ: Humana Press.

Jellinger, K. A. (2004). Infantile neuroaxonal dystrophy (Seitelberger disease). In J. A. Golden & B. N. Harding (Eds.), *Developmental neuropathology* (pp. 337–343). Basel: ISN Neuropath Press.

Jellinger, K. A. (2010). Basic mechanisms of neurodegeneration: A critical update. *Journal of Cellular and Molecular Medicine, 14*(3), 457–487.

Jellinger, K. A. (2012). Neuropathology of sporadic Parkinson's disease: Evaluation and changes of concepts. *Movement Disorders, 27*(1), 8–30.

Jellinger, K. A. (2013). Challenges in the neuropathological diagnosis of dementias. *International Journal of Neuropathology, 1*, 8–52.

Jellinger, K., Kienzl, E., Rumpelmair, G., Riederer, P., Stachelberger, H., Ben-Shachar, D., et al. (1992). Iron-melanin complex in substantia nigra of parkinsonian brains: An x-ray microanalysis. *Journal of Neurochemistry, 59*(3), 1168–1171.

Jeyaraju, D. V., Cisbani, G., & Pellegrini, L. (2009). Calcium regulation of mitochondria motility and morphology. *Biochimica et Biophysica Acta, 1787*(11), 1363–1373.

Jiang, D., Li, X., Williams, R., Patel, S., Men, L., Wang, Y., et al. (2009). Ternary complexes of iron, amyloid-beta, and nitrilotriacetic acid: Binding affinities, redox properties, and relevance to iron-induced oxidative stress in Alzheimer's disease. *Biochemistry, 48*(33), 7939–7947.

Johnson, C. J., Gilbert, P. U., Abrecht, M., Baldwin, K. L., Russell, R. E., Pedersen, J. A., et al. (2013). Low copper and high manganese levels in prion protein plaques. *Viruses, 5*(2), 654–662.

Jomova, K., & Valko, M. (2011). Advances in metal-induced oxidative stress and human disease. *Toxicology, 283*(2–3), 65–87.

Jomova, K., Vondrakova, D., Lawson, M., & Valko, M. (2010). Metals, oxidative stress and neurodegenerative disorders. *Molecular and Cellular Biochemistry, 345*(1–2), 91–104.

Karbowski, M., & Neutzner, A. (2012). Neurodegeneration as a consequence of failed mitochondrial maintenance. *Acta Neuropathologica, 123*(2), 157–171.

Kawahara, M. (2005). Effects of aluminum on the nervous system and its possible link with neurodegenerative diseases. *Journal of Alzheimer's Disease, 8*(2), 171–182, discussion 209–115.

Kawahara, M., & Kato-Negishi, M. (2011). Link between aluminum and the pathogenesis of Alzheimer's disease: The integration of the aluminum and amyloid cascade hypotheses. *International Journal of Alzheimer's Disease, 2011*, 276393.

Ke, Y., & Qian, Z. M. (2007). Brain iron metabolism: Neurobiology and neurochemistry. *Progress in Neurobiology, 83*(3), 149–173.

Kell, D. B. (2010). Towards a unifying, systems biology understanding of large-scale cellular death and destruction caused by poorly liganded iron: Parkinson's, Huntington's, Alzheimer's, prions, bactericides, chemical toxicology and others as examples. *Archives of Toxicology, 84*(11), 825–889.

Keogh, M. J., Jonas, P., Coulthard, A., Chinnery, P. F., & Burn, J. (2012). Neuroferritinopathy: A new inborn error of iron metabolism. *Neurogenetics, 13*(1), 93–96.

Khalid, M., Aoun, R. A., & Mathews, T. A. (2011). Altered striatal dopamine release following a sub-acute exposure to manganese. *Journal of Neuroscience Methods, 202*(2), 182–191.

Khalil, M., Langkammer, C., Ropele, S., Petrovic, K., Wallner-Blazek, M., Loitfelder, M., et al. (2011). Determinants of brain iron in multiple sclerosis: A quantitative 3T MRI study. *Neurology*, 77(18), 1691–1697.

Kienzl, E., Jellinger, K., Stachelberger, H., & Linert, W. (1999). Iron as catalyst for oxidative stress in the pathogenesis of Parkinson's disease?. *Life Sciences, 65*(18–19), 1973–1976.

Koeppen, A. H., Kuntzsch, E. C., Bjork, S. T., Ramirez, R., Mazurkiewicz, J. E., & Feustel, P. J. (2013). Friedreich ataxia: metal dysmetabolism in dorsal root ganglia. *Acta Neuropathologica Communications, 1*, 26. http://dx.doi.org/10.1186/2051-5960-1181-1126.

Kovacs, G. G., Botond, G., & Budka, H. (2010). Protein coding of neurodegenerative dementias: The neuropathological basis of biomarker diagnostics. *Acta Neuropathologica, 119*(4), 389–408.

Kozlowski, H., Janicka-Klos, A., Brasun, J., Gaggelli, E., Valensin, D., & Valensin, G. (2009). Copper, iron, and zinc ions homeostasis and their role in neurodegenerative disorders (metal uptake, transport, distribution and regulation). *Coordination Chemistry Reviews, 253*, 2665–2685.

Kruer, M. C., Hiken, M., Gregory, A., Malandrini, A., Clark, D., Hogarth, P., et al. (2011). Novel histopathologic findings in molecularly-confirmed pantothenate kinase-associated neurodegeneration. *Brain, 134*(Pt 4), 947–958.

Kumar, V., Bal, A., & Gill, K. D. (2009). Aluminium-induced oxidative DNA damage recognition and cell-cycle disruption in different regions of rat brain. *Toxicology, 264*(3), 137–144.

Kurisaki, H., Yomono, H., Murayama, S., & Hebisawa, A. (2002). Multiple system atrophy with a-/hypo-ceruloplasminemia: Distribution of iron in brains of 2 autopsy cases. *Rinshō Shinkeigaku, 42*(4), 293–298.

Langkammer, C., Krebs, N., Goessler, W., Scheurer, E., Ebner, F., Yen, K., et al. (2010). Quantitative MR imaging of brain iron: A postmortem validation study. *Radiology, 257*(2), 455–462.

Lassmann, H., Bruck, W., & Lucchinetti, C. F. (2007). The immunopathology of multiple sclerosis: An overview. *Brain Pathology, 17*(2), 210–218.

Lassmann, H., van Horssen, J., & Mahad, D. (2012). Progressive multiple sclerosis: Pathology and pathogenesis. *Nature Reviews Neurology, 8*(11), 647–656.

Lee, J. Y., Cho, E., Seo, J. W., Hwang, J. J., & Koh, J. Y. (2012). Alteration of the cerebral zinc pool in a mouse model of Alzheimer disease. *Journal of Neuropathology and Experimental Neurology*, *71*(3), 211–222.

Lehn, A., Boyle, R., Brown, H., Airey, C., & Mellick, G. (2012). Neuroferritinopathy. *Parkinsonism & Related Disorders*, *18*(8), 909–915.

Leskovjan, A. C., Kretlow, A., Lanzirotti, A., Barrea, R., Vogt, S., & Miller, L. M. (2011). Increased brain iron coincides with early plaque formation in a mouse model of Alzheimer's disease. *NeuroImage*, *55*(1), 32–38.

Levine, S. M., Bilgen, M., & Lynch, S. G. (2013). Iron accumulation in multiple sclerosis: An early pathogenic event. *Expert Review of Neurotherapeutics*, *13*(3), 247–250.

LeVine, S. M., & Chakrabarty, A. (2004). The role of iron in the pathogenesis of experimental allergic encephalomyelitis and multiple sclerosis. *Annals of the New York Academy of Sciences*, *1012*, 252–266.

Li, X., Jankovic, J., & Le, W. (2011). Iron chelation and neuroprotection in neurodegenerative diseases. *Journal of Neural Transmission*, *118*(3), 473–477.

Li, A., Paudel, R., Johnson, R., Courtney, R., Lees, A. J., Holton, J. L., et al. (2012). Pantothenate kinase-associated neurodegeneration is not a synucleinopathy. *Neuropathology and Applied Neurobiology*, *39*, 121–131.

Lin, M. T., & Beal, M. F. (2006). Mitochondrial dysfunction and oxidative stress in neurodegenerative diseases. *Nature*, *443*(7113), 787–795.

Linder, M. C. (2012). The relationship of copper to DNA damage and damage prevention in humans. *Mutation Research*, *733*(1–2), 83–91.

Linert, W., Herlinger, E., Jameson, R. F., Kienzl, E., Jellinger, K., & Youdim, M. B. (1996). Dopamine, 6-hydroxydopamine, iron, and dioxygen–their mutual interactions and possible implication in the development of Parkinson's disease. *Biochimica et Biophysica Acta*, *1316*(3), 160–168.

Liu, Y., & Connor, J. R. (2012). Iron and ER stress in neurodegenerative disease. *Biometals*, *25*(4), 837–845.

Liu, X., Sullivan, K. A., Madl, J. E., Legare, M., & Tjalkens, R. B. (2006). Manganese-induced neurotoxicity: The role of astroglial-derived nitric oxide in striatal interneuron degeneration. *Toxicological Sciences*, *91*(2), 521–531.

Liu, G., Zhang, C., Yin, J., Li, X., Cheng, F., Li, Y., et al. (2009). α-Synuclein is differentially expressed in mitochondria from different rat brain regions and dose-dependently down-regulates complex I activity. *Neuroscience Letters*, *454*(3), 187–192.

Loeb, V., Yakunin, E., Saada, A., & Sharon, R. (2010). The transgenic overexpression of alpha-synuclein and not its related pathology associates with complex I inhibition. *The Journal of Biological Chemistry*, *285*(10), 7334–7343.

Long-Smith, C. M., Sullivan, A. M., & Nolan, Y. M. (2009). The influence of microglia on the pathogenesis of Parkinson's disease. *Progress in Neurobiology*, *89*(3), 277–287.

Lopes, K. O., Sparks, D. L., & Streit, W. J. (2008). Microglial dystrophy in the aged and Alzheimer's disease brain is associated with ferritin immunoreactivity. *Glia*, *56*(10), 1048–1060.

Lothian, A., Hare, D. J., Grimm, R., Ryan, T. M., Masters, C. L., & Roberts, B. R. (2013). Metalloproteomics: principles, challenges and applications to neurodegeneration. *Frontiers in Aging Neuroscience*, *5*, 35.

Lovell, M. A., & Markesbery, W. R. (2006). Amyloid beta peptide, 4-hydroxynonenal and apoptosis. *Current Alzheimer Research*, *3*(4), 359–364.

Luk, K. C., Kehm, V., Carroll, J., Zhang, B., O'Brien, P., Trojanowski, J. Q., et al. (2012). Pathological alpha-synuclein transmission initiates Parkinson-like neurodegeneration in nontransgenic mice. *Science*, *338*(6109), 949–953.

Lumsden, A. L., Henshall, T. L., Dayan, S., Lardelli, M. T., & Richards, R. I. (2007). Huntingtin-deficient zebrafish exhibit defects in iron utilization and development. *Human Molecular Genetics*, *16*(16), 1905–1920.

Lutsenko, S., Bhattacharjee, A., & Hubbard, A. L. (2010). Copper handling machinery of the brain. *Metallomics, 2*(9), 596–608.
Lv, Z., Jiang, H., Xu, H., Song, N., & Xie, J. (2011). Increased iron levels correlate with the selective nigral dopaminergic neuron degeneration in Parkinson's disease. *Journal of Neural Transmission, 118*(3), 361–369.
Lyubartseva, G., & Lovell, M. A. (2012). A potential role for zinc alterations in the pathogenesis of Alzheimer's disease. *Biofactors, 38*(2), 98–106.
Mann, V. M., Cooper, J. M., Daniel, S. E., Srai, K., Jenner, P., Marsden, C. D., et al. (1994). Complex I, iron, and ferritin in Parkinson's disease substantia nigra. *Annals of Neurology, 36*(6), 876–881.
Manso, Y., Adlard, P. A., Carrasco, J., Vasak, M., & Hidalgo, J. (2011). Metallothionein and brain inflammation. *Journal of Biological Inorganic Chemistry, 16*(7), 1103–1113.
Martin, W. R., Wieler, M., & Gee, M. (2008). Midbrain iron content in early Parkinson disease: A potential biomarker of disease status. *Neurology, 70*(16 Pt 2), 1411–1417.
Massaad, C. A., & Klann, E. (2011). Reactive oxygen species in the regulation of synaptic plasticity and memory. *Antioxidants & Redox Signaling, 14*(10), 2013–2054.
Mastroberardino, P. G., Hoffman, E. K., Horowitz, M. P., Betarbet, R., Taylor, G., Cheng, D., et al. (2009). A novel transferrin/TfR2-mediated mitochondrial iron transport system is disrupted in Parkinson's disease. *Neurobiology of Disease, 34*(3), 417–431.
McNaught, K. S., Belizaire, R., Isacson, O., Jenner, P., & Olanow, C. W. (2003). Altered proteasomal function in sporadic Parkinson's disease. *Experimental Neurology, 179*(1), 38–46.
Mehta, V., Pei, W., Yang, G., Li, S., Swamy, E., Boster, A., et al. (2013). Iron is a sensitive biomarker for inflammation in multiple sclerosis lesions. *PLoS One, 8*(3), e57573.
Meloni, G., & Vasak, M. (2011). Redox activity of alpha-synuclein-Cu is silenced by Zn(7)-metallothionein-3. *Free Radical Biology & Medicine, 50*(11), 1471–1479.
Milatovic, D., Zaja-Milatovic, S., Gupta, R. C., Yu, Y., & Aschner, M. (2009). Oxidative damage and neurodegeneration in manganese-induced neurotoxicity. *Toxicology and Applied Pharmacology, 240*(2), 219–225.
Miller, Y., Ma, B., & Nussinov, R. (2010). Zinc ions promote Alzheimer Abeta aggregation via population shift of polymorphic states. *Proceedings of the National Academy of Sciences of the United States of America, 107*(21), 9490–9495.
Mills, E., Dong, X. P., Wang, F., & Xu, H. (2010). Mechanisms of brain iron transport: Insight into neurodegeneration and CNS disorders. *Future Medicinal Chemistry, 2*(1), 51–64.
Miu, A. C., & Benga, O. (2006). Aluminum and Alzheimer's disease: A new look. *Journal of Alzheimer's Disease, 10*(2–3), 179–201.
Miyajima, H., Adachi, J., Tatsuno, Y., Takahashi, Y., Fujimoto, M., Kaneko, E., et al. (1998). Increased very long-chain fatty acids in erythrocyte membranes of patients with aceruloplasminemia. *Neurology, 50*(1), 130–136.
Mizoroki, T., Meshitsuka, S., Maeda, S., Murayama, M., Sahara, N., & Takashima, A. (2007). Aluminum induces tau aggregation in vitro but not in vivo. *Journal of Alzheimer's Disease, 11*(4), 419–427.
Mocchegiani, E., Bertoni-Freddari, C., Marcellini, F., & Malavolta, M. (2005). Brain, aging and neurodegeneration: Role of zinc ion availability. *Progress in Neurobiology, 75*(6), 367–390.
Molina-Holgado, F., Hider, R. C., Gaeta, A., Williams, R., & Francis, P. (2007). Metals ions and neurodegeneration. *Biometals, 20*(3–4), 639–654.
Montine, K. S., Quinn, J. F., Zhang, J., Fessel, J. P., Roberts, L. J., 2nd., Morrow, J. D., et al. (2004). Isoprostanes and related products of lipid peroxidation in neurodegenerative diseases. *Chemistry and Physics of Lipids, 128*(1–2), 117–124.
Morris, D. R., & Levenson, C. W. (2012). Ion channels and zinc: Mechanisms of neurotoxicity and neurodegeneration. *Journal of Toxicology, 2012*, 785647.

Mot, A. I., Wedd, A. G., Sinclair, L., Brown, D. R., Collins, S. J., & Brazier, M. W. (2011). Metal attenuating therapies in neurodegenerative disease. *Expert Review of Neurotherapeutics*, *11*(12), 1717–1745.

Mounsey, R. B., & Teismann, P. (2012). Chelators in the treatment of iron accumulation in Parkinson's disease. *International Journal of Cell Biology*, *2012*, 983245.

Muhoberac, B. B., & Vidal, R. (2013). Abnormal iron homeostasis and neurodegeneration. *Frontiers in Aging Neuroscience*, *5*, 32.

Munch, G., Luth, H. J., Wong, A., Arendt, T., Hirsch, E., Ravid, R., et al. (2000). Crosslinking of alpha-synuclein by advanced glycation endproducts—An early pathophysiological step in Lewy body formation? *Journal of Chemical Neuroanatomy*, *20*(3–4), 253–257.

Nakahara, J., Maeda, M., Aiso, S., & Suzuki, N. (2012). Current concepts in multiple sclerosis: Autoimmunity versus oligodendrogliopathy. *Clinical Reviews in Allergy & Immunology*, *42*(1), 26–34.

Nubling, G., Bader, B., Levin, J., Hildebrandt, J., Kretzschmar, H., & Giese, A. (2012). Synergistic influence of phosphorylation and metal ions on tau oligomer formation and coaggregation with alpha-synuclein at the single molecule level. *Molecular Neurodegeneration*, 7, 35.

Nunez, M. T., Urrutia, P., Mena, N., Aguirre, P., Tapia, V., & Salazar, J. (2012). Iron toxicity in neurodegeneration. *Biometals*, *25*(4), 761–776.

Oakley, A. E., Collingwood, J. F., Dobson, J., Love, G., Perrott, H. R., Edwardson, J. A., et al. (2007). Individual dopaminergic neurons show raised iron levels in Parkinson disease. *Neurology*, *68*(21), 1820–1825.

Oettinghaus, B., Licci, M., Scorrano, L., & Frank, S. (2012). Less than perfect divorces: Dysregulated mitochondrial fission and neurodegeneration. *Acta Neuropathologica*, *123*(2), 189–203.

Olanow, C. W. (2004). Manganese-induced parkinsonism and Parkinson's disease. *Annals of the New York Academy of Sciences*, *1012*, 209–223.

Olivares, D., Huang, X., Branden, L., Greig, N. H., & Rogers, J. T. (2009). Physiological and pathological role of alpha-synuclein in Parkinson's disease through iron mediated oxidative stress; the role of a putative iron-responsive element. *International Journal of Molecular Sciences*, *10*(3), 1226–1260.

Orr, A. L., Li, S., Wang, C. E., Li, H., Wang, J., Rong, J., et al. (2008). N-terminal mutant huntingtin associates with mitochondria and impairs mitochondrial trafficking. *The Journal of Neuroscience*, *28*(11), 2783–2792.

Ortega, R., Cloetens, P., Deves, G., Carmona, A., & Bohic, S. (2007). Iron storage within dopamine neurovesicles revealed by chemical nano-imaging. *PLoS One*, *2*(9), e925.

Oteiza, P. I., Mackenzie, G. G., & Verstraeten, S. V. (2004). Metals in neurodegeneration: Involvement of oxidants and oxidant-sensitive transcription factors. *Molecular Aspects of Medicine*, *25*(1–2), 103–115.

Palmieri, B., & Sblendorio, V. (2007). Oxidative stress tests: Overview on reliability and use. Part I. *European Review for Medical and Pharmacological Sciences*, *11*(5), 309–342.

Pamphlett, R., & Kum Jew, S. (2013). Uptake of inorganic mercury by human locus ceruleus and corticomotor neurons: implications for amyotrophic lateral sclerosis. *Acta Neuropathologica Communications*, *1*, 13. http://dx.doi.org/10.1186/2051-5960-1181-1113.

Parihar, M. S., Parihar, A., Fujita, M., Hashimoto, M., & Ghafourifar, P. (2008). Mitochondrial association of alpha-synuclein causes oxidative stress. *Cellular and Molecular Life Sciences*, *65*(7–8), 1272–1284.

Pandolfo, M., & Pastore, A. (2009). The pathogenesis of Friedreich ataxia and the structure and function of frataxin. *Journal of Neurology*, *256*(Suppl 1), 9–17.

Parker, S. J., Koistinaho, J., White, A. R., & Kanninen, K. M. (2013). Biometals in rare neurodegenerative disorders of childhood. *Frontiers in Aging Neuroscience*, *5*, 14. http://dx.doi.org/10.3389/fnagi.2013.00014.

Pastore, A. (2013). Mechanisms and metal involvement in neurodegenerative diseases. In R. Ward, R. Crichton, & D. Dexter (Eds.), *Mechanisms and metal involvement in neurodegenerative diseases* (pp. 98–117). London, UK: The Royal Society of Chemistry.

Peneder, T. M., Scholze, P., Berger, M. L., Reither, H., Heinze, G., Bertl, J., et al. (2011). Chronic exposure to manganese decreases striatal dopamine turnover in human alpha-synuclein transgenic mice. *Neuroscience, 180*, 280–292.

Percy, M. E., Kruck, T. P., Pogue, A. I., & Lukiw, W. J. (2011). Towards the prevention of potential aluminum toxic effects and an effective treatment for Alzheimer's disease. *Journal of Inorganic Biochemistry, 105*(11), 1505–1512.

Perl, D. P., & Moalem, S. (2006). Aluminum and Alzheimer's disease, a personal perspective after 25 years. *Journal of Alzheimer's Disease, 9*(3 Suppl), 291–300.

Pivovarova, N. B., & Andrews, S. B. (2010). Calcium-dependent mitochondrial function and dysfunction in neurons. *The FEBS Journal, 277*(18), 3622–3636.

Post, J. I., Eibl, J. K., & Ross, G. M. (2008). Zinc induces motor neuron death via a selective inhibition of brain-derived neurotrophic factor activity. *Amyotrophic Lateral Sclerosis, 9*(3), 149–155.

Prabhakaran, K., Chapman, G. D., & Gunasekar, P. G. (2011). alpha-Synuclein overexpression enhances manganese-induced neurotoxicity through the NF-kappaB-mediated pathway. *Toxicology Mechanisms and Methods, 21*(6), 435–443.

Prabhakaran, K., Ghosh, D., Chapman, G. D., & Gunasekar, P. G. (2008). Molecular mechanism of manganese exposure-induced dopaminergic toxicity. *Brain Research Bulletin, 76*(4), 361–367.

Przedborski, S. (2010). Inflammation and Parkinson's disease pathogenesis. *Movement Disorders, 25*(Suppl 1), S55–S57.

Ramalingam, M., & Kim, S. J. (2012). Reactive oxygen/nitrogen species and their functional correlations in neurodegenerative diseases. *Journal of Neural Transmission, 119*(8), 891–910.

Reusche, E., Koch, V., Lindner, B., Harrison, A. P., & Friedrich, H. J. (2001). Alzheimer morphology is not increased in dialysis-associated encephalopathy and long-term hemodialysis. *Acta Neuropathologica, 101*(3), 211–216.

Reynolds, R., & Crichton, R. (2013). Multiple sclerosis. In R. Ward, R. Crichton, & D. Dexter (Eds.), *Mechanisms and metal involvement in neurodegenerative diseases* (pp. 163–185). London, UK: The Royal Society of Chemistry.

Riederer, P., Sofic, E., Rausch, W. D., Schmidt, B., Reynolds, G. P., Jellinger, K., et al. (1989). Transition metals, ferritin, glutathione, and ascorbic acid in parkinsonian brains. *Journal of Neurochemistry, 52*(2), 515–520.

Rivera-Mancia, S., Perez-Neri, I., Rios, C., Tristan-Lopez, L., Rivera-Espinosa, L., & Montes, S. (2010). The transition metals copper and iron in neurodegenerative diseases. *Chemico-Biological Interactions, 186*(2), 184–199.

Roberts, B. R., Tainer, J. A., Getzoff, E. D., Malencik, D. A., Anderson, S. R., Bomben, V. C., et al. (2007). Structural characterization of zinc-deficient human superoxide dismutase and implications for ALS. *Journal of Molecular Biology, 373*(4), 877–890.

Rodella, L. F., Ricci, F., Borsani, E., Stacchiotti, A., Foglio, E., Favero, G., et al. (2008). Aluminium exposure induces Alzheimer's disease-like histopathological alterations in mouse brain. *Histology and Histopathology, 23*(4), 433–439.

Roos, P. M., Vesterberg, O., & Nordberg, M. (2006). Metals in motor neuron diseases. *Experimental Biology and Medicine (Maywood, NJ), 231*(9), 1481–1487.

Ropele, S., de Graaf, W., Khalil, M., Wattjes, M. P., Langkammer, C., Rocca, M. A., et al. (2011). MRI assessment of iron deposition in multiple sclerosis. *Journal of Magnetic Resonance Imaging, 34*(1), 13–21.

Rosas, H. D., Chen, Y. I., Doros, G., Salat, D. H., Chen, N. K., Kwong, K. K., et al. (2012). Alterations in brain transition metals in Huntington disease: An evolving and intricate story. *Archives of Neurology, 69*(7), 887–893.

Roth, J. A. (2009). Are there common biochemical and molecular mechanisms controlling manganism and parkisonism. *Neuromolecular Medicine*, *11*(4), 281–296.

Rothstein, J. D. (2009). Current hypotheses for the underlying biology of amyotrophic lateral sclerosis. *Annals of Neurology*, *65*(Suppl 1), S3–S9.

Rouault, T. A. (2013). Iron metabolism in the CNS: implications for neurodegenerative diseases. *Nature Reviews Neuroscience*, *14*(8), 551–564.

Rowinska-Zyrek, M., Valensin, D., Luczkowski, M., & Kozlowski, H. (2013). Mechanisms and metal involvement in neurodegenerative diseases. In R. Ward, R. Crichton, & D. Dexter (Eds.), *Mechanisms and metal involvement in neurodegenerative diseases* (pp. 118–162). London, UK: The Royal Society of Chemistry.

Ruiperez, V., Darios, F., & Davletov, B. (2010). Alpha-synuclein, lipids and Parkinson's disease. *Progress in Lipid Research*, *49*(4), 420–428.

Salazar, J., Mena, N., Hunot, S., Prigent, A., Alvarez-Fischer, D., Arredondo, M., et al. (2008). Divalent metal transporter 1 (DMT1) contributes to neurodegeneration in animal models of Parkinson's disease. *Proceedings of the National Academy of Sciences of the United States of America*, *105*(47), 18578–18583.

Salvador, G. A. (2010). Iron in neuronal function and dysfunction. *Biofactors*, *36*(2), 103–110.

SanMartin, C. D., Paula-Lima, A. C., Hidalgo, C., & Nunez, M. T. (2012). Sub-lethal levels of amyloid beta-peptide oligomers decrease non-transferrin-bound iron uptake and do not potentiate iron toxicity in primary hippocampal neurons. *Biometals*, *25*(4), 805–813.

Santner, A., & Uversky, V. N. (2010). Metalloproteomics and metal toxicology of alpha-synuclein. *Metallomics*, *2*(6), 378–392.

Santos, R., Lefevre, S., Sliwa, D., Seguin, A., Camadro, J. M., & Lesuisse, E. (2010). Friedreich ataxia: Molecular mechanisms, redox considerations, and therapeutic opportunities. *Antioxidants & Redox Signaling*, *13*(5), 651–690.

Savelieff, M. G., Lee, S., Liu, Y., & Lim, M. H. (2013). Untangling amyloid-beta, tau, and metals in Alzheimer's disease. *ACS Chemical Biology*, *8*(5), 856–865.

Savory, J., Herman, M. M., & Ghribi, O. (2006). Mechanisms of aluminum-induced neurodegeneration in animals: Implications for Alzheimer's disease. *Journal of Alzheimer's Disease*, *10*(2–3), 135–144.

Schipper, H. M. (2012). Neurodegeneration with brain iron accumulation—Clinical syndromes and neuroimaging. *Biochimica et Biophysica Acta*, *1822*(3), 350–360.

Schipper, H. M., Song, W., Zukor, H., Hascalovici, J. R., & Zeligman, D. (2009). Heme oxygenase-1 and neurodegeneration: Expanding frontiers of engagement. *Journal of Neurochemistry*, *110*(2), 469–485.

Schneider, S. A., Hardy, J., & Bhatia, K. P. (2012). Syndromes of neurodegeneration with brain iron accumulation (NBIA): An update on clinical presentations, histological and genetic underpinnings, and treatment considerations. *Movement Disorders*, *27*(1), 42–53.

Schneider, S. A., Zorzi, G., & Nardocci, N. (2013). Pathophysiology and treatment of neurodegeneration with brain iron accumulation in the pediatric population. *Current Treatment Options in Neurology*, *15*(5), 652–667.

Schon, E. A., & Przedborski, S. (2011). Mitochondria: The next (neurode)generation. *Neuron*, *70*(6), 1033–1053.

Schrag, M., Mueller, C., Oyoyo, U., Smith, M. A., & Kirsch, W. M. (2011). Iron, zinc and copper in the Alzheimer's disease brain: A quantitative meta-analysis. Some insight on the influence of citation bias on scientific opinion. *Progress in Neurobiology*, *94*(3), 296–306.

Seguin, A., Sutak, R., Bulteau, A. L., Garcia-Serres, R., Oddou, J. L., Lefevre, S., et al. (2010). Evidence that yeast frataxin is not an iron storage protein in vivo. *Biochimica et Biophysica Acta*, *1802*(6), 531–538.

Shamoto-Nagai, M., Maruyama, W., Yi, H., Akao, Y., Tribl, F., Gerlach, M., et al. (2006). Neuromelanin induces oxidative stress in mitochondria through release of iron: Mechanism behind the inhibition of 26S proteasome. *Journal of Neural Transmission*, *113*(5), 633–644.

Shavali, S., Brown-Borg, H. M., Ebadi, M., & Porter, J. (2008). Mitochondrial localization of alpha-synuclein protein in alpha-synuclein overexpressing cells. *Neuroscience Letters*, *439*(2), 125–128.

Shcherbatykh, I., & Carpenter, D. O. (2007). The role of metals in the etiology of Alzheimer's disease. *Journal of Alzheimer's Disease*, *11*(2), 191–205.

Sheline, C. T., Zhu, J., Zhang, W., Shi, C., & Cai, A. L. (2013). Mitochondrial inhibitor models of Huntington's disease and Parkinson's disease induce zinc accumulation and are attenuated by inhibition of zinc neurotoxicity in vitro or in vivo. *Neurodegenerative Diseases*, *11*(1), 49–58.

Shimanouchi, T., Onishi, R., Kitaura, N., Umakoshi, H., & Kuboi, R. (2011). Copper-mediated growth of amyloid beta fibrils in the presence of oxidized and negatively charged liposomes. *Journal of Bioscience and Bioengineering*, *112*(6), 611–615.

Shin, R. W. (1997). Interaction of aluminum with paired helical filament tau is involved in neurofibrillary pathology of Alzheimer's disease. *Gerontology*, *43*(Suppl 1), 16–23.

Sian-Hulsmann, J., Mandel, S., Youdim, M. B., & Riederer, P. (2011). The relevance of iron in the pathogenesis of Parkinson's disease. *Journal of Neurochemistry*, *118*(6), 939–957.

Singh, B. K., Kumar, A., Ahmad, I., Kumar, V., Patel, D. K., Jain, S. K., et al. (2011). Oxidative stress in zinc-induced dopaminergic neurodegeneration: Implications of superoxide dismutase and heme oxygenase-1. *Free Radical Research*, *45*(10), 1207–1222.

Singh, N., Haldar, S., Tripathi, A. K., McElwee, M., Horback, K., Anbalagan, C., et al. (2013). Iron-mediated protein aggregation in prion disorders and Parkinson's disease. *Antioxid Redox Signal*, in print.

Smith, M. A., Casadesus, G., Joseph, J. A., & Perry, G. (2002). Amyloid-beta and tau serve antioxidant functions in the aging and Alzheimer brain. *Free Radical Biology & Medicine*, *33*(9), 1194–1199.

Smith, A. P., & Lee, N. M. (2007). Role of zinc in ALS. *Amyotrophic Lateral Sclerosis*, *8*(3), 131–143.

Smith, M. A., Zhu, X., Tabaton, M., Liu, G., McKeel, D. W., Jr., Cohen, M. L., et al. (2010). Increased iron and free radical generation in preclinical Alzheimer disease and mild cognitive impairment. *Journal of Alzheimer's Disease*, *19*(1), 363–372.

Spadoni, F., Stefani, A., Morello, M., Lavaroni, F., Giacomini, P., & Sancesario, G. (2000). Selective vulnerability of pallidal neurons in the early phases of manganese intoxication. *Experimental Brain Research*, *135*(4), 544–551.

Spillantini, M. G., Schmidt, M. L., Lee, V. M., Trojanowski, J. Q., Jakes, R., & Goedert, M. (1997). Alpha-synuclein in Lewy bodies. *Nature*, *388*(6645), 839–840.

Squitti, R. (2012a). Copper dysfunction in Alzheimer's disease: From meta-analysis of biochemical studies to new insight into genetics. *Journal of Trace Elements in Medicine and Biology*, *26*(2–3), 93–96.

Squitti, R. (2012b). Metals in Alzheimer's disease: A systemic perspective. *Frontiers in Bioscience*, *17*, 451–472.

Squitti, R., & Polimanti, R. (2013). Copper phenotype in Alzheimer's disease: Dissecting the pathway. *American Journal of Neurodegenerative Disease*, *2*(2), 46–56.

Stankiewicz, J., Panter, S. S., Neema, M., Arora, A., Batt, C. E., & Bakshi, R. (2007). Iron in chronic brain disorders: Imaging and neurotherapeutic implications. *Neurotherapeutics*, *4*(3), 371–386.

Sultana, R., & Butterfield, D. A. (2010). Role of oxidative stress in the progression of Alzheimer's disease. *Journal of Alzheimer's Disease*, *19*(1), 341–353.

Taba, P. (2013). Metals and movement disorders. *Current Opinion in Neurology*, *26*(4), 435–441.

Tabner, B. J., El-Agnaf, O. M., German, M. J., Fullwood, N. J., & Allsop, D. (2005). Protein aggregation, metals and oxidative stress in neurodegenerative diseases. *Biochemical Society Transactions*, *33*(Pt 5), 1082–1086.

Takanashi, M., Mochizuki, H., Yokomizo, K., Hattori, N., Mori, H., Yamamura, Y., et al. (2001). Iron accumulation in the substantia nigra of autosomal recessive juvenile parkinsonism (ARJP). *Parkinsonism & Related Disorders*, 7(4), 311–314.

Tamagno, E., Guglielmotto, M., Aragno, M., Borghi, R., Autelli, R., Giliberto, L., et al. (2008). Oxidative stress activates a positive feedback between the gamma- and beta-secretase cleavages of the beta-amyloid precursor protein. *Journal of Neurochemistry*, *104*(3), 683–695.

Telpoukhovskaia, M. A., & Orvig, C. (2013). Werner coordination chemistry and neurodegeneration. *Chemical Society Reviews*, *42*(4), 1836–1846.

Thomas, M., & Jankovic, J. (2004). Neurodegenerative disease and iron storage in the brain. *Current Opinion in Neurology*, *17*(4), 437–442.

Thompson, K. J., Shoham, S., & Connor, J. R. (2001). Iron and neurodegenerative disorders. *Brain Research Bulletin*, *55*(2), 155–164.

Tiffany-Castiglioni, E., Hong, S., & Qian, Y. (2011). Copper handling by astrocytes: Insights into neurodegenerative diseases. *International Journal of Developmental Neuroscience*, *29*(8), 811–818.

Tougu, V., Karafin, A., Zovo, K., Chung, R. S., Howells, C., West, A. K., et al. (2009). Zn(II)- and Cu(II)-induced non-fibrillar aggregates of amyloid-beta (1–42) peptide are transformed to amyloid fibrils, both spontaneously and under the influence of metal chelators. *Journal of Neurochemistry*, *110*(6), 1784–1795.

Trushina, E., Dyer, R. B., Badger, J. D., 2nd., Ure, D., Eide, L., Tran, D. D., et al. (2004). Mutant huntingtin impairs axonal trafficking in mammalian neurons in vivo and in vitro. *Molecular and Cellular Biology*, *24*(18), 8195–8209.

Uversky, V. N. (2007). Neuropathology, biochemistry, and biophysics of alpha-synuclein aggregation. *Journal of Neurochemistry*, *103*(1), 17–37.

Valiente-Gabioud, A. A., Torres-Monserrat, V., Molina-Rubino, L., Binolfi, A., Griesinger, C., & Fernandez, C. O. (2012). Structural basis behind the interaction of Zn(2)(+) with the protein alpha-synuclein and the Abeta peptide: A comparative analysis. *Journal of Inorganic Biochemistry*, *117*, 334–341.

Valko, M., Morris, H., & Cronin, M. T. (2005). Metals, toxicity and oxidative stress. *Current Medicinal Chemistry*, *12*(10), 1161–1208.

Verina, T., Kiihl, S. F., Schneider, J. S., & Guilarte, T. R. (2011). Manganese exposure induces microglia activation and dystrophy in the substantia nigra of non-human primates. *Neurotoxicology*, *32*(2), 215–226.

Vidal, R., Delisle, M. B., Rascol, o., & Ghetti, B. (2011). Hereditary ferritinopathies. In D. W. Dickson & R. O. Weller (Eds.), *Neurodegeneration: The molecular pathology of dementia and movement disorders* (2nd ed., pp. 461–466). Oxford: Blackwell.

Vonk, W. I., & Klomp, L. W. (2008). Role of transition metals in the pathogenesis of amyotrophic lateral sclerosis. *Biochemical Society Transactions*, *36*(Pt 6), 1322–1328.

Wallis, L. I., Paley, M. N., Graham, J. M., Grunewald, R. A., Wignall, E. L., Joy, H. M., et al. (2008). MRI assessment of basal ganglia iron deposition in Parkinson's disease. *Journal of Magnetic Resonance Imaging*, *28*(5), 1061–1067.

Walsh, A. J., Lebel, R. M., Eissa, A., Blevins, G., Catz, I., Lu, J. Q., et al. (2013). Multiple sclerosis: Validation of MR imaging for quantification and detection of iron. *Radiology*, *267*, 531–542.

Walton, J. R. (2010). Evidence for participation of aluminum in neurofibrillary tangle formation and growth in Alzheimer's disease. *Journal of Alzheimer's Disease*, *22*(1), 65–72.

Walton, J. R. (2012). Aluminum disruption of calcium homeostasis and signal transduction resembles change that occurs in aging and Alzheimer's disease. *Journal of Alzheimer's Disease*, *29*(2), 255–273.

Walton, J. R. (2013). Aluminum involvement in the progression of Alzheimer's disease. *Journal of Alzheimer's Disease, 35*(1), 7–43.
Walton, J. R., & Wang, M. X. (2009). APP expression, distribution and accumulation are altered by aluminum in a rodent model for Alzheimer's disease. *Journal of Inorganic Biochemistry, 103*(11), 1548–1554.
Wang, C. Y., Wang, T., Zheng, W., Zhao, B. L., Danscher, G., Chen, Y. H., et al. (2010). Zinc overload enhances APP cleavage and Abeta deposition in the Alzheimer mouse brain. *PLoS One, 5*(12), e15349.
Ward, R., Crichton, R., & Dexter, D. (Eds.), (2013). *Mechanisms and metal involvement in neurodegenerative diseases*. London, UK: The Royal Society of Chemistry.
Watt, N. T., Taylor, D. R., Kerrigan, T. L., Griffiths, H. H., Rushworth, J. V., Whitehouse, I. J., et al. (2012). Prion protein facilitates uptake of zinc into neuronal cells. *Nature Communications, 3*, 1134.
Wegrzynowicz, M., Holt, H. K., Friedman, D. B., & Bowman, A. B. (2012). Changes in the striatal proteome of YAC128Q mice exhibit gene-environment interactions between mutant huntingtin and manganese. *Journal of Proteome Research, 11*(2), 1118–1132.
White, A. R., Barnham, K. J., Huang, X., Voltakis, I., Beyreuther, K., Masters, C. L., et al. (2004). Iron inhibits neurotoxicity induced by trace copper and biological reductants. *Journal of Biological Inorganic Chemistry, 9*(3), 269–280.
Williams, R., Rohr, A. M., Wang, W. T., Choi, I. Y., Lee, P., Berman, N. E., et al. (2011). Iron deposition is independent of cellular inflammation in a cerebral model of multiple sclerosis. *BMC Neuroscience, 12*, 59.
Wu, Z., Du, Y., Xue, H., Wu, Y., & Zhou, B. (2012). Aluminum induces neurodegeneration and its toxicity arises from increased iron accumulation and reactive oxygen species (ROS) production. *Neurobiology of Aging33*(1), 199 e191-112.
Wypijewska, A., Galazka-Friedman, J., Bauminger, E. R., Wszolek, Z. K., Schweitzer, K. J., Dickson, D. W., et al. (2010). Iron and reactive oxygen species activity in parkinsonian substantia nigra. *Parkinsonism & Related Disorders, 16*(5), 329–333.
Yokel, R. A. (2006). Blood–brain barrier flux of aluminum, manganese, iron and other metals suspected to contribute to metal-induced neurodegeneration. *Journal of Alzheimer's Disease, 10*(2–3), 223–253.
Youdim, M. B. (2008). Brain iron deficiency and excess; cognitive impairment and neurodegeneration with involvement of striatum and hippocampus. *Neurotoxicity Research, 14*(1), 45–56.
Yumoto, S., Kakimi, S., Ohsaki, A., & Ishikawa, A. (2009). Demonstration of aluminum in amyloid fibers in the cores of senile plaques in the brains of patients with Alzheimer's disease. *Journal of Inorganic Biochemistry, 103*(11), 1579–1584.
Zatta, P., Lucchini, R., van Rensburg, S. J., & Taylor, A. (2003). The role of metals in neurodegenerative processes: Aluminum, manganese, and zinc. *Brain Research Bulletin, 62*(1), 15–28.
Zecca, L., Youdim, M. B., Riederer, P., Connor, J. R., & Crichton, R. R. (2004). Iron, brain ageing and neurodegenerative disorders. *Nature Reviews Neuroscience, 5*(11), 863–873.
Zhang, X., Haaf, M., Todorich, B., Grosstephan, E., Schieremberg, H., Surguladze, N., et al. (2005). Cytokine toxicity to oligodendrocyte precursors is mediated by iron. *Glia, 52*(3), 199–208.
Zhang, Q. L., Jia, L., Jiao, X., Guo, W. L., Ji, J. W., Yang, H. L., et al. (2012). APP/PS1 transgenic mice treated with aluminum: An update of Alzheimer's disease model. *International Journal of Immunopathology and Pharmacology, 25*(1), 49–58.
Zheng, W., & Monnot, A. D. (2012). Regulation of brain iron and copper homeostasis by brain barrier systems: Implication in neurodegenerative diseases. *Pharmacology and Therapeutics, 133*(2), 177–188.

Zhou, Z. D., Lan, Y. H., Tan, E. K., & Lim, T. M. (2010). Iron species-mediated dopamine oxidation, proteasome inhibition, and dopaminergic cell demise: Implications for iron-related dopaminergic neuron degeneration. *Free Radical Biology & Medicine, 49*(12), 1856–1871.

Zhu, X., Raina, A. K., Lee, H. G., Casadesus, G., Smith, M. A., & Perry, G. (2004). Oxidative stress signalling in Alzheimer's disease. *Brain Research, 1000*(1–2), 32–39.

Zivadinov, R., Weinstock-Guttman, B., & Pirko, I. (2011). Iron deposition and inflammation in multiple sclerosis. Which one comes first? *BMC Neuroscience, 12*, 60.

Zucca, F. A., Bellei, C., Giannelli, S., Terreni, M. R., Gallorini, M., Rizzio, E., et al. (2006). Neuromelanin and iron in human locus coeruleus and substantia nigra during aging: Consequences for neuronal vulnerability. *Journal of Neural Transmission, 113*(6), 757–767.

Zucconi, G. G., Cipriani, S., Scattoni, R., Balgkouranidou, I., Hawkins, D. P., & Ragnarsdottir, K. V. (2007). Copper deficiency elicits glial and neuronal response typical of neurodegenerative disorders. *Neuropathology and Applied Neurobiology, 33*(2), 212–225.

CHAPTER TWO

Pantothenate Kinase-Associated Neurodegeneration (PKAN) and *PLA2G6*-Associated Neurodegeneration (PLAN): Review of Two Major Neurodegeneration with Brain Iron Accumulation (NBIA) Phenotypes

Manju A. Kurian[*,†,1], **Susan J. Hayflick**[‡,§,¶]
[*]Neurosciences Unit, UCL-Institute of Child Health, London, United Kingdom
[†]Department of Neurology, Great Ormond Street Hospital, London, United Kingdom
[‡]Department of Molecular & Medical Genetics, Oregon Health & Science University, Portland, OR, USA
[§]Department of Pediatrics, Oregon Health & Science University, Portland, OR, USA
[¶]Department of Neurology, Oregon Health & Science University, Portland, OR, USA
[1]Corresponding author: e-mail address: manju.kurian@ucl.ac.uk

Contents

International Review of Neurobiology, Volume 110
ISSN 0074-7742
http://dx.doi.org/10.1016/B978-0-12-410502-7.00003-X

Abstract

Neurodegeneration with brain iron accumulation (NBIA) comprises a heterogeneous group of disorders characterized by the presence of radiologically discernible high brain iron, particularly within the basal ganglia. A number of childhood NBIA syndromes are described, of which two of the major subtypes are pantothenate kinase-associated neurodegeneration (PKAN) and *PLA2G6*-associated neurodegeneration (PLAN). PKAN and PLAN are autosomal recessive NBIA disorders due to mutations in *PANK2* and *PLA2G6*, respectively. Presentation is usually in childhood, with features of neurological regression and motor dysfunction. In both PKAN and PLAN, a number of classical and atypical phenotypes are reported. In this chapter, we describe the clinical, radiological, and genetic features of these two disorders and also discuss the pathophysiological mechanisms postulated to play a role in disease pathogenesis.

1. INTRODUCTION

Neurodegeneration with brain iron accumulation (NBIA) comprises a heterogeneous group of disorders characterized clinically by progressive motor dysfunction, with evidence of radiologically discernible brain iron, particularly within the basal ganglia. A number of NBIA phenotypes are reported (Gregory & Hayflick, 2013), including pantothenate kinase-associated neurodegeneration (PKAN) (Gregory & Hayflick, 2002), *PLA2G6*-associated neurodegeneration (PLAN) (Gregory, Kurian, Maher, Hogarth, & Hayflick, 2008), mitochondrial membrane protein-associated neurodegeneration (MPAN), and the newly described beta propeller-associated neurodegeneration (BPAN). Most of the major NBIA subtypes of childhood (PKAN/PLAN/MPAN) are autosomal recessive disorders, except BPAN, which shows an X-linked pattern of inheritance. In this overview, we will focus on the clinical phenotypes, radiological features, pathological findings, disease mechanisms, and management strategies in PKAN and PLAN. Together, these two NBIA subtypes account for approximately two-thirds of patients with childhood onset disease. Increasingly, adult phenotypes caused by these genes are recognized (Schneider & Bhatia, 2010).

2. PANTOTHENATE KINASE-ASSOCIATED NEURODEGENERATION

First described as a clinical syndrome in 1924 (Hallervorden, 1924), PKAN was originally known as Hallervorden–Spatz syndrome (HSS)

(Dooling, Schoene, & Richardson, 1974), but in view of the unethical activities of these German neuropathologists before and during the second World War (Shevell, 2003), the syndrome was renamed PKAN, the new acronym clearly reflecting the underlying genetic cause of this disease. PKAN is a major form of NBIA accounting for approximately 50% of childhood NBIA. It has an estimated prevalence of 1–3/million (Gregory, Polster, & Hayflick, 2009) with a general population carrier frequency of 1/275–500. PKAN is typically characterized by the onset of progressive neurological symptoms in association with a characteristic pattern of basal ganglia iron deposition. The allelic disorder, hypobetalipoproteinaemia, acanthocytosis, retinitis pigmentosa, and pallidal degeneration (HARP) syndrome, is also caused by *PANK2* mutations and is considered part of the PKAN phenotypic spectrum (Ching, Westaway, Gitschier, Higgins, & Hayflick, 2002; Gregory et al., 2009; Houlden et al., 2003).

2.1. Clinical presentation

2.1.1 Onset of clinical symptoms

Classic PKAN: 75% of PKAN cases have a classic phenotype (childhood onset disease and faster disease progression), whereas the remainder of patients present with "atypical PKAN" (disease onset in the second or third decade of life, slower progression of disease). Classical PKAN will usually clinically present before age 6 years (mean age 3.4 years) in the majority of patients, but there is some variability (6 months–12 years). Some affected children often have a history of nonspecific features prior to presentation, including clumsiness, dyspraxia, and motor/global neurodevelopmental delay (Gregory & Hayflick, 2002). Early clinical features are commonly gait abnormalities and postural instability, which result from a combination of lower-limb spasticity, dystonia, and rigidity. Visual symptoms may be the presenting feature of PKAN. Toe walking and upper body dystonia are less common presenting signs.

Atypical PKAN: In contrast, patients with atypical PKAN present at a mean age of 14 years (range 1–28 years) (Hartig et al., 2006; Hayflick et al., 2003), often with speech difficulties (palilalia or dysarthria), mild gait disturbance with subtle dystonia, or neuropsychiatric features.

2.1.2 Clinical disease features

Classic PKAN: The clinical features of classic PKAN are remarkably homogeneous between patients. In classical PKAN, dystonia is always reported and it is the most common extrapyramidal feature. Dystonia can affect any part of the body but is often very prominent in the limbs and face. The majority of PKAN patients have oromandibular dystonia and dysarthria (Hartig et al., 2006). Cranial dystonia may lead to recurrent tongue trauma, in some cases requiring full-mouth dental extraction (Gregory & Hayflick, 2002). Limb dystonia can lead to long bone fracture, where extreme bone stress (from severe dystonia) and osteopenia (from reduced mobility) are contributory risk factors. Other motor phenotypes such as choreoathetosis, rigidity, and parkinsonism are rarely described. Pyramidal tract features are also reported, resulting in upper motor neuron signs of hypertonicity, hyperreflexia, and spasticity. Seizures are rarely reported in this group of patients.

Two-thirds of patients with classical PKAN (Hayflick et al., 2003) demonstrate a pigmentary retinopathy, which can lead to significant visual impairment (Hayflick et al., 2003). Clinical symptoms include nyctalopia with subsequent progressive loss of peripheral visual fields and sometimes eventual blindness. Abnormal vertical saccades and saccadic pursuits are reported. Sectoral iris paralysis and partial loss of the pupillary ruff in keeping with bilateral Adie's pupil (Egan et al., 2005) are also reported. Early changes on funduscopic examination include a flecked appearance to the retina, and later in the disease course, there is evidence of bone spicule formation, prominent choroidal vasculature, and "bull's-eye" annular maculopathy. The pigmentary retinopathy is thought to occur early in the disease course and requires a full ophthalmologic diagnostic evaluation including electroretinogram (ERG) and visual field testing for accurate diagnosis. Of note, individuals with a normal ophthalmologic examination at initial diagnosis generally tend not to develop retinopathy later.

As well as motor and visual symptoms, patients also develop cognition dysfunction with progressive disease, although there is much variability in severity (Freeman et al., 2007). Age of PKAN disease onset had a strong inverse correlation with intellectual impairment (Freeman et al., 2007). More recent data suggest that due to the practical difficulties in performing cognitive testing in those with PKAN (because the severity of their motor impairments limits what cognitive assessments can be performed), cognitive decline may be overestimated in PKAN patients (Mahoney, Selway, & Lin, 2011). Acanthocytosis is evident in 8% of affected individuals.

Atypical PKAN: There are clearly defined differences in the main clinical features observed in typical PKAN compared to atypical PKAN. In atypical

PKAN, one-third of patients have neuropsychiatric manifestations including behavioral difficulties and vocal and motor tics (Scarano, Pellecchia, Filla, & Barone, 2002), obsessions, obsessive–compulsive disorder, aggression, change in personality, emotional lability, impulsivity, depression, a frontotemporal-like dementia early in their disease (Pellecchia et al., 2005), and, rarely, psychotic symptoms (del Valle-López, Pérez-García, Sanguino-Andrés, & González-Pablos, 2011; Pellecchia et al., 2005). Speech abnormalities are common and include palilalia, tachylalia/tachylogia, and dysarthria (Benke & Butterworth, 2001; Benke, Hohenstein, Poewe, & Butterworth, 2000). Like in classical PKAN, many atypical PKAN patients develop motor features later in their disease course. Indeed, retrospective history taking reveals that some have an early history of clumsiness in childhood or adolescence. Dystonia is the most common extrapyramidal feature in atypical PKAN, although it is generally perceived to be less severe than that seen in classical PKAN (Hayflick et al., 2003). Other motor symptoms are also reported in atypical PKAN, including parkinsonism (Zhou et al., 2001), adult-onset pure akinesia (Molinuevo, Marti, Blesa, & Tolosa, 2003), focal dystonias (Zhou et al., 2001), corticospinal tract signs, freezing of gait (Guimaraes & Santos, 1999), and an essential tremor-like syndrome (Yamashita et al., 2004). Retinitis pigmentosa is only rarely described in atypical disease, although recent literature suggests that subclinical retinal changes may be more common than previously thought in the atypical PKAN subgroup. Optic atrophy is not associated with atypical phenotype. Similar to classical PKAN, cognitive impairment is also reported in late-onset PKAN phenotype (Freeman et al., 2007).

2.1.3 Disease course

PKAN is a disorder of neuroregression, and lost skills are not regained. The rate of disease progression seems to be faster in patients with earlier disease onset. Patients with PKAN show a pattern of stepwise decline, with periods of relative clinical stability combined with episodic neurological deterioration, cognitive decline, and loss of motor skills. The reason for this stepwise pattern of regression is unclear, and there seems to be no correlation with intercurrent infection or illness (Gregory & Hayflick, 2013). Loss of ambulation is seen in the majority of patients with classical disease within 10–15 years of diagnosis (Hartig et al., 2006; Hayflick et al., 2003). Secondary complications are commonly encountered including gut-related symptoms, such as gastroesophageal reflux, dysphagia, and constipation. Premature death does occur, but advances in medical care have resulted in a greater number of PKAN patients surviving into adulthood. Death is usually secondary to (i) cardiorespiratory complications (chest infections and aspiration pneumonia) and (ii) complications from malnutrition (such as immunodeficiency) and rarely associated with (iii) status dystonicus. Atypical PKAN appears to be less aggressive than classic disease, and most individuals remain ambulant into adulthood, with loss of ambulation occurring over a longer timescale, usually within 15–40 years of disease onset.

2.2. Neuroimaging features in PKAN (Fig. 2.1)

As for all forms of NBIA, brain MRI is the neuroimaging modality of choice for the diagnostic evaluation of suspected PKAN. The characteristic feature on neuroimaging is the "eye-of-the-tiger" sign defined as a central area of hyperintensity within a hypointense globus pallidus on coronal or axial T2-weighted imaging. Some patients with PKAN may also have additional hypointensity indicative of iron deposition within the substantia nigra (McNeill et al., 2008). The PKAN "eye-of-the-tiger signature" is highly

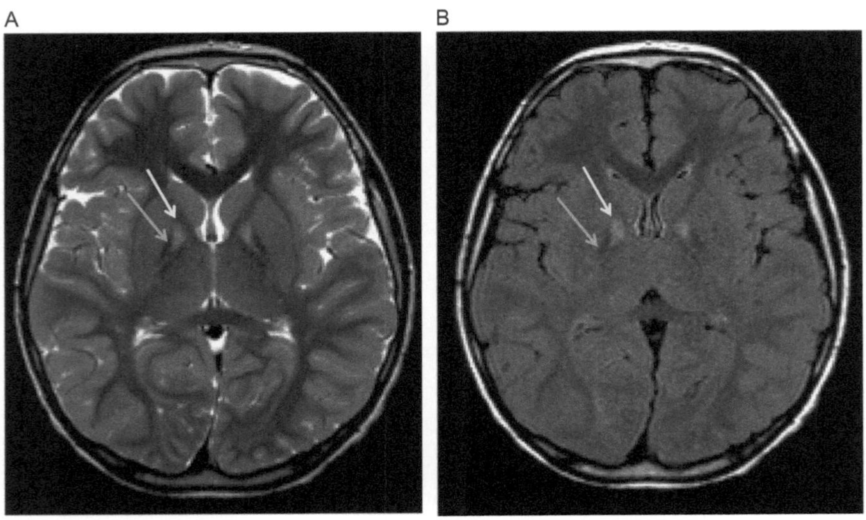

Figure 2.1 Radiological features of PKAN on brain MRI. Axial T2 (A) and axial FLAIR (B) images indicating the "eye-of-the-tiger" sign with medial globus pallidus hyperintensity (yellow arrow) surrounded by a region of hypointensity (blue arrow). (For interpretation of the references to color in this figure legend, the reader is referred to the online version of this chapter.)

predictive for at least one mutation in *PANK2* (Ching et al., 2002; Guimaraes & Santos, 1999). In addition, MRI has accurately predicted PKAN in presymptomatic siblings (Hayflick et al., 2001). Although the majority of patients with *PANK2* mutations have the classical "eye-of-the-tiger" sign, it is not universally present in every patient for a number of reasons: (i) imaging may have been undertaken too early in the disease course (Chiapparini et al., 2011); (ii) in more advanced disease, the region of hyperintensity is replaced by a more uniform hypointensity from increasing iron accumulation (Baumeister, Auer, Hortnagel, Freisinger, & Meitinger, 2005; Delgado et al., 2012); and (iii) specific *PANK2* mutations may be associated with features of NBIA on neuroimaging without classical "eye-of-the-tiger," such as is evident in a subset of the cohort of PKAN patients from the Dominican Republic (Delgado et al., 2012). Of note, appearances similar to the "eye-of-the-tiger" sign are also observed in other NBIA disorders including MPAN, where MRI indicates iron within the globus pallidus and hyperintense streaking of the medial medullary lamina (Hogarth et al., 2013). "Eye-of-the-tiger" may also be seen in multiple system atrophy (Strecker et al., 2007), neuroferritinopathy (McNeill et al., 2008), and progressive supranuclear palsy (Hartig et al., 2006).

2.3. Neuropathological features of PKAN

Before the modern era of high-resolution MR imaging, NBIA disorders were diagnosed on postmortem examination, which showed iron-rich rust-brown pigmentation within the globus pallidus and substantia nigra (Hallervorden, 1924). More recently, post-gene discovery, there have been studies providing further insight into the neuropathologic features of PKAN (Kruer et al., 2011). The globus pallidus is the most affected structure, with depletion of viable neurons within the globus pallidus interna. Iron, mainly as coarse granular hemosiderin deposits, is distributed in a perivascular pattern. Spheroids (swollen axons) are also seen (Koeppen & Dickson, 2001) in the pallidonigral system and also in the cerebrum (Swaiman, 2001). Large spheroid structures (degenerating neurons) stain for ubiquitin and smaller axonal spheroids are positive for amyloid precursor protein detected by immunoreactivity, with less anti-ubiquitin staining (Kruer et al., 2011; Malandrini et al., 1995). Li et al. (2012) report postmortem findings from a single case (20-year-old male) with significant tau pathology (neurofibrillary tangles and neuropil threads).

2.4. Molecular genetics of PKAN

In 2001, Zhou et al. (2001) demonstrated that PKAN was caused by mutations in the *PANK2* gene. PKAN is inherited in an autosomal recessive manner, and for genetic counseling purposes, recurrence risk at conception for each subsequent sibling of an affected individual with PKAN has (i) 1 in 4 chance of being affected with PKAN, (ii) 1 in 2 chance of being an asymptomatic carrier, and (iii) 1 in 4 chance of being an unaffected non-carrier. From a practical perspective, carrier testing for at-risk relatives, prenatal testing for future at-risk pregnancies, and preimplantation genetic diagnosis are all possible if both disease-causing mutations have been identified in an affected family member.

To date, mutations have been identified in all coding regions of the *PANK2* gene (Hartig et al., 2006; Hayflick et al., 2003). A wide variety of mutations have been reported but most *PANK2* mutations are missense variants distributed across the conserved domains of *PANK2*. Globally, the c.1561G>A missense mutation is the most common cause of PKAN, and many PKAN patients with this mutation have a shared haplotype, suggested of an ancestral founder effect. Mutation founder effects have been reported in specific populations including in the Netherlands (Rump et al., 2005) and also in a small isolated community from the Dominican Republic (c.680A>G, p.Tyr227Cys), where there is a significant increase in c.680A>G carrier frequency (Delgado et al., 2012).

Other more commonly identified mutations include c.1351C>T and c.1583C>T. Many rarer *PANK2* variants have also been identified, and mutations "private" to individual families have been reported. In about 5–10% of cases, only one mutated allele can be detected. Some of these cases are resolved with gene dosage analysis by multiplex ligation-dependent probe amplification (MLPA) or exon-level array comparative genomic hybridization (aCGH), where a second mutation is detected as an intragenic deletion or duplication. Mutations in the promoter, regulatory, or intronic regions (which cannot be detected on standard diagnostic testing) may also account for some of these unresolved "single mutation" cases.

Genotype–phenotype correlations for PKAN are still emerging, and certain patterns are already reported. Patients with 2 null mutations (resulting in absent PANK2 enzyme) consistently have the classical early-onset phenotype. Homozygotes of c.1561G>A missense mutation have classic PKAN. A recent study of *PANK2* mutations confirmed that p.Gly521Arg leads to a protein that is misfolded and devoid of activity (Zhang, Rock, & Jackowski,

2006). Homozygosity of other alleles seems not to be so clearly predictive of phenotype. Intrafamilial disease variation in families with multiple affected members is more evident in families with atypical rather than classical PKAN.

2.5. Pathophysiological disease mechanisms in PKAN

PANK2 encodes a predicted 50.5 kDa protein that is a functional pantothenate kinase (Zhou et al., 2001). PKAN is attributed to loss of function of pantothenate kinase 2, one of the four human pantothenate kinase proteins.

Pantothenate kinase is an essential regulatory enzyme in coenzyme A (CoA) biosynthesis, catalyzing the phosphorylation of pantothenate (vitamin B5), N-pantothenoyl-cysteine, and pantetheine (Fig. 2.2). CoA has a multitude of roles in adenosine triphosphate synthesis as well as fatty acid and neurotransmitter metabolism. The PANK2 enzyme localizes to mitochondria in both the human and mouse brain (Brunetti et al., 2012). *PANK2* mutations are postulated to cause mitochondrial dysfunction,

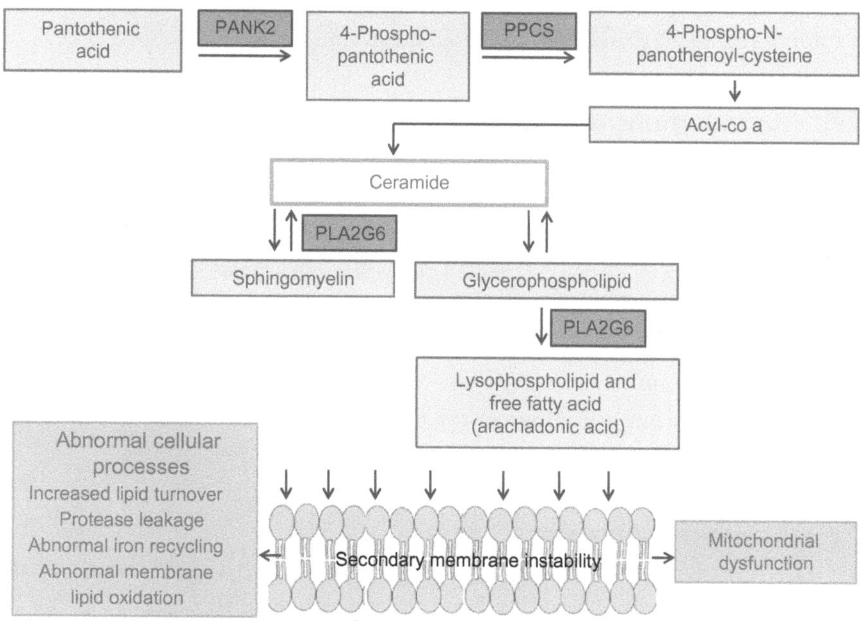

Figure 2.2 PLA2G6 and PANK2 enzymes: biochemical pathway and implicated cellular processes in PLAN and PKAN. (For color version of this figure, the reader is referred to the online version of this chapter.)

although the precise mechanisms are yet to be elucidated. Recent data from Pank2-defective neurons derived from knockout mice show altered mitochondrial membrane potential, swollen mitochondria at the ultrastructural level, and defective respiration (Brunetti et al., 2012). PANK2 dysfunction leads to neurotoxic accumulation of its substrates cysteine and pantetheine (Yang, Campbell, & Bondy, 2000; Yoon, Koh, Floyd, & Park, 2000). Cysteine is a potent iron chelator, and high local cysteine levels may lead to secondary iron accumulation and secondary oxidative stress-induced neuronal injury.

The lack of a robust mammalian model of the disease (that accurately recapitulates the human phenotype) is an ongoing limiting factor in further elucidation of the disease mechanisms in PKAN. A *Drosophila melanogaster* model is one of the best animal models to date, and the *PANK2* (fumble)$^{-/-}$ fly has reduced coordination and an impaired ability to climb, with recovery of the motor phenotype with pantetheine therapy (Rana et al., 2010). Despite promising data from knockout murine neurons (Brunetti et al., 2012), the knockout mouse model of PKAN develops retinal degeneration and azoospermia (Kuo et al., 2005) but no motor features nor brain iron accumulation. Other PANK isoforms (1, 3, and 4) are suspected to compensate for PANK2 loss in these animal models.

2.6. Management of PKAN treatment

2.6.1 Clinical assessment

1. Detailed neurological examination
2. Neurodevelopmental assessment by a multidisciplinary team including physiotherapists, occupational therapists, and speech and language therapists
3. Ophthalmologic assessment and ERG
4. MRI brain scan
5. Medical genetics consultation for counseling
6. Regular monitoring of height, weight, and nutritional status.

2.6.2 Treatment strategies

To date, the majority of treatment strategies are supportive and focus on palliation and symptom control, rather than disease modification.

2.6.3 Spasticity and dystonia

1. Intramuscular focal botulinum toxin to specific muscle groups
2. Oral baclofen and intrathecal baclofen in severe cases

3. Trihexyphenidyl and benzodiazepines
4. Deep brain stimulation (DBS).

Data regarding surgical intervention with DBS are still emerging. A cohort of patients treated with DBS showed improved motor function with gains in writing, speech, walking, and global measures of motor skills (Castelnau et al., 2005). Long-term data on this study are eagerly awaited. In addition, Mahoney et al. (2011) report improvement in cognitive function post-DBS in a cohort of seven children with PKAN. A number of additional single-case reports with varying follow-up times and anecdotal reports from PKAN families also support the notion that DBS can provide benefit in some cases (Isaac, Wright, Bhattacharyya, Baxter, & Rowe, 2008; Krause et al., 2006; Lim et al., 2012; Mikata, Yehya, Darwish, Karam, & Comair, 2009; Shields, Sharma, Gale, & Eskandar, 2007). The largest study to date by Timmermann et al. (2010) reports 23 PKAN patients undergoing DBS treatment (from 16 different DBS centers) and found improvement in both dystonia (particularly in the most severely affected patients) and the quality of life up to 15 months post-DBS. Further data on DBS treatment will no doubt inform clinical practice for PKAN patients in the future.

2.6.4 Multidisciplinary team input

1. Physiotherapy and occupational therapy to optimize joint mobility, minimize contractures, maintain posture, and maximize motor function
2. Adaptive aids for ambulation and mobility (walkers and wheelchairs)
3. Speech and language therapy (for dysarthria) and communication devices
4. Swallow assessment for safety of swallow
5. Dietetic input to maintain adequate caloric requirements and prevent malnutrition
6. Treatment of constipation and gastroesophageal reflux
7. Prompt PEG referral (as needed) to support any feeding difficulties
8. Vision support
9. Appropriate educational setting (and statementing of needs as appropriate)
10. Dental extraction or bite blocking if orolingual dystonia leads to recurrent tongue biting
11. Prompt recognition and treatment of painful factors that may exacerbate the movement disorder, such as occult GI bleeding, urinary tract infections, pressure sores from immobility, and bone fractures.

2.6.5 *New experimental therapies under consideration (see Chapter 7)*

2.6.5.1 Iron chelation (see Chapter 7)

The potential for iron chelation using deferiprone (an iron chelator that is able to cross the blood–brain barrier) to modify disease and improve clinical symptoms is highly topical at present with the start of the Treat Iron-Related Childhood-Onset Neurodegeneration (TIRCON) trial. To date, there are mixed reports in the literature. Abbruzzese et al. (2011) reported good tolerability of deferiprone, with reduction of radiologically discernible brain iron, and clinical improvement in some PKAN patients. Another pilot phase II trial showed that deferiprone was tolerated well in the nine PKAN patients who completed the study, with statistically significant reduction of iron in the pallida by MRI evaluation but disappointingly without clinical improvement of symptoms. Zorzi et al. (2011) postulated that a longer trial period may be necessary to produce clinical benefit.

2.6.5.2 Pantothenate

In PKAN patients who have residual PANK2 activity, the possibility of using high-dose pantothenate therapy has been considered. Pantothenate is well tolerated with no known toxicity. The effect of pantothenate supplementation in PKAN is currently unknown although patients with atypical PKAN have anecdotally reported improvement in motor symptoms, speech, cognition, and well-being while on treatment.

3. *PLA2G6*-ASSOCIATED NEURODEGENERATION

In 1952, Seitelberger (1952) first described an infantile-onset disorder characterized by neurological regression and lipid storage in the brain. Originally coined as Seitelberger disease, he described a disorder later to be known as infantile neuroaxonal dystrophy (INAD). More than 50 years later, the disease-causing gene was identified (*PLA2G6*) (Morgan et al., 2006), and over time, it is clear that a number of different, yet related phenotypes are caused by mutations in this gene. The newly termed PLAN is a second major NBIA phenotype and comprises a continuum of three overlapping phenotypes:

- Classic INAD
- Atypical neuroaxonal dystrophy (atypical NAD), including Karak syndrome
- *PLA2G6*-related dystonia-parkinsonism.

Disease prevalence is not established but PLAN is rare, with an estimated prevalence rate of approximately 1/million (Gregory et al., 2009).

3.1. Clinical features of PLAN

3.1.1 Disease onset

Classical INAD: INAD usually manifests between 6 months and 3 years of age. Infants are usually born following an unremarkable pregnancy and have a normal neurodevelopmental course in early infancy. At presentation, the majority of children display neurological regression with loss of previously acquired skills. Some present following an intercurrent illness (Kurian et al., 2008). Gait disturbance and loss of ambulation are often seen in the early stages of disease, as well as truncal hypotonia and strabismus.

Atypical NAD: Patients with atypical NAD have a slightly later age of disease onset in early childhood but it can also be as late as the end of the 2nd decade. Like in INAD, presentation is often associated with gait impairment or ataxia, but a number of patients also present with social communication difficulties, displaying speech difficulties, and autistic traits (Gregory, Westaway, et al., 2008). Indeed, these nonspecific features may be the only symptoms present for a significant period of time before the onset of motor symptoms.

PLA2G6-related dystonia-parkinsonism: *PLA2G6*-related dystonia-parkinsonism has a wide age range of presentation (4–30 years) (Bower, Bushara, Dempsey, Das, & Tuite, 2011; Paisan-Ruiz et al., 2009; Yoshino et al., 2010), although most presented in early adulthood (second and early third decade). Patients presenting with this condition in childhood showed features similar to those seen in atypical NAD (gait abnormalities and speech difficulties), whereas young adults seem to present with gait disturbance or neuropsychiatric symptoms.

3.1.2 Disease clinical features

Classical INAD: Clinical presentation is fairly homogenous in this PLAN subtype. Truncal hypotonia is often seen early in the disease course. Upper motor neuron signs are commonly reported. Initially patients show hyperreflexia and hypertonicity, and later in the disease course, there is evidence of spastic tetraparesis, with symmetrical pyramidal tract signs, areflexia, and contractures on clinical examination. Visual features are also commonly reported, and strabismus and nystagmus are often seen early in the disease course. As disease progresses, optic nerve pallor and then optic atrophy are reported in the majority of cases. Seizures are uncommon in INAD and usually a late manifestation of disease (Nardocci et al., 1999; Wu et al., 2009).

Atypical NAD: Patients with atypical NAD are less homogenous in clinical presentation than those with INAD. Over the disease course, patients mainly develop a predominantly extrapyramidal phenotype with prominent dystonia and dysarthria. Neuropsychiatric disturbances are also common (Gregory, Westaway, et al., 2008) and include hyperactivity, impulsivity, poor attention, and periods of emotional lability. Visual features are similar to those seen in INAD. Spastic tetraparesis may be evident but is usually a feature at end-stage disease, and in contrast to INAD, it is rarely preceded by early truncal hypotonia.

PLA2G6-related dystonia-parkinsonism: To date, only a few cases of this adult-onset PLAN subtype have been described in the literature. Dystonia-parkinsonism is universally seen developing in late teenage years or early twenties and is associated with rapid decline in cognitive function. Psychiatric features are commonly reported and may precede or occur at the same time as motor dysfunction. Parkinsonism manifests with progressive bradykinesia, resting coarse tremor, rigidity, and postural instability. Hand/foot dystonia is commonly seen but some patients will also have a more generalized dystonia. Although it has been reported that patients respond to L-dopa, this appears to be a temporary effect, and patients rapidly develop progressive motor symptoms of medication-related dyskinesia.

3.1.3 Disease course

Classical INAD: Of all the PLAN phenotypes, disease progression is the fastest in INAD. Severe neurological regression leads to spasticity, contractures, progressive cognitive dysfunction, and visual impairment. By end-stage disease children are in a vegetative state. Death is common from the end of the 1st decade and usually is a consequence of secondary complications such as intercurrent respiratory illnesses or aspiration pneumonia secondary to bulbar dysfunction. Appropriate supportive care can improve longevity in these children.

Atypical NAD: Affected patients often will have fairly stable course during early childhood with neurological deterioration in midchildhood (Nardocci et al., 1999). Atypical cases are rare, and only a handful of cases are reported in the literature, including those with Karak syndrome (Mubaidin et al., 2003). Consequently, little is currently known about the life span in atypical NAD, although with its less severe presentation, and course, it is typically longer than that described in classic disease.

PLA2G6-related dystonia-parkinsonism: The rarity of cases precludes an accurate picture of long-term prognosis. Following onset of symptoms, motor and cognitive decline progress rapidly in affected patients.

3.1.4 Electrophysiological investigations in PLAN

Infantile PLAN is associated with a number of abnormalities on electrophysiological investigation that can aid diagnosis (Carrilho et al., 2008). High-amplitude fast activity is often seen on electroencephalography (EEG). Fast rhythms of EEG have not been reported in atypical NAD, although some children may develop epileptiform EEG changes. In INAD, denervation may be seen on electromyography (EMG), with a distal axonal-type sensorimotor neuropathy on nerve conduction studies (NCS). Visual evoked potentials show either absent or delayed reduced amplitude.

3.2. Neuroimaging features of PLAN (Fig. 2.3)

For INAD, cerebellar atrophy is a universal feature and often the earliest sign on MRI (Farina et al., 1999). Cerebellar gliosis is additionally seen in many cases but not in all INAD patients (Kurian et al., 2008). Secondary posterior corpus callosum abnormalities (vertical orientation, thinning, and elongation of the splenium) secondary to cerebellar atrophy are also reported. Brain iron accumulation within the (medial) globus pallidus (McNeill et al., 2008), dentate nuclei, and substantia nigra is also described, with increasing severity with

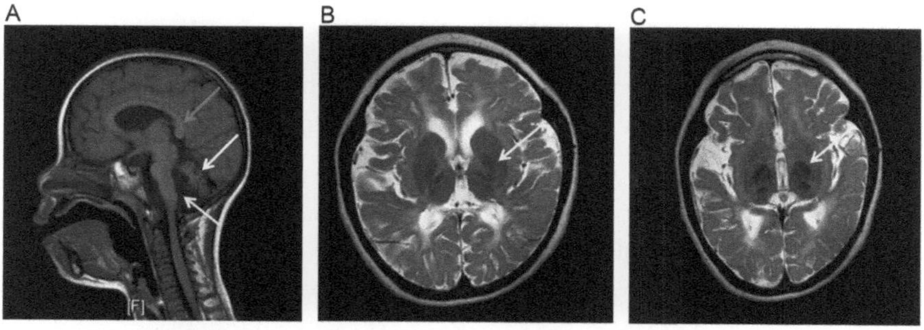

Figure 2.3 Radiological features of PLAN on brain MRI. Sagittal T1-weighted sequence (A) showing abnormal orientation of the posterior splenium of the corpus callosum (blue arrow), apparent claval hypertrophy (yellow arrow), and marked atrophy of the cerebellum (white arrow). T2-weighted sequences showing hypointensity in the globus pallidus (B) and substantia nigra (C), both indicated by yellow arrows. Note symmetrical white matter changes are evident on images (B) and (C), which can be reported in PLAN. (For interpretation of the references to color in this figure legend, the reader is referred to the online version of this chapter.)

age (Kurian et al., 2008). Clinical optic atrophy is also evident radiologically with reduced volume of the optic chiasm and optic nerves. Cerebral white matter changes such as high signal on T2-weighted sequences as well as atrophy have also been described. More recently, apparent "claval hypertrophy" has been identified as a consistent feature in INAD (Maawali et al., 2011).

In atypical NAD, prominent brain iron is the main neuroradiological feature with or without cerebellar atrophy. In patients with *PLA2G6*-related dystonia-parkinsonism, to date, reports only detail the presence of nonspecific changes such as cerebral atrophy.

3.3. Neuropathological features of PLAN

Electron microscopic examination of nerve structure in conjunctival, skin, muscle, rectal, or sural nerve tissue demonstrates the present of axonal spheroids in later-stage INAD. However, it is not a consistent finding in early disease or indeed in all INAD patients. The presence of peripheral spheroids has been reported in atypical PLAN but not in patients with *PLA2G6*-related dystonia-parkinsonism—the advent of molecular testing has negated the need for biopsy; therefore, it is still unclear whether this pathological hallmark is truly evident in the non-INAD PLAN subtypes.

Paisán-Ruiz et al. (2012) examined postmortem brain tissue from PLAN patients (age of death 8–36 years) and demonstrated widespread alpha-synuclein-positive Lewy pathology (diffuse neocortical type). In three-fifths of cases, there was hyperphosphorylated tau accumulation as threads, pretangles, and neurofibrillary tangles. The authors reported that in cases of *PLA2G6*-associated dystonia-parkinsonism, there was less tau involvement but still severe alpha-synuclein pathology. These data certainly suggest a link between the clinical and pathological features of PLAN and parkinsonian disorders.

3.4. Molecular genetic features of PLAN

In 2006, Morgan et al. (2006) identified *PLA2G6* as the causative gene for INAD and atypical NAD. PLAN is inherited in an autosomal recessive manner, with similar recurrence risks to those described in PKAN. Identification of 2 disease-causing mutations will therefore facilitate appropriate genetic counseling, prenatal testing, and also preimplantation diagnosis.

To date, over 70 disease-causing mutations in *PLA2G6* have been reported, including pathogenic missense variants, small exonic deletions, nonsense mutations, splice site mutations, and more recently copy number variants (Crompton et al., 2010). Common mutations have been identified in a number of reportedly unrelated families although the commonality of ethnic background in these families suggests a possible founder effect (such as the homozygous mutation c.1634A>C, p.Lys545Thr in a number of Pakistani PLAN families).

Limited genotype–phenotype correlations are evident although it appears that, similar to PKAN, all individuals with two null alleles present with infantile-onset PLAN. In addition, patients with atypical NAD tend to have two missense *PLA2G6* mutations. Furthermore, mutations in *PLA2G6*-related dystonia-parkinsonism do not impair the catalytic activity of the PLA2G6 enzyme (Engel, Jing, O'Brien, Sun, & Kotzbauer, 2010), in contrast to mutations studied for INAD.

3.5. Pathogenic disease mechanisms in PLAN

PLA2G6 encodes an 85 kDa calcium-independent phospholipase A_2 enzyme $iPLA_2$-VIA (which is one of several calcium-independent phospholipases), active in a tetrameric form. This class of enzyme catalyzes the hydrolysis of glycerophospholipids, generating a free fatty acid (usually arachidonic acid) and a lysophospholipid (Fig. 2.2). $iPLA_2$-VIA protein has proposed roles in phospholipid remodeling, arachidonic acid release, leukotriene and prostaglandin synthesis, and apoptosis (Balsinde & Balboa, 2005). The $iPLA_2$ enzymes regulate levels of phospholipids (Baburina & Jackowski, 1999) thereby playing an important role in cell membrane homeostasis. Defects in $iPLA_2$-VIA could lead to an imbalance of membrane phospholipids with secondary structural consequences to the cell. This may indeed contribute to the axonal pathology observed in INAD (Morgan et al., 2006).

Both knockout and knock-in murine models of PLAN have been developed (Wada, Kojo, & Seino, 2013), which show progressive motor dysfunction, hematopoietic abnormalities, and widespread axonal spheroids akin to those seen in human disease.

3.6. Management of PLAN

3.6.1 Clinical assessment

1. Detailed neurological examination
2. Neurodevelopmental assessment by a multidisciplinary team including physiotherapists, occupational therapists, and speech and language therapists
3. MRI brain scan

4. Ophthalmologic assessment and VER for optic atrophy
5. EEG
6. EMG/NCS
7. Medical genetics consultation for counseling
8. Consider biopsy if any doubts regarding diagnosis following genetic testing
9. Regular monitoring of height, weight, and nutritional status
10. Periodic assessment of vision and hearing
11. Psychiatry review if prominent psychiatric symptoms.

3.6.2 Multidisciplinary team input

1. Control of drooling and secretions with transdermal hyoscine patches or oral glycopyrrolate therapy
2. Supportive treatment for constipation (softeners and laxatives)
3. Feeding modifications (such as nasogastric or PEG feeding) as needed to prevent aspiration pneumonia and achieve adequate nutrition
4. A rehabilitation program with physiotherapy, occupational therapy, and orthopedic input to optimize joint mobility, minimize contractures, maintain posture, and maximize motor function.

3.6.3 New experimental therapies under consideration (see Chapter 7)

As previously discussed earlier in this chapter, the iron chelator, deferiprone is currently under investigation in PKAN. Promising results may lead the way for future trials in PLAN.

3.6.4 PLA2G6-dystonia parkinsonism

1. Consider treatment with dopaminergic agents such as L-dopa
2. Appropriate treatment of neuropsychiatric symptoms by a psychiatrist.

3.6.5 Treatment strategies

3.6.5.1 For INAD and atypical NAD

1. Routine pharmacological treatment of spasticity (e.g., with baclofen) and seizures (with standard antiepileptic agents)
2. Treatment of dystonia in atypical NAD: trial of oral or intrathecal baclofen. A single case of atypical NAD treated with DBS resulted in clinical benefit (L Cif—personal communication).

4. CONCLUSION

Both PKAN and PLAN are major NBIA subtypes. In both disorders, neurological regression in tandem with cognitive decline, progressive motor dysfunction (with pyramidal and extrapyramidal features), and additional visual and neuropsychiatric features are reported. Both disorders are associated with a number of distinct, yet overlapping phenotypes. Advances in molecular genetic techniques and the availability of genetic testing will no doubt lead to further detailed characterization of the subtypes and clinical spectrum of PLAN and PKAN. Future research is likely to provide insight into the epigenetic mechanisms that determine specific PLAN/PKAN phenotypes. Despite advances in understanding the clinical, molecular, and pathological features of PKAN and PLAN, treatment strategies remain palliative, with currently little disease-modifying therapies for these two conditions. Future research in developing novel therapeutic strategies should be prioritized, given the life-limiting nature of these disorders.

ACKNOWLEDGMENTS

We would like to thank the patients and families for their ongoing support for NBIA research, which has allowed detailed characterization of these phenotypes. MAK is funded by the Wellcome Trust and Great Ormond Street Hospital Children's Charity (GOSHCC); SJH is supported by the NBIA Disorders Association and Oregon Clinical and Translational Research Institute (UL1 RR024140 NCRR), a component of the National Institutes of Health (NIH).

REFERENCES

Abbruzzese, G., Cossu, G., Balocco, M., Marchese, R., Murgia, D., Melis, M., et al. (2011). A pilot trial of deferiprone for neurodegeneration with brain iron accumulation. *Haematologica, 96*, 1708–1711.

Baburina, I., & Jackowski, S. (1999). Cellular responses to excess phospholipid. *Journal of Biological Chemistry, 274*, 9400–9408.

Balsinde, J., & Balboa, M. A. (2005). Cellular regulation and proposed biological functions of group VIA calcium-independent phospholipase A2 in activated cells. *Cellular Signalling, 17*, 1052–1062.

Baumeister, F. A., Auer, D. P., Hortnagel, K., Freisinger, P., & Meitinger, T. (2005). The eye-of-the-tiger sign is not a reliable disease marker for Hallervorden–Spatz syndrome. *New England Journal of Medicine, 348*, 33–40.

Benke, T., & Butterworth, B. (2001). Palilalia and repetitive speech: Two case studies. *Brain and Language, 78*, 62–81.

Benke, T., Hohenstein, C., Poewe, W., & Butterworth, B. (2000). Repetitive speech phenomena in Parkinson's disease. *Journal of Neurology, Neurosurgery, and Psychiatry, 69*, 319–324.

Bower, M. A., Bushara, K., Dempsey, M. A., Das, S., & Tuite, P. J. (2011). Novel mutations in siblings with later-onset PLA2G6-associated neurodegeneration (PLAN). *Movement Disorders*, *26*, 1768–1769.

Brunetti, D., Dusi, S., Morbin, M., Uggetti, A., Moda, F., D'Amato, I., et al. (2012). Pantothenate kinase-associated neurodegeneration: Altered mitochondria membrane potential and defective respiration in Pank2 knock-out mouse model. *Human Molecular Genetics*, *21*, 5294–5305.

Carrilho, I., Santos, M., Guimarães, A., Teixeira, J., Chorão, R., Martins, M., et al. (2008). Infantile neuroaxonal dystrophy: What's most important for the diagnosis? *European Journal of Paediatric Neurology*, *12*, 491–500.

Castelnau, P., Cif, L., Valente, E. M., Vayssiere, N., Hemm, S., Gannau, A., et al. (2005). Pallidal stimulation improves pantothenate kinase-associated neurodegeneration. *Annals of Neurology*, *57*, 738–741.

Chiapparini, L., Savoiardo, M., D'Arrigo, S., Reale, C., Zorzi, G., Zibordi, F., et al. (2011). The "eye-of-the-tiger" sign may be absent in the early stages of classic pantothenate kinase associated neurodegeneration. *Neuropediatrics*, *42*, 159–162.

Ching, K. H., Westaway, S. K., Gitschier, J., Higgins, J. J., & Hayflick, S. J. (2002). HARP syndrome is allelic with pantothenate kinase-associated neurodegeneration. *Neurology*, *58*, 1673–1674.

Crompton, D., Rehal, P. K., MacPherson, L., Foster, K., Lunt, P., Hughes, I., et al. (2010). Multiplex ligation-dependent probe amplification (MLPA) analysis is an effective tool for the detection of novel intragenic PLA2G6 mutations: Implications for molecular diagnosis. *Molecular Genetics and Metabolism*, *100*, 207–212.

del Valle-López, P., Pérez-García, R., Sanguino-Andrés, R., & González-Pablos, E. (2011). Adult onset Hallervorden–Spatz disease with psychotic symptoms. *Actas Españolas de Psiquiatría*, *39*, 260–262.

Delgado, R. F., Sanchez, P. R., Speckter, H., Then, E. P., Jiminez, R., Oviedo, J., et al. (2012). Missense PANK2 mutation without "eye of the tiger" sign: MR findings in a large group of patients with pantothenate kinase-associated neurodegeneration (PKAN). *Journal of Magnetic Resonance Imaging*, *35*, 788–794.

Dooling, E. C., Schoene, W. C., & Richardson, E. P. (1974). Hallervorden–Spatz syndrome. *Archives of Neurology*, *30*, 70–83.

Egan, R. A., Weleber, R. G., Hogarth, P., Gregory, A., Coryell, J., Westaway, S. K., et al. (2005). Neuro-ophthalmologic and electroretinographic findings in pantothenate kinase-associated neurodegeneration (formerly Hallervorden–Spatz syndrome). *American Journal of Ophthalmology*, *140*, 267–274.

Engel, L. A., Jing, Z., O'Brien, D. E., Sun, M., & Kotzbauer, P. T. (2010). Catalytic function of PLA2G6 is impaired by mutations associated with infantile neuroaxonal dystrophy but not dystonia-parkinsonism. *PLoS One*, *5*, e12897.

Farina, L., Nardocci, N., Bruzzone, M. G., D'Incerti, L., Zorzi, G., Verga, L., et al. (1999). Infantile neuroaxonal dystrophy: Neuroradiological studies in 11 patients. *Neuroradiology*, *41*, 376–380.

Freeman, K., Gregory, A., Turner, A., Blasco, P., Hogarth, P., & Hayflick, S. (2007). Intellectual and adaptive behavior functioning in pantothenate kinase-associated neurodegeneration. *Journal of Intellectual Disability Research*, *51*, 417–426.

Gregory, A., & Hayflick, S. J. (2012). Pantothenate kinase-associated neurodegeneration. In R. A. Pagon, M. P. Adam, T. D. Bird, C. R. Dolan, C. T. Fong, & K. Stephens (Eds.), *GeneReviews™ [Internet]*Seattle, WA: University of Washington (updated 2013 January 31).

Gregory, A., & Hayflick, S. J. (2013). Neurodegeneration with brain iron accumulation disorders overview. In R. A. Pagon, M. P. Adam, T. D. Bird, C. R. Dolan, C. T. Fong, & K. Stephens (Eds.), *GeneReviews™ [Internet]*. Seattle, WA: University of Washington, Seattle.

Gregory, A., Kurian, M. A., Maher, E. R., Hogarth, P., & Hayflick, S. J. (2008). PLA2G6-associated neurodegeneration. In R. A. Pagon, M. P. Adam, T. D. Bird, C. R. Dolan, C. T. Fong, & K. Stephens (Eds.), *GeneReviews™ [Internet]*. Seattle, WA: University of Washington (updated 2012 April 19).

Gregory, A., Polster, B. J., & Hayflick, S. J. (2009). Clinical and genetic delineation of neurodegeneration with brain iron accumulation. *Journal of Medical Genetics*, *46*, 73–80.

Gregory, A., Westaway, S. K., Holm, I. E., Kotzbauer, P. T., Hogarth, P., Sonek, S., et al. (2008). Neurodegeneration associated with genetic defects in phospholipase A2. *Neurology*, *71*, 1402–1409.

Guimaraes, J., & Santos, J. V. (1999). Generalized freezing in Hallervorden–Spatz syndrome: Case report. *European Journal of Neurology*, *6*, 509–513.

Hallervorden, J. (1924). Uber eine familiare Erkrankung im extrapyramidalen System. *Deutsche Zeitschrift für Nervenheilkunde*, *81*, 204–210.

Hartig, M. B., Hortnagel, K., Garavaglia, B., Zorzi, G., Kmiec, T., Klopstock, T., et al. (2006). Genotypic and phenotypic spectrum of PANK2 mutations in patients with neurodegeneration with brain iron accumulation. *Annals of Neurology*, *59*, 248–256.

Hayflick, S. J., Penzien, J. M., Michl, W., Sharif, U. M., Rosman, N. P., & Wheeler, P. G. (2001). Cranial MRI changes may precede symptoms in Hallervorden–Spatz syndrome. *Pediatric Neurology*, *25*, 166–169.

Hayflick, S. J., Westaway, S. K., Levinson, B., Zhou, B., Johnson, M. A., Ching, K. H., et al. (2003). Genetic, clinical, and radiographic delineation of Hallervorden–Spatz syndrome. *New England Journal of Medicine*, *348*, 33–40.

Hogarth, P., Gregory, A., Kruer, M. C., Sanford, L., Wagoner, W., Natowicz, M. R., et al. (2013). New form of neurodegeneration with brain iron accumulation: Features associated with MPAN. *Neurology*, *80*, 268–275.

Houlden, H., Lincoln, S., Farrer, M., Cleland, P. G., Hardy, J., & Orrell, R. W. (2003). Compound heterozygous PANK2 mutations confirm HARP and Hallervorden–Spatz syndromes are allelic. *Neurology*, *61*, 1423–1426.

Isaac, C., Wright, I., Bhattacharyya, D., Baxter, P., & Rowe, J. (2008). Pallidal stimulation for pantothenate kinase-associated neurodegeneration dystonia. *Archives of Disease in Childhood*, *93*, 239–240.

Koeppen, A. H., & Dickson, A. C. (2001). Iron in the Hallervorden–Spatz syndrome. *Pediatric Neurology*, *25*, 148–155.

Krause, M., Fogel, W., Tronnier, V., Pohle, S., Hörtnagel, K., Thyen, U., et al. (2006). Long-term benefit to pallidal deep brain stimulation in a case of dystonia secondary to pantothenate kinase-associated neurodegeneration. *Movement Disorders*, *21*, 2255–2257.

Kruer, M. C., Hiken, M., Gregory, A., Malandrini, A., Clark, D., Hogarth, P., et al. (2011). Novel histopathologic findings in molecularly-confirmed pantothenate kinase-associated neurodegeneration. *Brain*, *134*, 947–958.

Kuo, Y. M., Duncan, J. L., Westaway, S. K., Yang, H., Nune, G., Xu, E. Y., et al. (2005). Deficiency of pantothenate kinase 2 (Pank2) in mice leads to retinal degeneration and azoospermia. *Human Molecular Genetics*, *14*, 49–57.

Kurian, M. A., Morgan, N. V., MacPherson, L., Foster, K., Peake, D., Gupta, R., et al. (2008). Phenotypic spectrum of neurodegeneration associated with mutations in the PLA2G6 gene (PLAN). *Neurology*, *70*, 1623–1629.

Li, A., Paudel, R., Johnson, R., Courtney, R., Lees, A. J., Holton, J. L., et al. (2012). Pantothenate kinase-associated neurodegeneration is not a synucleinopathy. *Neuropathology and Applied Neurobiology*, *39*, 121–131.

Lim, B. C., Ki, C. S., Cho, A., Hwang, H., Kim, K. J., Hwang, Y. S., et al. (2012). Pantothenate kinase-associated neurodegeneration in Korea: Recurrent R440P mutation in PANK2 and outcome of deep brain stimulation. *European Journal of Neurology*, *4*, 556–561.

Maawali, A. L., Yoon, G., Halliday, W., Clarke, A. T. R., Feigenbaum, A., Banwell, B., et al. (2011). Hypertrophy of the clava, a new MRI sign in patients with PLA2G6 mutations. In: *Poster presentation: American Society of Human genetics Meeting, October 2011.*

Mahoney, R., Selway, R., & Lin, J. P. (2011). Cognitive functioning in children with pantothenate kinase-associated neurodegeneration undergoing deep brain stimulation. *Developmental Medicine and Child Neurology*, *53*, 275–279.

Malandrini, A., Cavallaro, T., Fabrizi, G. M., Berti, G., Salvestroni, R., & Salvadori, C. (1995). Ultrastructure and immunoreactivity of dystrophic axons indicate a different pathogenesis of Hallervorden–Spatz disease and infantile neuroaxonal dystrophy. *Virchows Archiv*, *427*, 415–421.

McNeill, A., Birchall, D., Hayflick, S. J., Gregory, A., Schenk, J. F., Zimmerman, E. A., et al. (2008). T2* and FSE MRI distinguishes four subtypes of neurodegeneration with brain iron accumulation. *Neurology*, *70*, 1614–1619.

Mikata, M. A., Yehya, A., Darwish, H., Karam, P., & Comair, Y. (2009). Deep brain stimulation as a mode of treatment of early onset pantothenate kinase-associated neurodegeneration. *European Journal of Paediatric Neurology*, *13*, 61–64.

Molinuevo, J. L., Marti, M. J., Blesa, R., & Tolosa, E. (2003). Pure akinesia: An unusual phenotype of Hallervorden–Spatz syndrome. *Movement Disorders*, *18*, 1351–1353.

Morgan, N. V., Westaway, S. K., Morton, J. E., Gregory, A., Gissen, P., Sonek, S., et al. (2006). PLA2G6, encoding a phospholipase A2, is mutated in neurodegenerative disorders with high brain iron. *Nature Genetics*, *38*, 752–754.

Mubaidin, A., Roberts, E., Hampshire, D., Dehyyat, M., Shurbaji, A., Mubaidien, M., et al. (2003). Karak syndrome: A novel degenerative disorder of the basal ganglia and cerebellum. *Journal of Medical Genetics*, *40*, 543–546.

Nardocci, N., Zorzi, G., Farina, L., Binelli, S., Scaioli, W., Ciano, C., et al. (1999). Infantile neuroaxonal dystrophy: Clinical spectrum and diagnostic criteria. *Neurology*, *52*, 1472–1478.

Paisan-Ruiz, C., Bhatia, K. P., Li, A., Hernandez, D., Davis, M., Wood, N. W., et al. (2009). Characterization of PLA2G6 as a locus for dystonia-parkinsonism. *Annals of Neurology*, *65*, 19–23.

Paisán-Ruiz, C., Li, A., Schneider, S. A., Holton, J. L., Johnson, R., Kidd, D., et al. (2012). Widespread Lewy body and tau accumulation in childhood and adult onset dystonia-parkinsonism cases with PLA2G6 mutations. *Neurobiology of Aging*, *33*, 814–823.

Pellecchia, M. T., Valente, E. M., Cif, L., Salvi, S., Albanese, A., Scarano, V., et al. (2005). The diverse phenotype and genotype of pantothenate kinase-associated neurodegeneration. *Neurology*, *64*, 1810–1812.

Rana, A., Seinen, E., Siudeja, K., Muntendam, R., Srinivasan, B., van der Want, J. J., et al. (2010). Pantethine rescues a Drosophila model for pantothenate kinase-associated neurodegeneration. *Proceedings of the National Academy of Sciences of the United States of America*, *107*, 6988–6993.

Rump, P., Lemmink, H. H., Verschuuren-Bemelmans, C. C., Grootscholten, P. M., Fock, J. M., Hayflick, S. J., et al. (2005). A novel 3-bp deletion in the PANK2 gene of Dutch patients with pantothenate kinase-associated neurodegeneration: Evidence for a founder effect. *Neurogenetics*, *6*, 201–207.

Scarano, V., Pellecchia, M. T., Filla, A., & Barone, P. (2002). Hallervorden–Spatz syndrome resembling a typical Tourette syndrome. *Movement Disorders*, *17*, 618–620.

Schneider, S. A., & Bhatia, K. P. (2010). Rare causes of dystonia parkinsonism. *Current Neurology and Neuroscience Reports*, *10*, 431–439.

Seitelberger, F. (1952). Eine unbekannte Form von infantiler lipoid-Speicher Krankheit des Gehirns. In: *First international congress of neuropathology, Rome, Italy.*

Shevell, M. (2003). Hallervorden and history. *New England Journal of Medicine*, *348*, 3–4.

Shields, D. C., Sharma, N., Gale, J. T., & Eskandar, E. N. (2007). Pallidal stimulation for dystonia in pantothenate kinase-associated neurodegeneration. *Pediatric Neurology, 37*, 442–445.

Strecker, K., Hesse, S., Wegner, F., Sabri, O., Schwarz, J., & Schneider, J. P. (2007). Eye of the tiger sign in multiple systems atrophy. *European Journal of Neurology, 14*, e1–e2.

Swaiman, K. F. (2001). Hallervorden–Spatz syndrome. *Pediatric Neurology, 25*, 102–108.

Timmermann, L., Pauls, K. A., Wieland, K., Jech, R., Kurlemann, G., Sharma, N., et al. (2010). Dystonia in neurodegeneration with brain iron accumulation: Outcome of bilateral pallidal stimulation. *Brain, 133*, 701–712.

Wada, H., Kojo, S., & Seino, K. I. (2013). Mouse models of human INAD by Pla2g6 deficiency. *Histology and Histopathology, 28*, 965–969.

Wu, Y., Jiang, Y., Gao, Z., Wang, J., Yuan, Y., Xiong, H., et al. (2009). Clinical study and PLA2G6 mutation screening analysis in Chinese patients with infantile neuroaxonal dystrophy. *European Journal of Neurology, 16*, 240–245.

Yamashita, S., Maeda, Y., Ohmori, H., Uchida, Y., Hirano, T., Yonemura, K., et al. (2004). Pantothenate kinase-associated neurodegeneration initially presenting as postural tremor alone in a Japanese family with homozygous N245S substitutions in the pantothenate kinase gene. *Journal of Neurological Sciences, 225*, 129–133.

Yang, E. Y., Campbell, A., & Bondy, S. C. (2000). Configuration of thiols dictates their ability to promote iron-induced reactive oxygen species generation. *Redox Report, 5*, 371–375.

Yoon, S. J., Koh, Y. H., Floyd, R. A., & Park, J. W. (2000). Copper, zinc superoxide dismutase enhances DNA damage and mutagenicity induced by cysteine/iron. *Mutation Research, 448*, 97–104.

Yoshino, H., Tomiyama, H., Tachibana, N., Ogaki, K., Li, Y., Funayama, M., et al. (2010). Phenotypic spectrum of patients with PLA2G6 mutation and PARK14-linked parkinsonism. *Neurology, 75*, 1356–1361.

Zhang, Y. M., Rock, C. O., & Jackowski, S. (2006). Biochemical properties of human pantothenate kinase 2 isoforms and mutations linked to pantothenate kinase-associated neurodegeneration. *Journal of Biological Chemistry, 281*, 107–114.

Zhou, B., Westaway, S. K., Levinson, B., Johnson, M. A., Gitschier, J., & Hayflick, S. J. (2001). A novel pantothenate kinase gene (PANK2) is defective in Hallervorden–Spatz syndrome. *Nature Genetics, 28*, 345–349.

Zorzi, G., Zibordi, F., Chiapparini, L., Bertini, E., Russo, L., Piga, A., et al. (2011). Iron-related MRI images in patients with pantothenate kinase-associated neurodegeneration (PKAN) treated with deferiprone: Results of a phase II pilot trial. *Movement Disorders, 26*, 1756–1759.

CHAPTER THREE

Mitochondrial Membrane Protein-Associated Neurodegeneration (MPAN)

Monika Hartig[*,†], **Holger Prokisch**[*,†,‡], **Thomas Meitinger**[*,†,‡,§], **Thomas Klopstock**[†,§,¶,1]

[*]Institute of Human Genetics, Technische Universität München, Munich, Germany
[†]German Network for Mitochondrial Disorders (mitoNET), Munich, Germany
[‡]Institute of Human Genetics, Helmholtz Zentrum München, Munich, Germany
[§]DZNE – German Center for Neurodegenerative Diseases, Munich, Germany
[¶]Department of Neurology, Friedrich-Baur-Institute, Ludwig-Maximilians-University, Munich, Germany
[1]Corresponding author: e-mail address: tklopsto@med.lmu.de

Contents

Abstract

Neurodegeneration with brain iron accumulation (NBIA) is a group of rare and devastating disorders characterized by iron deposition in the brain. Mutations in *C19orf12* cause autosomal recessive inherited mitochondrial membrane protein-associated neurodegeneration (MPAN), which may account for up to 30% of NBIA cases. The C19orf12 gene product is an orphan mitochondrial membrane protein, and most mutations are predicted to cause loss of function.

From 67 MPAN cases so far reported, we describe here the clinical, radiological, and genetic features. Key clinical features are pyramidal and extrapyramidal signs, cognitive decline, neuropsychiatric abnormalities, optic atrophy, and motor axonal neuropathy. Magnetic resonance imaging shows the eponymous brain iron accumulation in globus pallidus and substantia nigra and in some cases a hyperintense streaking of the medial medullary lamina. The latter sign may discriminate MPAN from other NBIA subtypes. In two postmortem MPAN cases, neuropathology showed axonal spheroids, Lewy bodies, and hyperphosphorylated tau-containing inclusions.

International Review of Neurobiology, Volume 110
ISSN 0074-7742
http://dx.doi.org/10.1016/B978-0-12-410502-7.00004-1

1. INTRODUCTION

Neurodegeneration with brain iron accumulation (NBIA) is a group of disorders sharing the hallmark of iron deposition in the brain. Almost a century ago, Hallervorden and Spatz were the first to describe the existence of neurodegeneration with iron accumulation in the brain (Hallervorden, 1924). Later, the term Hallervorden–Spatz disease was abandoned because of the unethical activities of the authors in Nazi Germany. Clinical signs of the disorder include motor dysfunction, visual loss, psychiatric features, and cognitive decline.

The prevalence of NBIA is low (1–3 cases/1 million population). A major breakthrough in the elucidation of NBIA has been the identification of loss-of-function mutations in the PANK2 gene in patients with pantothenate kinase-associated neurodegeneration (pantothenate kinase-associated neurodegeneration (PKAN), NBIA 1, MIM# 234200) (Zhou et al., 2001). PANK2 encodes a pantothenate kinase, which is the key regulating enzyme in the CoA pathway and is located in the mitochondria (Hortnagel, Prokisch, & Meitinger, 2003). Pathogenic PANK2 mutations have been found in approximately 50% of NBIA patients (Hartig et al., 2006; Hayflick et al., 2003).

Up to now, eight further NBIA genes (*CP*, *FTL*, *PLA2G6*, *FA2H*, *ATP13A2*, *C2orf37*, *C19orf12*, *WDR45*) have been described (Alazami et al., 2008; Curtis et al., 2001; Haack et al., 2012; Harris et al., 1995; Hartig et al., 2011; Kruer et al., 2010; Morgan et al., 2006; Schneider et al., 2010). Mutations in the PLA2G6 gene were initially identified in patients with infantile neuroaxonal dystrophy (NBIA 2, MIM# 256600) and atypical neuroaxonal dystrophy (MIM# 610217). Later, PLA2G6 mutations were found in patients with dystonia–parkinsonism expanding the clinical spectrum of PLA2G6-associated neurodegeneration (PLAN, MIM# 610217).

Mutations in *FTL* (NBIA 3, neuroferritinopathy MIM# 606159), *CP* (aceruloplasminemia MIM# 604290), *FA2H* (FAHN, fatty acid hydroxylase-associated neurodegeneration MIM# 612319), *ATP13A2* (Kufor–Rakeb syndrome MIM# 606693), *C2orf37* (Woodhouse–Sakati syndrome, MIM# 241080), and WDR45 (NBIA5, BPAN, beta-propeller protein-associated neurodegeneration MIM# 300894) appear to be rare, accounting for less than 5% of NBIA patients. For about 30% of patients, the responsible genes have not been identified yet.

The percentage of NBIA patients carrying mutations in *C19orf12* vary in the literature, ranging from 5% to 30% (Hartig et al., 2011; Hogarth et al., 2013; Panteghini et al., 2012). The corresponding phenotype was named mitochondrial membrane protein-associated neurodegeneration (MPAN, NBIA 4, MIM# 614298) due to the subcellular localization of the C19orf12 protein.

Although the identification of genes underlying different forms of NBIA has provided important insights into the nature of this complex disorder, the pivotal underlying pathology is still not understood. Moreover, insights gained from gene identification have not led to development of gene-specific treatments.

This chapter focuses on the clinical, radiological, and genetic aspects of MPAN.

2. CLINICAL FINDINGS

Until now, 67 MPAN patients have been described in the literature (Deschauer et al., 2012; Dezfouli et al., 2013; Dogu et al., 2012; Goldman et al., 2013; Hartig et al., 2011; Hogarth et al., 2013; Panteghini et al., 2012; Schottmann, Stenzel, Lutzkendorf, Schuelke, & Knierim, 2013; Schulte et al., 2013). A compilation of all these cases shows that MPAN leads to a distinctive phenotype with prominent pyramidal and extrapyramidal signs, cognitive decline, neuropsychiatric abnormalities, optic atrophy, and motor axonal neuropathy (Table 3.1). There are clues that can be helpful in distinguishing between different forms of NBIA. While pyramidal, extrapyramidal, cognitive, and psychiatric manifestations are common in most NBIA subtypes, optic atrophy is found predominantly in PLAN, FAHN, and MPAN, and motor axonal neuropathy (lower motor neuron signs) is the most distinctive feature of MPAN.

In detail, the most frequent symptoms and signs in the described 67 MPAN cases were upper motor neuron signs (spastic paresis and pyramidal signs) in 87.9%, cognitive decline in 85.9%, dysarthria in 82.3%, optic atrophy in 75.0%, dystonia in 66.7%, psychiatric abnormalities in 64.3%, lower motor neuron signs (muscle atrophy, fasciculations, and neurogenic electromyography changes) in 56.3%, dysphagia in 55.6%, and parkinsonism in 44.6% (Table 3.1).

Age at onset differs widely between 3 and 30 years, the mean age at onset being around 10 years. Accordingly, there is great variability in terms of

Table 3.1 Summary of clinical findings in MPAN patients

Symptom	No. of positive cases	No. of total cases	%
Upper motor neuron signs	58	66	87.9
Cognitive decline	55	64	85.9
Dysarthria	51	62	82.3
Optic atrophy	45	60	75.0
Dystonia	42	63	66.7
Psychiatric abnormalities	36	56	64.3
Lower motor neuron signs	27	48	56.3
Dysphagia	15	27	55.6
Parkinsonism	29	65	44.6

Compilation of patients from references (Deschauer et al., 2012; Dezfouli et al., 2013; Dogu et al., 2012; Goldman et al., 2013; Hartig et al., 2011; Hogarth et al., 2013; Panteghini et al., 2012; Schottmann et al., 2013; Schulte et al., 2013).

disease progression. In our own series of 24 cases, all but one individual were still able to walk at age 18 years, but 7 out of 8 patients presenting above that age lost independent ambulation and needed a wheelchair by the mean (±SD) age of 21.7±4.4 years (n=7; range 18–31 years). Compared to the most prevalent NBIA subtype, PKAN, MPAN patients were on average older at age of onset and the disease progressed more slowly.

3. NEUROIMAGING

The imaging hallmark in all NBIA subtypes is the eponymous brain iron accumulation, mostly in globus pallidus and substantia nigra (Fig. 3.1). This can be easily pictured by T2-weighted magnetic resonance imaging (MRI). In the two large case series (Hartig et al., 2011; Hogarth et al., 2013), all 47 MPAN cases showed increased iron deposition in the globus pallidus, and all but two also had iron deposition in the substantia nigra. Noteworthy, one of these 47 patients showed an "eye of the tiger" sign, which is otherwise considered pathognomonic for PKAN. Hogarth et al. observed a hyperintense streaking of the medial medullary lamina between the globus pallidus interna and externa in 5 out of 23 subjects that might be mistaken for an "eye of the tiger" sign (Fig. 3.1). In addition to iron accumulation, generalized brain atrophy and/or cerebellar atrophy is found in a fraction of MPAN patients.

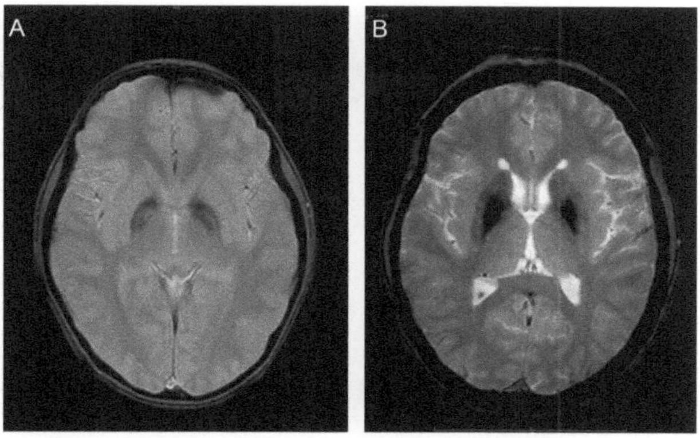

Figure 3.1 Axial T2-weighted magnetic resonance imaging (1.5 T) shows bilateral hypointensity of the globus pallidus in two MPAN patients (A and B). The patient on the left shows additional hyperintense streaking of the medial medullary lamina (A).

4. NEUROPATHOLOGY

The number of reported autopsy studies of MPAN is very limited; only two postmortem examinations are available. Hartig et al. (2011) reported one case who presented with clumsiness and fatigue at the age of 6 years. During the further course of the disease optic atrophy, spastic tetraparesis, ataxia, marked dysarthria, axonal motor neuropathy, and cognitive decline were observed. Death occurred at the age of 23. The case reported by Hogarth et al. presented at a relatively advanced age of 30 years (Hogarth et al., 2013). Dementia was the main clinical finding. The patient died at age 41. Although the scarcity of cases is a limitation, the conformity in the pathological findings of both published cases is intriguing. The presentation is unique and combines iron accumulation, spheroids, Lewy bodies, and tau pathology.

MPAN's neuropathology involves various regions of the brain including the basal ganglia (globus pallidus, substantia nigra, and corpus striatum), the archicortex (in particular the hippocampus), the neocortex, the pons, and the spinal cord. The main site of iron deposition is the globus pallidus followed by the substantia nigra. In other regions, no or minimal signs of iron are detected. The iron is deposited in neurons, astrocytes, and perivascular macrophages.

Like in PKAN, axonal spheroids have been detected in globus pallidus, putamen, and caudate nucleus and at a lower density in the thalamus, internal capsule, brain stem, cerebral cortex, dentate nucleus, and spinal cord. Spheroids show strong immunoreactivity for ubiquitin and faint immunoreactivity for amyloid precursor protein (APP).

Alpha-synuclein-positive Lewy body pathology was present to a higher degree than seen in typical cases of Lewy body disease and included the occurrence of Lewy bodies and Lewy neurites. Lewy bodies and/or Lewy neurites were found in the neocortex, for instance, in the frontal cortex, the hippocampus, brain stem including the substantia nigra and pons, basal ganglia (in particular the globus pallidus), and spinal cord. In the substantia nigra, the occurrence of Lewy bodies and Lewy neurites was accompanied by almost complete neuronal loss.

In addition, the C19orf12 defect also leads to occurrence of hyperphosphorylated tau-containing neuronal inclusions mainly in the hippocampus but also in other brain regions. The localization of the tau protein around pyramidal cell nuclei shows a resemblance to tauopathies. No neurofibrillary tangles were observed in the globus pallidus or the substantia nigra, but tau-positive extracellular structures were detected in the hippocampus.

Further loss of myelin was observed in pyramidal tracts of the spinal cord and most notably in the optic tract.

5. GENETICS

MPAN is an autosomal recessive inherited disorder, which is caused by mutations in the C19orf12 gene. This is a genomically small gene located at chromosome 19q12 (16 kb) with three exons and encodes two isoforms with two alternative first exons (NM_001031726.2 and NM_031448.3) (Fig. 3.2). While the first exon of the shorter isoform (141 amino acids) is not protein-coding, the longer isoform contains a start codon in the exon 1, resulting in a protein eleven amino acids longer (152 amino acids). The C19orf12 proteins are highly conserved in evolution and contain a transmembrane domain. Mitochondrial localization was shown by polyclonal antibodies on cellular fractions, immune fluorescence, and *in vitro* import experiments. The function of this orphan protein is still unclear. *C19orf12* expression occurs mainly in the brain, blood cells, and adipocytes. The expression of *C19orf12* in adipocytes and its coregulation with genes involved in fatty acid metabolism indicate a role of C19orf12 in lipid metabolism.

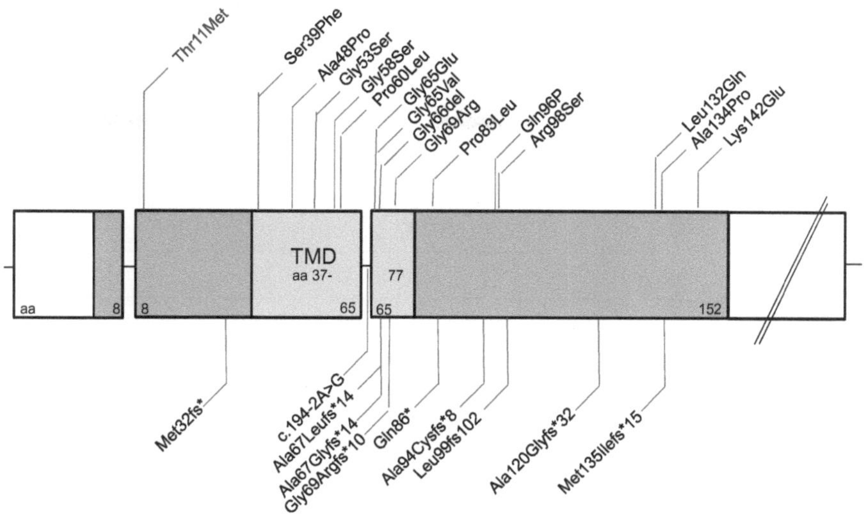

Figure 3.2 Gene structure and identified disease alleles. Gene structure of the larger isoform of *C19orf12* (NM_001031726.2) with the identified mutations. The predicted transmembrane domain is marked in yellow. (For interpretation of the references to color in this figure legend, the reader is referred to the online version of this chapter.)

Twenty-eight different mutations have been described in *C19orf12* in 55 published families (67 cases) including frameshift mutations, missense mutations, nonsense mutations, and splice-site mutations (Table 3.2 and Fig. 3.2). The 11-base pair deletion mutation c.204_214del11 was first identified in a Polish family with three affected siblings (Hartig et al., 2011). The variant is predicted to truncate the C19orf12 protein and has since been identified in 31 other families originating from Poland or other Eastern Europe countries. Haplotype analysis suggests that the 11 bp deletion derives from a common founder at least 50–100 generations ago (Hartig et al., 2011). The second frequent mutation is the p.Thr11Met mutation, which was found in eight families from Turkey, Iran, Poland, and Russia with German origin (Deschauer et al., 2012; Dezfouli et al., 2013; Dogu et al., 2012; Hartig et al., 2011; Schulte et al., 2013). Seven further mutations were identified in two or more families. The remaining mutations ($n = 19$) were private for individual families.

Mutations predicted to delete part of the protein have been found in 42 from 55 families, indicating loss of function in consequence for most disease alleles. This was confirmed by the absence of C19orf12 protein in fibroblast cell lines from patients with the common deletion shown by immunoblot analysis (Hartig et al., 2011). Most missense mutations

Table 3.2 Mutations identified in *C19orf12*

DNA mutation	Protein alteration	Reference
Missense mutations		
c.32C>T	p.Thr11Met	Deschauer et al. (2012), Dezfouli et al. (2013), Dogu et al. (2012), Hartig et al. (2011), Schulte et al. (2013)
c.53A>G	p.Asp18Gly	Schulte et al. (2013)
c.116C>T	p.Ser39Phe	Hogarth et al. (2013)
c.142G>C	p.Ala48Pro	Hogarth et al. (2013)
c.157G>A	p.Gly53Arg	Hartig et al. (2011), Hogarth et al. (2013)
c.172G>A	p.Gly58Ser	Panteghini et al. (2012)
c.179C>T	p.Pro60Leu	Hogarth et al. (2013)
c.194G>A	p.Gly65Glu	Hartig et al. (2011), Hogarth et al. (2013)
c.194G>T	p.Gly65Val	Hogarth et al. (2013)
c.197_199del3	p.Gly66del	Deschauer et al. (2012)
c.205G>A	p.Gly69Arg	Goldman et al. (2013), Hartig et al. (2011), Hogarth et al. (2013)
c.248C>T	p.Pro83Leu	Hogarth et al. (2013)
c.287A>C	p.Gln96Pro	Panteghini et al. (2012)
c.294G>C	p.Arg98Ser	Hogarth et al. (2013)
c.395T>A	p.Leu132Gln	Schulte et al. (2013)
c.400G>C	p.Ala134Pro	Schulte et al. (2013)
c.424A>G	p.Lys142Glu	Goldman et al. (2013), Hartig et al. (2011), Schulte et al. (2013)
Nonsense mutations and frameshift mutations		
c.94delA	p.Met32fs*	Hogarth et al. (2013)
c.177_178insG	p.Leu60Alafs10*	Schottmann et al. (2013)
c.191insG	p.Ala67Glyfs*14	Dezfouli et al. (2013)
c.194delG	p.Ala67Leufs*6	Hogarth et al. (2013)
c.204_214del11	p.Gly69Argfs*10	Deschauer et al. (2012), Goldman et al. (2013), Hartig et al. (2011), Hogarth et al. (2013), Schulte et al. (2013)

Table 3.2 Mutations identified in *C19orf12*—cont'd

DNA mutation	Protein alteration	Reference
c.278dupC	p.Ala94Cysfs*8	Hogarth et al. (2013)
c.297insGCTC	p.Leu99fs102	Panteghini et al. (2012)
c.358dupG	p.Ala120Glyfs*32	Hogarth et al. (2013)
c.404insT	p.Met135Ilefs*15	Dezfouli et al. (2013)
c.256C>T	p.Gln86*	Hogarth et al. (2013)
Splice-site mutations		
c.194-2A>G	p.?	Hogarth et al. (2013)

($n=9$) are clustered in the transmembrane domain ($n=8$) or at the C-terminal end of the protein ($n=6$) (Fig. 3.2). The mutation p.Thr11Met is the only missense mutation, which is located at the N-terminal part of the protein. It affects only the longer isoform of the protein, indicating the important function of this isoform. Since the variant p.Lys142Glu is found in most cases in association with the variant p.Gly69Arg, it was suggested that both variants affect the same allele (Hogarth et al., 2013). Therefore, the pathogenicity of the variant p.Lys142Glu has been a subject of debate. However, we have identified two patients where we found both variants separately in combination with other mutations supporting their pathogenicity (unpublished results).

Seven of clinically typical MPAN patients had a mutation in just one of the two copies of the *c19orf12* gene. Four out of five of these mutations were frameshift or nonsense mutations located at the C-terminal part of the protein (p.G69R, p.Q86X, p.A94CfsX8, p.A120GfsX32, and p.L99fs102) (Hogarth et al., 2013; Panteghini et al., 2012). Given that in the case of autosomal recessive inheritance a pathogenic alteration should affect both alleles, incomplete mutation detection might be one possible explanation. In order to increase sensitivity of genetic testing, deletion screening and sequencing of promoter regions has been performed in some of these cases, but no second mutation was identified (Hogarth et al., 2013; Panteghini et al., 2012; Hartig, unpublished). In one family with only one heterozygous mutation detected (p.A120GfsX32), family history pointed to possible autosomal-dominant inheritance. In this family, the father of the patient showed typical MPAN pathology with widespread abundant Lewy bodies and both axonal spheroids and increased iron in globus pallidus and substantia nigra (Hogarth et al., 2013).

Some genotype–phenotype correlations can be observed in MPAN cases. The p.Thr11Met variant was found in six patients from five families in a homozygous state. Age of onset in this group was at the mean of 25 years (range 21–28) compared to the mean of 11 years of all published MPAN cases. Twenty-four patients with two frameshift or nonsense mutations presented with a mean age of onset of 9.7 years.

6. CONCLUSION

From the clinical perspective, NBIA as a group is diagnosed by MR imaging. The clinical and radiological phenotype of the MPAN subtype is distinctive in some aspects. Considering the high proportion and the small size of the gene with only three exons, C19orf12 should always be analyzed in NBIA patients who do not show the "eye of the tiger" sign. Especially in NBIA patients with optic atrophy, motor axonal neuropathy, psychiatric findings, and cognitive decline, genetic testing of C19orf12 should be performed in order to confirm the clinical diagnosis and provide genetic counseling to the families.

Despite the vast increase in knowledge that has been gained in the past 15 years—mainly driven by the identification of novel NBIA genes—much remains to be learned about the pathophysiological basis of NBIA. In particular, the work on NBIA has unfortunately not led to a causative treatment yet. Currently, a large international consortium funded by the European Commission Seventh Framework Programme (TIRCON, treat iron-related childhood-onset neurodegeneration) tries to overcome the obstacles of rare disease research and to build a framework for improved research and care for NBIA patients.

ACKNOWLEDGMENTS

For this study, the authors acknowledge funding from the European Commission Seventh Framework Programme (FP7/2007-2013, HEALTH-F2-2011, grant agreement No. 277984, TIRCON) and from the German Federal Ministry of Education and Research (BMBF, grant numbers 01GM1113A, 01GM1207, and 01GM1113C) for the German and European networks for mitochondrial disorders (mitoNET and GENOMIT).

Disclosure: TK has been a principal investigator or investigator on industry-sponsored trials funded by Santhera Pharmaceuticals Ltd (idebenone in LHON and idebenone in Friedreich ataxia) and by H. Lundbeck A/S (carbamylated erythropoietin in Friedreich ataxia). He has received research support from government entities (German Research Foundation; German Federal Ministry of Education and Research; European Commission 7th Framework Programme) and from commercial entities (Santhera Pharmaceuticals Ltd; Actelion Pharmaceuticals Ltd; H. Lundbeck A/S). He has been serving on scientific advisory boards for commercial entities (Santhera Pharmaceuticals Ltd; Actelion Pharmaceuticals Ltd) and for nonprofit entities (Center for Rare Diseases Bonn, Germany; Hoffnungsbaum e.V., Germany). He has received speaker honoraria and travel costs from commercial entities (Dr. Willmar Schwabe GmbH & Co. KG; Eisai Co., Ltd; Santhera Pharmaceuticals Ltd; Actelion Pharmaceuticals Ltd; Boehringer Ingelheim Pharma GmbH & Co. KG, GlaxoSmithKline GmbH & Co. KG). He has been doing consultancies for Gerson Lehrman Group, United States, and FinTech Global Capital, Japan. He has been serving as a section editor for BMC Medical Genetics from 2011.

The other authors have nothing to disclose.

REFERENCES

Alazami, A. M., Al-Saif, A., Al-Semari, A., Bohlega, S., Zlitni, S., Alzahrani, F., et al. (2008). Mutations in C2orf37, encoding a nucleolar protein, cause hypogonadism, alopecia, diabetes mellitus, mental retardation, and extrapyramidal syndrome. *American Journal of Human Genetics*, *83*(6), 684–691.

Curtis, A. R., Fey, C., Morris, C. M., Bindoff, L. A., Ince, P. G., Chinnery, P. F., et al. (2001). Mutation in the gene encoding ferritin light polypeptide causes dominant adult-onset basal ganglia disease. *Nature Genetics*, *28*(4), 350–354.

Deschauer, M., Gaul, C., Behrmann, C., Prokisch, H., Zierz, S., & Haack, T. B. (2012). C19orf12 mutations in neurodegeneration with brain iron accumulation mimicking juvenile amyotrophic lateral sclerosis. *Journal of Neurology*, *259*(11), 2434–2439.

Dezfouli, M. A., Alavi, A., Rohani, M., Rezvani, M., Nekuie, T., Klotzle, B., et al. (2013). PANK2 and C19orf12 mutations are common causes of neurodegeneration with brain iron accumulation. *Movement Disorders: official journal of the Movement Disorder Society*, *28*(2), 228–232.

Dogu, O., Krebs, C., Kaleagasi, H., Demirtas, Z., Oksuz, N., Walker, R., et al. (2012). Rapid disease progression in adult-onset mitochondrial membrane protein-associated neurodegeneration. *Clinical Genetics*, *84*(4), 350–355.

Goldman, J. G., Eichenseer, S. R., Berry-Kravis, E., Zimnowodzki, S., Gregory, A., Hogarth, P., et al. (2013). Clinical features of neurodegeneration with brain iron accumulation due to a C19orf12 gene mutation. *Movement Disorders: Official Journal of the Movement Disorder Society*, *28*(10), 1462–1463.

Haack, T. B., Hogarth, P., Kruer, M. C., Gregory, A., Wieland, T., Schwarzmayr, T., et al. (2012). Exome sequencing reveals de novo WDR45 mutations causing a phenotypically distinct, X-linked dominant form of NBIA. *American Journal of Human Genetics*, *91*(6), 1144–1149.

Hallervorden, J. (1924). Uber eine familiare Erkrankung im extrapyramidalen System. *Deutsche Zeitschrift für Nervenheilkunde*, *81*, 204–210.

Harris, Z. L., Takahashi, Y., Miyajima, H., Serizawa, M., MacGillivray, R. T., & Gitlin, J. D. (1995). Aceruloplasminemia: Molecular characterization of this disorder of iron metabolism. *Proceedings of the National Academy of Sciences of the United States of America*, *92*(7), 2539–2543.

Hartig, M. B., Hortnagel, K., Garavaglia, B., Zorzi, G., Kmiec, T., Klopstock, T., et al. (2006). Genotypic and phenotypic spectrum of PANK2 mutations in patients with neurodegeneration with brain iron accumulation. *Annals of Neurology*, *59*(2), 248–256.

Hartig, M. B., Iuso, A., Haack, T., Kmiec, T., Jurkiewicz, E., Heim, K., et al. (2011). Absence of an orphan mitochondrial protein, c19orf12, causes a distinct clinical subtype of neurodegeneration with brain iron accumulation. *American Journal of Human Genetics*, *89*(4), 543–550.

Hayflick, S. J., Westaway, S. K., Levinson, B., Zhou, B., Johnson, M. A., Ching, K. H., et al. (2003). Genetic, clinical, and radiographic delineation of Hallervorden–Spatz syndrome. *The New England Journal of Medicine*, *348*(1), 33–40.

Hogarth, P., Gregory, A., Kruer, M. C., Sanford, L., Wagoner, W., Natowicz, M. R., et al. (2013). New NBIA subtype: Genetic, clinical, pathologic, and radiographic features of MPAN. *Neurology*, *80*(3), 268–275.

Hortnagel, K., Prokisch, H., & Meitinger, T. (2003). An isoform of hPANK2, deficient in pantothenate kinase-associated neurodegeneration, localizes to mitochondria. *Human Molecular Genetics*, *12*(3), 321–327.

Kruer, M. C., Paisan-Ruiz, C., Boddaert, N., Yoon, M. Y., Hama, H., Gregory, A., et al. (2010). Defective FA2H leads to a novel form of neurodegeneration with brain iron accumulation (NBIA). *Annals of Neurology*, *68*(5), 611–618.

Morgan, N. V., Westaway, S. K., Morton, J. E., Gregory, A., Gissen, P., Sonek, S., et al. (2006). PLA2G6, encoding a phospholipase A2, is mutated in neurodegenerative disorders with high brain iron. *Nature Genetics*, *38*(7), 752–754.

Panteghini, C., Zorzi, G., Venco, P., Dusi, S., Reale, C., Brunetti, D., et al. (2012). C19orf12 and FA2H mutations are rare in Italian patients with neurodegeneration with brain iron accumulation. *Seminars in Pediatric Neurology*, *19*(2), 75–81.

Schneider, S. A., Paisan-Ruiz, C., Quinn, N. P., Lees, A. J., Houlden, H., Hardy, J., et al. (2010). ATP13A2 mutations (PARK9) cause neurodegeneration with brain iron accumulation. *Movement Disorders: Official Journal of the Movement Disorder Society*, *25*(8), 979–984.

Schottmann, G., Stenzel, W., Lutzkendorf, S., Schuelke, M., & Knierim, E. (2013). A novel frameshift mutation of C19ORF12 causes NBIA4 with cerebellar atrophy and manifests with severe peripheral motor axonal neuropathy. *Clinical Genetics*, http://dx.doi.org/10.1111/cge.12137.

Schulte, E. C., Claussen, M. C., Jochim, A., Haack, T., Hartig, M., Hempel, M., et al. (2013). Mitochondrial membrane protein associated neurodegenration: A novel variant of neurodegeneration with brain iron accumulation. *Movement Disorders: Official Journal of the Movement Disorder Society*, *28*(2), 224–227.

Zhou, B., Westaway, S. K., Levinson, B., Johnson, M. A., Gitschier, J., & Hayflick, S. J. (2001). A novel pantothenate kinase gene (PANK2) is defective in Hallervorden–Spatz syndrome. *Nature Genetics*, *28*(4), 345–349.

CHAPTER FOUR

BPAN: The Only X-Linked Dominant NBIA Disorder

Tobias B. Haack[*,†], **Penny Hogarth**[‡,§], **Allison Gregory**[‡], **Holger Prokisch**[*,†,1], **Susan J. Hayflick**[‡,§,¶]

[*]Institute of Human Genetics, Technische Universität München, Munich, Germany
[†]Institute of Human Genetics, Helmholtz Zentrum München, Munich, Germany
[‡]Department of Molecular and Medical Genetics, Oregon Health and Science University, Portland, OR, USA
[§]Department of Neurology, Oregon Health and Science University, Portland, OR, USA
[¶]Department of Paediatrics, Oregon Health and Science University, Portland, OR, USA
[1]Corresponding author: e-mail address: Prokisch@helmholtz-muenchen.de

Contents

Abstract

Beta-propeller protein-associated neurodegeneration (BPAN) is the most recently identified subtype of neurodegeneration with brain iron accumulation (NBIA), being unique with respect to the underlying disease genetics, the associated clinical presentation, and the suggested pathomechanism. Mutations in X-chromosomal *WDR45* arise *de novo*; however, the dominant pattern of inheritance is unusual for an X-linked disorder and additional mechanisms such as X-inactivation or somatic mosaicism are likely to contribute to the phenotype that is indistinguishable between males and females. The course of the disease is two-staged with developmental delay and intellectual disability in childhood and a second phase of rapid neurological deterioration characterized by parkinsonism and dementia occurring in adolescence or early adulthood. At this time, neuroimaging findings are characteristic and provide excellent diagnostic guidance. There is increasing evidence that human WDR45 deficiency impairs autophagy, thereby raising the possibility that this rare disorder will offer insights into more common neurodegenerative disorders such as Parkinson or Alzheimer disease.

International Review of Neurobiology, Volume 110
ISSN 0074-7742
http://dx.doi.org/10.1016/B978-0-12-410502-7.00005-3

1. INTRODUCTION

Beta-propeller protein-associated neurodegeneration (BPAN) is a recognizable NBIA disorder based on the clinical features, the MRI pattern, and the natural history of disease. The recent discovery of causative mutations in *WDR45* has provided a beginning to investigate the biological basis of this rare disorder. *WDR45* localizes to the X chromosome; hence, the genetic factors that underlie this disorder and its manifestation are complex, multifaceted, and yet to be fully understood. Nevertheless, most clinicians can recognize patients likely to have BPAN and confirm the diagnosis with clinical genetic testing.

2. CLINICAL FEATURES

The phenotype of BPAN was first recognized by Hogarth, Gregory, and Hayflick in 2002. Several idiopathic NBIA patients were examined and found to have a similar clinical picture and pattern of disease progression that differed from that observed in other NBIA disorders. The typical pattern is normal birth history with a diagnosis in early childhood of developmental delay and intellectual disability. In adolescence or early adulthood, patients undergo a period of neurological regression with onset of parkinsonism and cognitive decline from their already abnormal baseline. The regression typically prompts a brain MRI, which shows iron accumulation in the basal ganglia. Patients continue to deteriorate with worsening parkinsonism, leading to death by middle age.

Children with BPAN may be recognizable prior to the period of clinical deterioration. In addition to manifesting global developmental delay, these children usually show expressive language delay disproportionate to their other disabilities. Epilepsy is frequently part of the clinical picture, with complex partial, atonic, absence, or generalized tonic–clonic seizures all observed and with some children exhibiting multiple seizure types. Rett-like behaviors and stereotypies are common; many children who are eventually diagnosed with BPAN carry an initial diagnosis of atypical Rett syndrome. Some overlap with atypical Angelman syndrome is evident, as well. Individuals with features of Rett or Angelman syndromes and lacking a genetic or chromosomal abnormality to explain their disease should be tested for BPAN. Other features of BPAN that are evident in childhood include dysfunctional sleep and ocular defects.

Regression in BPAN usually occurs in adolescence or early adulthood and begins with bradykinesia, freezing of gait, and rigidity. The onset of parkinsonism may be subtle, but over time, clear signs and symptoms are evident. The parkinsonism in BPAN is responsive to L-DOPA; however, the duration of benefit lasts usually only a few years and is limited by the development of disabling dyskinesias. With disease progression, patients typically have progressive cognitive decline with loss of limited expressive language skills, as well. Further details of the clinical phenotype and natural history can be found in a recently published paper (Hayflick et al., 2013). Brain MRI done after the onset of regression will show iron accumulation in the basal ganglia and prompt suspicion of one of the NBIA disorders.

3. BRAIN IMAGING

The natural history of BPAN is unusual, and the pattern of abnormalities seen on brain MRI is also distinctive. Iron appears as hypointense signal on T2-weighted imaging. In BPAN, iron is abundant in the substantia nigra and globus pallidus once there are clinical manifestations of parkinsonism (Fig. 4.1). On T1-weighted imaging, the substantia nigra/cerebral peduncles show a hyperintense "halo," which is very distinctive and can be very helpful in guiding diagnostic genetic testing. These changes are found after clinical deterioration. The early MRI in BPAN either is normal or shows only nonspecific cerebral atrophy.

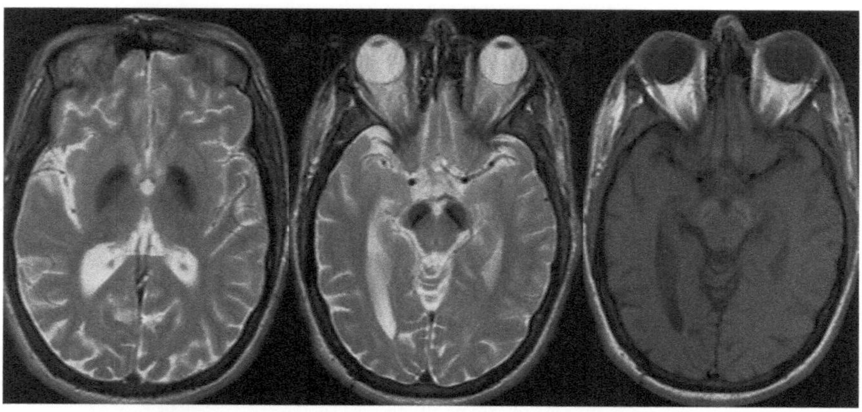

Figure 4.1 Brain MRI in BPAN showing hypointense signal on T2-weighted sequence representing iron in globus pallidus (A) and substantia nigra (B) and T1-weighted sequence showing the hyperintense "halo" in substantia nigra.

4. HISTOPATHOLOGY

Macroscopic neuropathological findings of a female BPAN patient included mild cerebellar atrophy, thinned cerebral peduncles, and dark gray-brown aspects of the substantia nigra and the globus pallidus. Microscopic work-up confirmed marked iron deposition in the substantia nigra and globus pallidus in co-occurrence with siderophages, axonal spheroids, and severe neuronal loss. While Lewy bodies and amyloid-beta plaques were absent, numerous tau-positive tangles were observed in various brain regions. For details, see Hayflick et al. (2013).

5. DIAGNOSIS

BPAN should be suspected when the characteristic natural history, clinical features, and brain MRI changes are observed. Clinical genetic testing is available to confirm a suspected diagnosis. While the BPAN phenotype represents the spectrum of disease as we currently recognize it, there are compelling reasons to expect this spectrum to broaden and to include patients with more mild and more severe manifestations. The basis for this derives from the interesting genetics of BPAN, including X chromosome linkage, somatic cell mosaicism, and skewing of X chromosome inactivation.

6. GENETICS

X-chromosomal *WDR45 de novo* mutations were identified in an exome sequencing study investigating a group of 14 NBIA index cases selected for distinct clinical presentation (Haack et al., 2012). Data analysis based on a recessive model of inheritance failed to identify mutations in known NBIA disease genes or obvious candidates. In contrast, considering dominant-acting mutations as a potential pathomechanism, a search for heterozygous mutations detected pathogenic mutations in *WDR45* in 13 individuals. Subsequent screening of unresolved NBIA cases revealed *WDR45* mutations in an additional 10 patients (Hayflick et al., 2013). Another study applied exome sequencing to investigate two index cases together with healthy family members enabling a search for *de novo* variation. *WDR45* mutations were detected in both individuals, and subsequent mutation screening revealed pathogenic mutations in an additional three index cases (Saitsu et al., 2013). Together, a total of 28 simplex cases have been reported to date.

Only two patients share the same mutation. Of the 27 different disease alleles, apart from three missense mutations affecting evolutionarily highly conserved amino acid residues, the remaining alleles are predicted loss-of-function mutations. With *WDR45* located on the X chromosome, one would expect to see mainly male affected in case of a recessive pattern or no gender bias in a dominant model of inheritance. In agreement with a dominant model, all mutations are known or suspected to be *de novo*. However, 25 out of 28 patients were females, suggesting that *WDR45* mutations are lethal in most males. Consistently, in at least one male patient, the mutation occurred postzygotic leading to somatic mosaicism (Haack et al., 2012). Somatic mutations in males and females together with skewing of X chromosome inactivation in females could on one hand explain the similar phenotype in both genders and would on the other hand predict a much broader range of phenotypes still found to be associated with *WDR45* mutations.

7. BIOLOGY

WDR45 belongs to the large family of WD40 proteins that share a conserved motif of about 40 amino acids terminating with tryptophan-aspartate (W–D) (Li & Roberts, 2001). WD40 proteins promote protein–protein interactions and play a role in various cellular processes including cell cycle control, translation regulation, signal transduction, and autophagy. WDR45 has a beta-propeller tertiary structure and a conserved domain for interaction with phospholipids. It has been shown to physically interact with the autophagy factors ATG2A and ATG2B (Behrends, Sowa, Gygi, & Harper, 2010). Knockdown studies and investigation of the *WDR45* ortholog *epg6* in *Caenorhabditis elegans* suggested a crucial role in the formation of the autophagosome with *epg6* silencing leading to an accumulation of early autophagic structures (Lu et al., 2011). Saitsu et al. showed that patient-derived lymphoblast cell lines have a decrease in autophagic flux compared to controls and accumulate early autophagic structures. While an involvement of autophagy makes perfectly sense also with respect to pathomechanisms discussed in more common neurodegenerative disorders such as Parkinson or Alzheimer disease, other functions of WDR45 yet to be characterized might also contribute to the disease phenotype. More extensive investigations of patients' brain samples or animal models might be valuable tools to address these questions. These future studies might also provide clues why involvement of the very basic pathway autophagy via *WDR45* mutations predominantly affects the brain, unlike recessive mutations in another autophagy-related factor, *EPG5*, which causes a multisystemic disorder, Vici syndrome, characterized by callosal agenesis, cataracts, cardiomyopathy, combined immunodeficiency, and hypopigmentation (Cullup et al., 2013). The role of defective autophagy in NBIA might also be substantiated by assigning the molecular genetic basis of so far idiopathic NBIA cases to this pathway.

REFERENCES

Behrends, C., Sowa, M. E., Gygi, S. P., & Harper, J. W. (2010). Network organization of the human autophagy system. *Nature*, *466*(7302), 68–76.

Cullup, T., Kho, A. L., Dionisi-Vici, C., Brandmeier, B., Smith, F., Urry, Z., et al. (2013). Recessive mutations in EPG5 cause Vici syndrome, a multisystem disorder with defective autophagy. *Nature Genetics*, *45*(1), 83–87.

Haack, T. B., Hogarth, P., Kruer, M. C., Gregory, A., Wieland, T., Schwarzmayr, T., et al. (2012). Exome sequencing reveals de novo WDR45 mutations causing a phenotypically distinct, X-linked dominant form of NBIA. *American Journal of Human Genetics*, *91*(6), 1144–1149.

Hayflick, S. J., Kruer, M. C., Gregory, A., Haack, T. B., Kurian, M. A., Houlden, H. H., et al. (2013). Beta-propeller protein-associated neurodegeneration: A new X-linked dominant disorder with brain iron accumulation. *Brain*, *136*, 1708–1717.

Li, D., & Roberts, R. (2001). WD-repeat proteins: Structure characteristics, biological function, and their involvement in human diseases. *Cellular and Molecular Life Sciences*, *58*(14), 2085–2097.

Lu, Q., Yang, P., Huang, X., Hu, W., Guo, B., Wu, F., et al. (2011). The WD40 repeat PtdIns(3)P-binding protein EPG-6 regulates progression of omegasomes to autophagosomes. *Developmental Cell*, *21*(2), 343–357.

Saitsu, H., Nishimura, T., Muramatsu, K., Kodera, H., Kumada, S., Sugai, K., et al. (2013). De novo mutations in the autophagy gene WDR45 cause static encephalopathy of childhood with neurodegeneration in adulthood. *Nature Genetics*, *45*, 445–449.

CHAPTER FIVE

Neuroferritinopathy

Michael J. Keogh*, Christopher M. Morris†, Patrick F. Chinnery*,1
*Wellcome Centre for Mitochondrial Research, Institute of Genetic Medicine, International Centre for Life, Newcastle University, Newcastle upon Tyne, United Kingdom
†Medical Toxicology Centre, Wolfson Unit of Clinical Pharmacology, Newcastle University, Newcastle Upon Tyne, United Kingdom
[1]Corresponding author: e-mail address: patrick.chinnery@newcastle.ac.uk

Contents

International Review of Neurobiology, Volume 110
ISSN 0074-7742
http://dx.doi.org/10.1016/B978-0-12-410502-7.00006-5

Abstract

Neuroferritinopathy remains the only autosomal dominant syndrome of neurodegeneration with brain iron accumulation (NBIA). While the majority of identified cases appear to be part of an extended pedigree in the northeast of England, patients are increasingly being identified across the globe. Since its discovery in 2001, there have been significant developments in our understanding of the pathological, radiological, and clinical aspects of the condition, though several key pathomechanistic questions, and crucially treatment paradigms, remain unaddressed. This chapter summarizes the genetic etiology, pathological, radiological, and clinical data from all published data to date and suggested potential new avenues for therapy.

1. INTRODUCTION

Neuroferritinopathy (MIM 606159, also called hereditary ferritinopathy and previously neurodegeneration with brain iron accumulation type 2, NBIA2) is the only known autosomal dominant syndrome of neurodegeneration with brain iron accumulation (NBIA). Originally discovered in a large pedigree in the northeast of England, there are now over 70 reported cases in the United Kingdom, France, Portugal, the United States, Japan, and Australia.

Clinically, the condition is characterized by a slowly progressive adult-onset movement disorder often associated with a heterogeneous cognitive and neuropsychiatric phenotype, and all reported cases occur due to mutations in the *FTL1* gene. Mutations within this gene result in disruption of the ferritin light-chain protein that results in an abnormal configuration of the ferritin molecule, resulting in an accumulation of iron, predominantly within the basal ganglia. The condition provides a direct link between abnormal iron metabolism and neurodegeneration. At present, there is no proven treatment for neuroferritinopathy and treatment is mainly symptomatic.

1.1. Genetics

The *FTL1* gene is 1577 bases in length and is located on chromosome 7 of the human genome. The gene contains four exons and encodes a single 175-amino acid protein, the ferritin light chain. *FTL1* is however only 1 of 3 ferritin genes in the human genome. The ferritin heavy-chain gene (*FTH*) is encoded on chromosome 5, and the 183-amino acid ferritin heavy-chain protein combines with the light chain to form cytosolic ferritin as a 24 subunit complex (Friedman, Arosio, Finazzi, Koziorowski, & Galazka-Friedman, 2011). Secondly, also on chromosome 5 lies the *FTMT* gene that encodes the precursor of the mitochondrial ferritin protein (FtMt). The structure and function of the three ferritins are discussed further in the biochemistry section.

To date (April 2013), there have been seven reported pathogenic mutations in the *FTL1* gene that have been described as causing neuroferritinopathy (Curtis et al., 2001; Devos et al., 2009; Kubota et al., 2009; Maciel et al., 2005; Mancuso et al., 2005; Ohta et al., 2008; Vidal et al., 2004). Six of the seven mutations occur within exon 4, with the exception being that described by Maciel et al. in exon 3 (Maciel et al., 2005). All pathogenic mutations within exon 4 are base insertions or duplications ranging from 1 to 16 bases in length causing a frameshift in the *FTL* gene. The single reported case of a mutation in exon 3 is the only single-base substitution reported (Table 5.1; Fig. 5.1).

To date, the most widely described and studied genotype is that of the c460*Ins*A mutation (Curtis et al., 2001). Based on the data from this large

Table 5.1 A summary of the described pathogenic mutations in the *FTL1* gene (April 2013)

Exon	Exonic position	Mutation	Helix disrupted	Reference
3	474G>A	Guanine to adenine base substitution	–	Maciel et al. (2005)
4	461InsC	An insertion of cytosine at bases 461 resulting in a frameshift	D and E	Mancuso et al. (2005)
4	461–462 4 bp dup	A four-base duplication at base 461/462 resulting in a frameshift	D and E	Kubota et al. (2009)
4	460InsA	A single-base insertion at base 460 resulting in a frameshift	D and E	Curtis et al. (2001)
4	469–484 dup	A 16-base duplication of bases 469–484 resulting in a frameshift	E	Ohta et al. (2008)
4	498InsTC	Insertion of thymine and cytosine resulting in a frame shift	E	Vidal et al. (2004)
4	458InsA	Insertion of an adenine at position 458 resulting in a frameshift	D and E	Devos et al. (2009)

Exonic position refers to the base number within the exon at which the mutation occurs. Helix disrupted: The helix in the α-chain of the FTL protein that is disrupted.

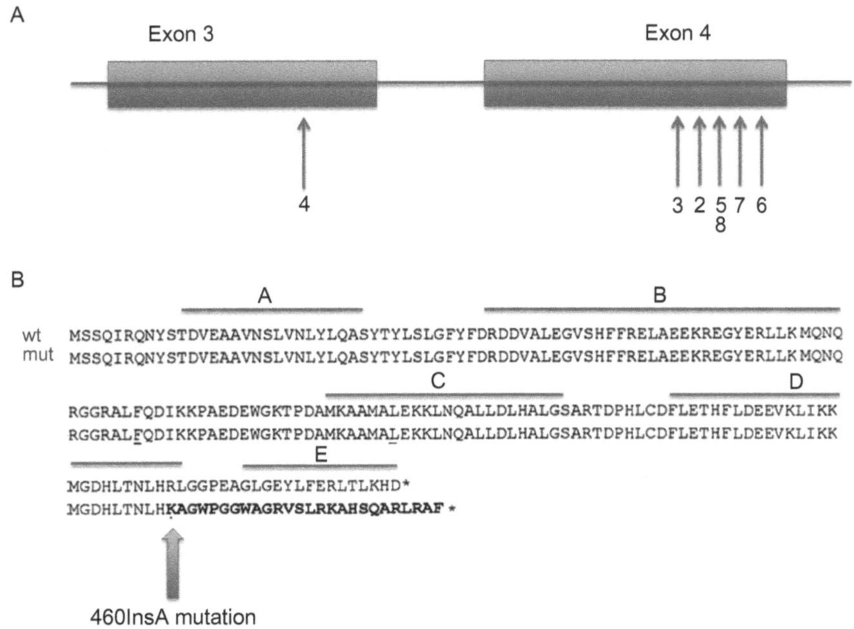

Figure 5.1 (A) Schematic of exons 3 and 4 of the *FTL1*gene. Numbers correspond to references describing the original mutations seen at each point. (B) The amino acid sequence of the *FTL1* gene. The 460*Ins*A mutation replaces the C-terminal 22-amino acid chain with a different 26-amino acid chain extending the peptide by four amino acids. (For interpretation of the references to color in this figure legend, the reader is referred to the web version of this article.)

pedigree, it is believed that this mutation has 100% penetrance. The number of cases from different genotypes is relatively low, though incomplete penetrance has been suggested in one small pedigree carrying the 461*Ins*C mutation (Mancuso et al., 2005).

1.2. Established mutations

The base insertions and duplications in exon 4 all extend the normal 22 C-terminal amino acid sequence of the ferritin light chain by varying degrees (Crompton et al., 2002). In the case of the most frequently seen mutation, the 460*Ins*A mutation, this is from 22 to 26 amino acids, elongating the total protein from 175 to 179 amino acids in length (Fig. 5.1). The location of the frameshift mutations in exon 4 results in specific alterations in the ferritin protein, with insertions close to the 5′ end of exon 4 (Curtis et al., 2001; Devos et al., 2009; Kubota et al., 2009; Mancuso et al., 2005), altering the structure of part of helix D and all of helix E though mutations in the 3′

end of exon 4 (Ohta et al., 2008; Vidal et al., 2004) only disrupt helix E (Kubota et al., 2009). In the case of the single-base substitution in exon 3 described by Maciel et al. (2005), a substitution of alanine to threonine occurs in the amino acid chain; however, this also appears to result in alterations to the protein tertiary structure in the same C-terminal region as the mutations described in exon 4 (Maciel et al., 2005).

2. BIOCHEMISTRY

Ferritins are vital proteins within the human body. While they possess a diverse array of functions, their primary role is to provide a dynamic storage system of iron that is able to respond to cellular demands while also protecting the intracellular environment from potential oxidative damage (Friedman et al., 2011).

2.1. The function of ferritin

The major biochemical reaction with which ferritin is involved is the oxidation of iron in its Fe (II) state to the Fe (III) state. Iron in its oxidized state can subsequently be deposited within the ferritin cavity in a ferric oxyhydroxide form. The benefit of this process for the intracellular environment is that "free" iron in its Fe(II) state can become involved in Fenton-like reactions, transforming molecules like hydrogen peroxide (H_2O_2) into potentially toxic free radicals ($Fe^{2+} + H_2O_2 \rightarrow Fe^{3+} + OH\cdot + OH^-$) (Arosio & Levi, 2010).

The ferroxidase reaction, which takes place on the H chain of ferritin, though not fully understood, enables Fe (II) to be oxidized to the Fe (III) safely without the production of free radicals. While H_2O_2 is produced as an intermediate, the ferroxidase reaction is able to use it as an oxidant preventing free-radical formation (Zhao et al., 2003). This ability of ferritin to oxidize Fe (II) while preventing the formation of free radicals and store iron in a "safe" form likely explains its presence in almost all living systems.

2.2. Cytosolic ferritin

Two types of ferritin exist: cytosolic ferritin, which is a mixture of ferritin heavy- and light-chain proteins, and mitochondrial ferritin encoded by the mitochondrial ferritin chain gene *FTMT*.

Cytosolic ferritin is an intracellular spherical heteropolymer consisting of 24 subunits (Friedman et al., 2011). Each subunit (H or L type) contains essentially the same 3D structure: a bundle of four helices (A–D) and a C-terminus containing a fifth short E helix that is involved in stabilizing interactions around the fourfold symmetry axes (Harrison & Arosio, 1996). Together, a mixture of the heavy chain (FTH) and light chain (FTL) form the ferritin spherical shell in the cytosol, and it is within this shell that up to 4000 iron atoms (Harrison & Arosio, 1996), though on average 2000–3000, are stored (Chasteen & Harrison, 1999). Iron is able to translocate from within this cavity into the cytoplasm via six hydrophilic channels.

Although both FTH production and FTL production are tightly controlled by the same iron-dependent post transcriptional control (reviewed extensively elsewhere (Mikulits, Schranzhofer, Beug, & Mullner, 1999)), transcribed FTH and FTL within the cell rarely form homopolymers and generally coassemble to form a large array of isoferritins (H_2L_{22}, H_4L_{20}, etc.) with their constitution varying markedly between different tissues due to the functionality of the different chains. H chains possess ferroxidase activity with direct enzymatic activity occurring on the chain, with some prokaryotic cells only possessing the H chain form (Arosio, Ingrassia, & Cavadini, 2009). However, such high numbers of catalytic sites are probably in excess of cellular requirements and paradoxically result in functional inefficiency. Higher eukaryotes therefore possess the L chain, which while possessing no catalytic activity itself contains acidic groups on the cavity surface that facilitate iron hydrolysis and mineralization (Santambrogio et al., 1993). It is believed that even a low relative proportion of L to H chains can significantly improve overall iron sequestration, iron transfer from ferroxidase sites to the core, and therefore overall efficiency of ferritin (Arosio et al., 1991). The subunit composition of ferritin found in brain is of a high H to L chain ratio, reflecting high ferroxidase activity and high antioxidant capacity when functioning correctly. In comparison, tissues that have a high iron storage capacity such as the spleen and liver have a higher L chain ratio (Harrison & Arosio, 1996).

The second role of ferritin is its role in the dynamic storage of iron, being able to both sequester and release iron during times of low and high cellular demand, respectively. While the process of iron release is not entirely understood, a channel-gated pathway release (in which maintaining the geometry of the ferritin chains is a vital factor in the ability to control iron (Jin, Takagi, Pancorbo, & Theil, 2001)) and lysosomal proteolysis pathways have both been suggested (Kidane, Sauble, & Linder, 2006). Under stressful intracellular conditions, ferritin is also able to release large quantities of iron by lysosomal autophagy (Larson, Howie, & So, 2004) and proteasomal pathways (Mehlhase, Sandig, Pantopoulos, & Grune, 2005).

2.3. Ferritin function in neuroferritinopathy

Transgenic animal models containing mutations in *FTH* and *FTL* have directly enabled the study of the biological consequences of ferritin mutations. Studies in which mutations have been incorporated into the *FTH* gene have shown that its deletion in transgenic knockout mice resulted in death in the early stages of development *in utero* (Ferreira et al., 2000), though mice carrying heterozygous mutations appear without overt features of iron accumulation, but with significantly increased oxidative stress within the brain (Thompson et al., 2003). In addition, overexpression of *FTH* in a transgenic cell model resulted in increased ability to withstand induced oxidative stress (Cozzi et al., 2000), while in a second experiment, overexpression of either wild type or mutant *FTH* was able to suppress apoptosis (Cozzi et al., 2003).

In contrast, over- or underexpression of *FTL* appears to cause little pathological or functional effects in transgenic mice (Cozzi et al., 2004; Cremonesi et al., 2004), supporting the theory that pathological mutations in the *FTL* gene causing neuroferritinopathy must significantly alter its biological function to induce a neurodegenerative process.

2.4. The effect of FTL mutations

As described in Section 1.2, mutations in the *FTL* gene primarily result in distortion of the C-terminal region of the FTL protein, altering either D or E helical loop (Kubota et al., 2009). These regions are essential in enabling the correct conformation and subsequent stability of the ferritin molecule (Jappelli, Luzzago, Tataseo, Pernice, & Cesareni, 1992). Molecular analysis has revealed that these changes subsequently prevent the C-terminal region from being buried within the central core leaving it exposed within the ferritin cavity (Cozzi et al., 2006). While not completely understood, it has been suggested that failure of the E helices may result in ineffective formation of the hydrophobic channels found in the ferritin shell that facilitate the passage and incorporation of iron (Bou-Abdallah, Biasiotto, Arosio, & Chasteen, 2004; Cozzi et al., 2006). Similar mutations in the *FTH* gene that distort the E helix also significantly alter the ability of ferritin to incorporate iron (Ingrassia, Gerardi, Biasiotto, & Arosio, 2006).

Further evidence supporting a functional impairment of ferritin with *FTL* mutations has been shown in cell models of neuroferritinopathy expressing the 460*Ins*A mutation (Cozzi et al., 2006). In this model, cellular iron levels were significantly increased compared to controls, despite increased expression of *FTH* and *FTL*. Increasing the expression of *FTH* is normally able to decrease cellular iron contents, even resulting in intracellular iron deficiency. However, high intracellular iron content despite increased expression of *FTH* in the 460*Ins*A model suggests that the mutation results in a reduced ability to sequester and regulate iron contents (Cozzi et al., 2006).

Exactly how this functional impairment of ferritin leads to neurodegeneration remains incompletely understood, though it has been hypothesized that an increased ability to induce proteasomal activation may be a central process. Ferritin molecules containing mutant *FTL* undergo an increased rate of degradation secondary to inducing higher rates of proteasomal activity (Cozzi et al., 2004). Cells carrying *FTL* or *FTH* mutations are also more susceptible to oxidative stress (Cozzi et al., 2000, 2004), leading to the possibility that *FTL* mutations increase the activity of proteasomal pathways or shift the balance away from protein production to degradation while simultaneously increasing the free iron pool. An alternative suggestion is that physical iron aggregates themselves are a key mechanism of cellular impairment by providing a physical interruption to normal intracellular processes (Barbeito et al., 2009; Vidal et al., 2008). This however is unlikely to be a key pathological process as subsequent cellular modeling has shown that only a minority of cells have iron aggregates while undergoing apoptosis (Cozzi et al., 2010).

3. PATHOLOGY

The early neuropathological features of NF echo the temporal pattern with which iron is naturally deposited within the normal aging brain. Several studies have demonstrated that iron flux within the brain is unidirectional, with little evidence for any significant iron efflux particularly within the basal ganglia (Dwork et al., 1990; Morris, Keith, Edwardson, & Pullen, 1992; Taylor, Crowe, & Morgan, 1991), and may suggest why the basal ganglia are most markedly affected.

3.1. Macroscopic

The classical macroscopic neuropathological features of neuroferritinopathy are cystic cavitation of the basal ganglia in association with significant iron deposition in the surrounding regions (Curtis et al., 2001; Mancuso et al.,

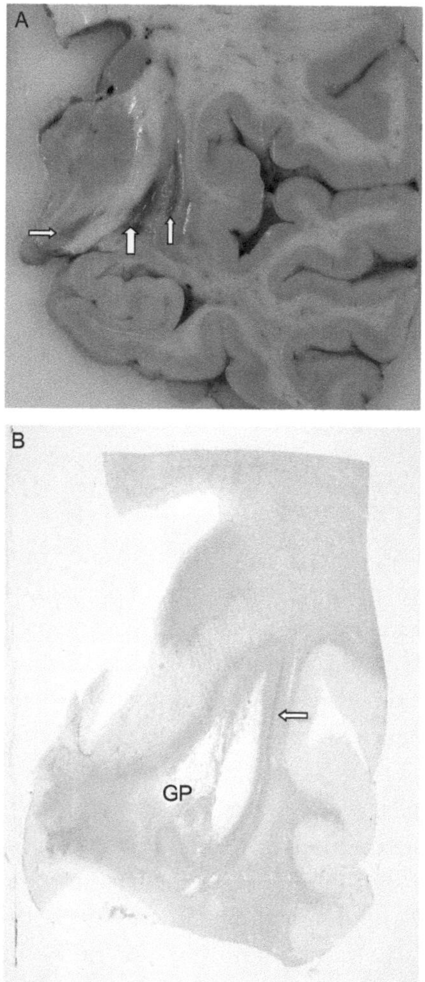

Figure 5.2 Basal ganglia pathology in neuroferritinopathy. (A) Macroscopic picture of the basal ganglia showing the presence of cystic cavities in the globus pallidus (broad vertical arrow) and putamen (narrow vertical arrow), with discoloration of the substantia nigra zona reticulata (horizontal arrow). The cerebral cortex appears unremarkable though a slight discoloration can be seen in the lower layers. Picture reduced to 80%. (B) Perls' stain with DAB intensification in the same case as A showing the cavities in the basal ganglia and the rim of iron staining (arrow) surrounding the putamen. GP, globus pallidus. (For interpretation of the references to color in this figure legend, the reader is referred to the online version of this chapter.)

2005; Vidal et al., 2004; Fig. 5.2). It appears that the putamen, particularly in its posteromedial regions, and the globus pallidus are the two nuclei, which are particularly susceptible to this cavitation (Mancuso et al., 2005 and personal experience). It is not clear however whether cavitation begins directly within the putamen or whether it actually begins in the internal medullary lamina between the putamen and globus pallidus (unpublished observations). Cavitation of the dentate has also been described along with cavitation within the head of the caudate nucleus, while changes in the substantia nigra consistent with cavitation have been seen on radiological imaging (McNeill et al., 2012). A further macroscopic pathological feature that has been described is that of cerebellar cortical damage (Mancuso et al., 2005). This feature appears particularly prominent in the 461*Ins*C and 498*Ins*TC genotypes as described by Vidal and Mancuso, respectively (Mancuso et al., 2005; Vidal et al., 2004), whereas, in comparison, no overt macroscopic cerebellar changes have been described in the 460*Ins*A genotype.

3.2. Microscopic examination

The most striking and characteristic microscopic abnormalities of neuroferritinopathy are intranuclear, intracytoplasmic, and occasionally extracellular spherical inclusions of up to 50 μm in diameter (Curtis et al., 2001; Mancuso et al., 2005; Vidal et al., 2004). These inclusions are most prominent in and surrounding the caudate, putamen, and globus pallidus (Vidal et al., 2004) together with the dentate, red nucleus, and motor cortex (unpublished data). The inclusion bodies in these regions were found predominantly within glial cells or extracellularly, though inclusions could also be seen in the neurons themselves (Vidal et al., 2004). It is important to note that all inclusions do not stain positive for ferritin (Curtis et al. (2001) and unpublished data) but predominantly stain positive with Perls' stain (for iron in the Fe^{3+} state) or Turnbull's stain (for iron in the Fe^{2+} state) (Mancuso et al., 2005; Fig. 5.3).

Microscopic pathological changes are also seen outside the basal ganglia nuclei. In cases of cavitation, surrounding rims of glial cells can be seen that stain positive for iron, together with robust astrocytosis with glial fibrillary acid protein staining (unpublished). In addition, white matter tracts adjacent to the anterior commissure and lateral aspects of the internal capsule appear to show marked vacuolation of glial nuclei (Mancuso et al., 2005). The pathological features of neuroferritinopathy do not appear to affect all of the basal ganglia nuclei, with the thalamus unaffected in almost all cases.

Nonpathognomonic features that have been described within the basal ganglia and deep subcortical regions are axonal swelling and ubiquitinated axonal torpedoes (Curtis et al., 2001), particularly within the putamen

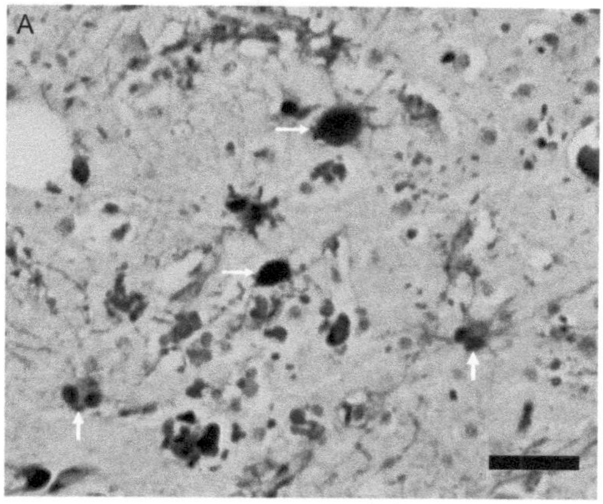

Figure 5.3 Glial pathology in neuroferritinopathy. A Perls' staining of the globus pallidus demonstrating the presence of iron-positive inclusions of various sizes in the cytoplasm of glia (vertical arrows) and free in the neuropil. Some of the larger inclusions show the presence of glial processes (horizontal arrow), suggesting that the inclusions increase in size and eventually engulf the cell. Scale bar = 30 μm. (For interpretation of the references to color in this figure legend, the reader is referred to the online version of this chapter.)

(Mancuso et al., 2005) and the pericystic region of the globus pallidus together with white matter tracts in the same region (Curtis et al., 2001). Such axonal swellings and swollen nuclei do not appear to stain positive for Aβ, amyloid precursor protein, or tau, though are often weakly positive for ubiquitin (unpublished data and Mancuso et al., 2005). There is also well-described neuronal loss in the basal ganglia, which is unsurprisingly most marked within the regions affected by the greatest degree of cavitation. It should be noted that there have been no descriptions of significant Lewy body formation, tau deposition, or neurofibrillary tangles beyond that which would be expected for the patient's age.

Despite only around half of patients clinically showing features of cerebellar dysfunction, this appears to be a region of significant abnormal histopathology, noted in patients with the 461*Ins*C and 498*Ins*TC mutations. Significant loss of Purkinje cells and granule cells has been described, particularly in the vermis (Mancuso et al., 2005), together with infrequent inclusion bodies within the nucleus of both cell types and the cytoplasm of dendrites (Vidal et al., 2004). Mancuso et al. also describe that the white matter between the dentate nucleus and the cerebellar cortex demonstrated a total loss of myelin in the presence of relatively well-preserved axons and the region also showed a significantly increased number of glia but significantly decreased numbers of oligodendrocytes (Mancuso et al., 2005). With regard to the dentate nucleus, Mancuso et al. noted significant quantities of birefringent staining in the dentate nucleus that was thought to represent large quantities of nonerythrocytically derived iron (Mancuso et al., 2005). In our experience, we have seen significant iron deposition, predominantly within glial cells not only within the body of the dentate nucleus but also in the surrounding rim and areas of demyelination in the white matter (unpublished data) (Figs 5.2 and 5.3).

The degree of pathology seen in the cerebral cortex appears variable, ranging from no obvious attributable pathology (Mancuso et al., 2005) to the presence of a mild degree of inclusion body deposition in perineuronal satellite cells throughout the cortex, with the exception of cortical cell layers I and II (Vidal et al., 2004). A transgenic mouse model of the 498*Ins*TC genotype developed by Vidal et al. suggested more prominent cortical intracellular inclusions than that which has been appreciated in humans, though a predominance of basal ganglia pathology remained (Vidal et al., 2008).

3.3. Ultrastructure

Electron microscopy of tissue has revealed that the inclusion bodies appear in the majority of cells to occupy a significant proportion of the nucleoplasm. This often localizes to the nucleus, displacing the chromatin to the edges of the cell creating the appearance of a thin intracellular rim (Mancuso et al., 2005). However, glia cells in particular may show cytoplasmic inclusion bodies that do not localize to the nucleus (unpublished data).

3.4. Nonneuronal pathology

Abnormal iron deposition has been described in several nonneural tissues and has resulted in some authors suggesting that neuroferritinopathy should in fact be called hereditary ferritinopathy (Vidal et al., 2004). The degree of extraneuronal tissue pathology has mainly been studied through transgenic animal modeling, though some data from patients do exist. Vidal et al. studied skin, muscle, and renal sections from a patient with the FTL498*Ins*TC insertion and showed that numerous fibroblasts in the papillary dermis showed intranuclear bodies similar to those seen in the brain (Vidal et al., 2004). They also showed that renal tubules, but not glomeruli, showed the presence of ferritin bodies in the tubular epithelium, while in muscle, inclusion bodies were present in endothelial cells of muscle capillaries, but not muscle cells (Vidal et al., 2004). An examination of nonneuronal pathology from a larger cohort of patients from the same pedigree who possessed the 498*Ins*TC mutation showed a significant heterogeneity of non-neuronal pathology (Ory-Magne et al., 2009). One patient showed ferritin inclusions in the liver, three showed inclusions in the skin, and one showed renal inclusions within the glomerulus, tubules, and fibroblasts (Ory-Magne et al., 2009). Mancuso et al.'s examination of a single patient with the C insertion at nt646–647 did not show any abnormalities in skeletal muscle, choroid plexus, peripheral nerve, or ependymal cells, which was the extent of their examination (Mancuso et al., 2005). While the most common c460*Ins*A mutation has often been reported to have no extraneuronal tissue involvement, we have observed characteristic features of neuroferritinopathy in hepatic tissues of two patients where tissue was available. Some authors have also suggested that there may be a more prominent involvement of muscle and peripheral nerve, though a molecular diagnosis of NF was not proven in the described case (Schroder, 2005).

3.5. Animal models

Transgenic animal models offer a unique opportunity to study the distribution of pathological changes in extraneuronal tissues and this has been performed in animals harboring 498*Ins*TC mutation (Vidal et al., 2008). These studies showed extensive inclusions in a number of tissues, including the skeletal and cardiac muscle, liver, kidney, gastrointestinal tract, adrenal glands, and skin, and in a more extensive extraneural assessment, characteristic pathological changes could be seen in numerous systems (Table 5.2).

3.6. Mitochondrial pathology

Muscle biopsies from patients carrying the 460*Ins*A mutation have shown a range of abnormalities of mitochondrial function (Chinnery et al., 2007). In a cohort of seven patients undergoing muscle biopsies, normal mitochondrial function was detected in a single patient. Two had a combined defect involving multiple respiratory chain complexes (complex I and III, and III and IV), and three had an isolated defect involving complex I (Chinnery et al., 2007). In addition, immunohistochemical analysis of brain tissue from several regions of a patient with the 646InsC genotype showed that neurons in the dentate nucleus and in the subthalamic nucleus displayed decreased immunoreactivity for COX I and COX II and to a lesser degree COX IV, although only COX II appeared to be more selectively decreased in the anterior putamen (Mancuso et al., 2005). Testing of respiratory chain

Table 5.2 Extent of nonneurological pathological changes consistent with excessive accumulation of ferritin and iron in a transgenic animal model of neuroferritinopathy harboring the 498InsTC mutation (Vidal et al., 2008)

Organ system	Cells
Cardiovascular	Cardiac muscle cells
Gastrointestinal	Hepatocytes, striated muscle of the tongue; mucosa, muscularis mucosae, submucosa, and muscularis externa layers of the gastrointestinal tract
Endocrine	Adrenal glands
Renal	Cuboidal cells of the convolute tubules of the kidneys; connective tissue of the urinary bladder
Skin	Fibroblasts of the reticular layer of the skin
Hematological	Endothelial cells of blood vessels of the spleen and mononuclear cells of lymphoid nodules and
Muscle	Striated and smooth muscle
Reproductive	Epithelium of uterine glands in females; seminiferous tubes of the testis, intratesticular genital ducts, and epididymis in males
Respiratory	Alveolar epithelial cells and smooth muscle cells

function from cerebellar tissue from the same patient possibly showed a generalized decrease in respiratory chain function though it was also speculated that it may have been due to postmortem changes (Mancuso et al., 2005).

4. CLINICAL PHENOTYPE

The phenotype of neuroferritinopathy can be broadly described as an adult-onset extrapyramidal movement disorder. It is however apparent that additional features such as cognitive impairment, cerebellar ataxia, and psychiatric disorders are also frequently described. Despite limited numbers of available cases in the literature, it appears that there are likely to be specific genotypic–phenotypic correlations within the disorder (Table 5.3). However, even within the most widely described genotypes such as the 460*Ins*A mutation, a marked degree of clinical heterogeneity can be seen, and the phenotype is constantly expanding. While there have only been a small number of cases described, there do not appear to be any significant sex differences with men and women equally affected.

Table 5.3 Summary of the clinical features described in published case reports of all known genotypes

Demographic data					**Clinical features**								
Genotype	**Ethnic group**	**Study**	**Number of cases (M:F)**	**Age onset**	**Chorea**	**Dystonia**	**Orofacial/ orolingual dyskinesia**	**Tremor**	**Bradykinesia**	**Ataxia**	**Dysarthria/ dysphagia**	**Neuropsychiatric /cognitive**
460InsA	British Caucasian	Chinnery et al.	20:20	39.4	70%	82.5%	65.5%	5%	45%	0%	+	41.7%
	Australian (of British ancestry)	Lehn et al.	5:0	47	+	+	+	+	+			
	American Caucasian	Ondo et al.	1:0	47	–	–	+	–	–	–	+	–
	British Caucasian	Wills et al.	1:0	37	+	+	+	+ (Palatal)	–	+	+	+
498–499InsTC	French	Ory-Magne et al.	7:4	36.8	+	+	+	+	+	+	+	+
458dupA	French	Devos et al.	3:1	24–44	–	+	+	–	+	+	+	+
474G>A	Portuguese	Maciel et al.	1:0	19	–	–	–	–	+	+		+
469–484dup	Japanese	Ohta et al.	1:1	35	–	–	–	+	+		+	+

Continued

Table 5.3 Summary of the clinical features described in published case reports of all known genotypes—cont'd

Demographic data					**Clinical features**							
Genotype	**Ethnic group**	**Study**	**Number of cases (M:F)**	**Age onset**	**Chorea**	**Dystonia**	**Orofacial/ orolingual dyskinesia**	**Tremor**	**Bradykinesia**	**Ataxia**	**Dysarthria/ dysphagia**	**Neuropsychiatric /cognitive**
461–462_4bp_dup	Japanese	Kubota et al.	7:0	52	+	+	+	+		−	−	+
646InsC	American (French-Canadian and Dutch ancestry)	Mancuso et al.	1:1	56	+	+	+	−	−	+	+	+

4.1. 460*Ins*A genotype

The largest study of the clinical features of any cohort of patients with neuroferritinopathy was by Chinnery et al. who defined the clinical phenotype of 40 patients (20 females and 20 males) harboring the 460*Ins*A mutation (Chinnery et al., 2007). The average age of onset of symptoms in this cohort was 39.4 years of age (sd 13.3, range 13–64). The predominant presenting features were either a focal onset chorea (50%) or dystonia (43%) with 76% beginning in the legs and 24% in the arms. The remaining 7% of patients presented with Parkinsonism (3 cases). Several patients also described more subtle movement disorders such as writer's cramp and blepharospasm early in the disease course together with 5% of cases having palatal tremor.

During follow-up (which ranged from 1 to 36 years), the presenting asymmetrical movement disorder remained the predominant phenotype (63% of cases), and the majority of patients developed either dystonia (83%) or chorea (70%) (Table 5.4). There were also atypical movement disorders that developed in a significant proportion of patients, specifically an oromandibular dyskinesia (65%) and an action-specific dystonia (63%) in which coexistent contraction of the platysma and frontalis muscles occurred giving the patients a startled appearance. Thirty-five percent of patients also eventually developed hypomimia and bradykinesia. Most patients remained ambulant 20 years after presentation. Cranial nerve and bulbar features were relatively uncommon with only 5% of patients having documented broken pursuit and 3% a limitation of upgaze. A functional dysphagia was noted in 40%, which was able to be managed by a soft diet (Table 5.4).

Cognitive features were noted in 4 of 9 patients who underwent a formal neuropsychometric assessment in this cohort. The results showed mild defects in verbal fluency in all patients, and two patients had features consistent with a subcortical dementia. Recently, Keogh et al. performed an analysis of 12 patients who had undergone cognitive assessment (Addenbrooke's cognitive examination or neuropsychometry) (Keogh, Singh, & Chinnery, 2013). Eighty-three percent of patients had an abnormality on one of these assessments, with abnormalities ranging from defects in verbal learning to frank psychiatric abnormalities including psychosis. Importantly, neuropsychiatric symptoms began within 5 years of the onset of motor symptoms in all patients suggesting they may be a prevalent but early feature of neuroferritinopathy. Recently, Lehn, Boyle, Brown, Airey, & Mellick (2012) have described the first five cases of neuroferritinopathy in Australia, with the 460*Ins*A mutation identified in two

Table 5.4 Clinical features of 40 symptomatic adult patients with the 460InsA genotype (Chinnery et al., 2007)

Clinical feature		**Number**	**Percentage (%)**
Presenting feature	Chorea	20/40	50
	Dystonia	17/40	42.5
	Parkinsonism	3/40	7.5
	Asymmetry	25/40	62.4
Speech and swallow	Dysarthria	31/40	77.5
	Dysphonia	19/40	47.5
	Orolingual dyskinesia	26/40	65
	Dysphagia	16/40	40
Ocular	Abnormal eye movements	3/40	7.5
Motor	Dystonia	33/40	82.5
	Chorea	28/40	70
	Bradykinesia	14/40	35.5
	Hyperreflexia	7/40	17.5
	Spasticity	0/40	0
	Ataxia	0/40	0
	Tremor	0/40	0
	Positive Babinski	0/40	0

pedigrees of patients originally of British descent. Their cohort presented with a similar array of features such as an orofacial dystonia and upper limb chorea, a rigid akinetic syndrome, upper limb dystonia, and upper limb chorea together with dysarthria. A similar slow progression in patients during follow-up was noted with progression of their motor disorder together with increasing severity or the development of dysarthria and dysphagia.

Finally, there have been isolated case reports by several authors describing single cases. Wills et al. described a case of a 37-year-old man who presented with oral and lingual dyskinetic movements, which progressed over the next 12 years to a phenotype that evolved a severe frontal dementia, dysphagia, generalized dystonia, and chorea together with atypical features of an eyelid apraxia and a 1 Hz palatal tremor (Wills, Sawle, Guilbert, & Curtis, 2002). Ondo et al. also reported a case of a patient aged 49 who presented with facial dyskinesias and vocal tics in the absence of any significant demonstrable features to date (Ondo, Adam, Jankovic, & Chinnery, 2010).

4.2. 458dupA genotype

Four patients from a French pedigree were described by Devos et al. with the 458*dup*A mutation (Devos et al., 2009). The clinical features were largely similar to those described by Chinnery et al. with the 460*Ins*A mutation. Patients between the ages of 24 and 44 presented with focal dystonia and chorea. Early onset of orofacial, laryngeal, or pharyngeal occurred, with all four patients exhibiting such symptoms within 2 years. Within 10 years, all patients either were anarthric or had a severe dysphagia and all required percutaneous feeding.

There were however some stark contrasts to the 460*Ins*A phenotype in two main areas: (1) All four patients had a severe cerebellar syndrome unlike cases in the 460*Ins*A group, and (2) the rate of clinical progression was far more rapid, with a progressive akinetic rigid syndrome in three patients and severe cognitive and behavioral symptoms in all four patients. Several additional and unusual features were also reported such as central sleep apnea, one of whom needed noninvasive ventilation at home and one patient developed a cardiomyopathy.

4.3. 474G>A genotype

There is a single case report of this genotype (Maciel et al., 2005), though two other patients are known to possess the mutation in the family but have not yet been described. The single described patient aged 13 developed a progressive gait disorder together with repeated episodes of acute psychosis over the next 4 years requiring admission to psychiatric institutions. Another patient aged 20 has the following clinical features: mild cerebellar ataxia, cogwheel rigidity, mild bradykinesia, and bilateral upgoing plantars. A marked clinical deterioration resulting in an akinetic rigid syndrome and severe ataxia was noted with neuroleptic medication and resolved upon its withdrawal.

4.4. 498InsTC genotype

Seven patients were described in a French pedigree by Ory-Magne et al. (2009). The mean age of onset was 36.8 years of age (range of 20–50 years old) and patients presented with either a postural tremor (4 of 7 patients), which has not been described in the 460*Ins*A group, or a cerebellar ataxia in combination with dysarthria (3 of 7 patients). Five patients subsequently developed a mixture of chorea and dystonia, and all patients eventually developed a cerebellar syndrome together with dysarthria. No patients exhibited any evidence of autonomic nervous system involvement, and with time, five patients developed Parkinsonism and four developed pyramidal signs. Of note was that for patients who were followed up for more than 10 years, all exhibited a combination of cerebellar dysfunction, Parkinsonism, abnormal involuntary movements, and subcortical cognitive impairment with predominantly frontal features.

4.5. 469–484dup genotype

There is only a single case report of NF caused by a 15-base duplication described in a Japanese man (Ohta et al., 2008). This patient developed an unusual clinical phenotype with a symmetrical upper limb tremor in his teens. By his early forties, he was aphonic and had a subcortical dementia (IQ = 66), and examination revealed a generalized hypotonia, micrographia, and hyporeflexia. Extrapyramidal features of rigidity, chorea dystonia, and spasticity were not observed. Brief clinical information is also given about the patient's mother who also developed a tremor in her teens, a gait disorder before her thirties, and subsequent cognitive impairment and akinetic mutism and died at age 64 prior to the era of molecular genetic testing.

4.6. 461–462_4bp_dup genotype

Kubota et al. described seven patients with a four-base duplication in a Japanese pedigree (Kubota et al., 2009). The mean age of onset was 52 ($n = 6$, range 41–60). The clinical phenotype was well described in three patients. The first presented in his early forties with unilateral upper and lower limb dystonia and chorea together with an orofacial dyskinesia. He subsequently developed depression, emotional lability, and suicidal ideation requiring hospitalization. The second patient developed unilateral choreoathetosis and orofacial dyskinesia and examination revealed generalized hypotonia at the age of 50. His asymmetrical movement disorder progressed but a strong asymmetry remained, and the patient died 15 years later. The third patient, aged 51, presented with asymmetrical chorea, which over a year progressed to involve all four limbs. Five years later, he had a fine postural tremor, various involuntary movements of his limbs and face, dysarthria and dysphagia, and moderate cognitive impairment (IQ = 83).

4.7. 461InsC

There have been two reported cases by Mancuso et al. of patients with this genotype (Mancuso et al., 2005). Both patients came from a family in the United States who were originally of French-Canadian and Dutch descent. One patient, aged 63, presented with a short history of an unsteady gait, dysarthria, and involuntary movements of all four limbs before developing depression in the context of preserved cognition. Clinical examination revealed abnormal saccadic pursuit, dysarthria, facial chorea, and an asymmetrical choreoathetosis. He also had severe ataxia with a positive Romberg's test. He died of pneumonia 5 years later. The second patient described from the pedigree was the proband's sister. She developed a stiff festinant gait at age 49, resulting in several falls. She also developed emotional lability, bilateral blepharospasm, oromandibular dystonia, dysarthria, hypophonia, dysphagia, sialorrhea, impaired dexterity of her hands, and micrographia. Examination revealed blepharospasm and facial, cervical, trunk, and limb dystonia. There was no evidence of any cognitive impairment, Parkinsonism, or ataxia. Table 5.3 describes the major clinical features in all genotypes.

5. RADIOLOGICAL IMAGING—GENERAL PRINCIPLES

MRI imaging has become a pivotal investigation in the investigation of all NBIA disorders. The technique has the ability to reflect underlying pathological changes of iron deposition, gliosis, cavitation, and atrophy (Curtis et al., 2001; Vidal et al., 2004) and may also offer the ability to differentiate between the various disorders(McNeill et al., 2008) and provide a tool to monitor the natural history and the effects of treatment.

With normal aging, there is a progressive accumulation of iron within the brain and in particular within the basal ganglia (Morris et al., 1992). This deposition can be readily seen with MRI imaging techniques (Haacke et al., 2005; Vymazal et al., 1995). Iron deposition is most accurately depicted on standard clinical MRI imaging as hypointensity or signal loss on T_2* sequences, though the more widely utilized T_2 fast spin echo (FSE) sequences are also of significant clinical value in detecting iron deposition (McNeill et al., 2008; Ohta & Takiyama, 2012). Hyperintensity of T_2 signal on the same sequences can also reflect "loose" tissue and gliosis (Koyama & Yagishita, 2006). Iron deposition with T_1-weighted imaging is often isointense and so rarely used to detect such features (Gregory, Polster, & Hayflick, 2009; Ohta & Takiyama, 2012). Though not yet widely utilized in clinical practice, T_2 relaxation time (Chavhan, Babyn, Thomas, Shroff, & Haacke, 2009) is probably an even more accurate assessment of tissue iron content within the brain than standard T_2 or $T_2{}^*$ imaging (Szumowski et al., 2010), and a variety of techniques and methods have been described to assess its deposition with these sequences (Brass, Chen, Mulkern, & Bakshi, 2006; Gelman et al., 1999; Hikita et al., 2005). The ability of $T_2{}^*$ to detect iron deposition particularly in cases where cavitation is yet to become apparent highlights the importance of specifically requesting this sequence when NBIA disorders are a possible diagnosis. In addition, techniques such as susceptibility weight imaging may become promising tools for detecting regional iron variation though they have not yet been utilized in patients with neuroferritinopathy or routinely in clinical practice (Lotfipour et al., 2012).

The noniron deposition related neuroradiological features of neuroferritinopathy such as cavitation, gliosis, and atrophy are also readily visible with standard T_1, T_2, and $T_2{}^*$ imaging. Cavitation is demonstrated by areas of low T_1 and high T_2 signal in comparison to the cerebrospinal fluid (CSF). The early stages of cavity formation are likely to be signified by T_2 hyperintensity and T_1 hypointensity (Keogh, Jonas, Coulthard, Chinnery, & Burn, 2012; Vidal et al., 2004), reflecting the early microcavitation seen on postmortem analysis (Vidal et al., 2004). Many cases of cavitation are also surrounded by a rim of T_2 hypointensity that is likely to reflect iron deposition itself (McNeill et al., 2008). Therefore, regions of the brain in which iron deposition is seen are typified by hypointensity on T_2-weighted imaging, with gliosis and cavitated regions showing T_2 hyperintensity. More general features such as generalized cortical atrophy and in several cases specific cerebellar atrophy have been identified with MRI imaging, and such findings have been shown to correlate with significant cognitive or cerebellar clinical features, respectively (Gregory et al., 2009; Mancuso et al., 2005; Ohta et al., 2008; Vidal et al., 2004).

5.1. Regional MRI findings

The normal features of the basal ganglia on standard T_2-weighted MRI show that the caudate, lentiform, and thalami are slightly hyperintense when compared to white matter but are of the same signal intensity as the cerebral cortex. In the adult brain, the globus pallidus and substantia nigra become slightly hypointense due to the normal iron deposition seen with age, often with the pars compacta being slightly hyperintense compared with the hypointensity of the adjacent structures of the pars reticulata and the red nuclei (Anderson, Costantino, & Stratford, 2004).

In the presymptomatic phase of the disease in patients carrying the c460*Ins*A mutation, Keogh et al. (2012) showed that iron deposition could been seen from very early childhood, with one patient aged 6 showing T_2^* hypointensity in the substantia nigra, globus pallidus, and motor cortex. Two other presymptomatic subjects who were aged between 17 and 36 showed more widespread iron deposition affecting the red nucleus, thalamus, and caudate with increasing involvement of nuclei correlating to age. There was no evidence of any abnormalities with T_1 imaging nor any evidence of T_2 or T_2^* hyperintensity in any individual (Keogh et al., 2012). These findings suggest that iron deposition may be seen decades before the onset of disease and progress from early involvement of the substantia nigra, globus pallidus, and motor cortex to eventually involve the other nuclei of the basal ganglia. Over time, the lesions become gradually more hypointense reflecting progressive iron deposition (Chinnery et al., 2007) and may eventually show T_2 hyperintensity reflecting gliosis or cavitation.

The most comprehensive analysis of the neuroradiological features of neuroferritinopathy was performed by McNeill et al. (2008). They assessed MRI brain scans of 21 symptomatic patients carrying the c460*Ins*A mutation at a mean age of 52 years old. T_2^* MRI imaging showed widespread hypointensity in the cerebral cortex (71%), globus pallidus (38%), putamen (28%), caudate nuclei (14%), thalamus (19%), substantia nigra (81%), and dentate nuclei (95%). In 52% of patients, the globus pallidus and putamen were involved by a confluent area of hyperintensity with a rim of peripheral hypointensity reflecting the likely central cavitation surrounded by a rim of iron deposition, and 48% showed only hypointensity of the basal ganglia nuclei, leading the authors to conclude that patients could be defined into either of these groups based on these findings (McNeill et al., 2008; Fig. 5.4). Two cases of neuroferritinopathy also showed the radiological "eye of the tiger sign" that was previously thought to be pathognomonic for pantothenate kinase-associated neurodegeneration (Hayflick, 2006) though it has also been seen in multisystem atrophy in addition to neuroferritinopathy (Strecker et al., 2007).

T_2—FSE MRI images showed similar results to that of the T_2^* sequences with 9% of patients showing small areas of hyperintensity within the globus pallidus and putamen, while 38% had areas of increased signal in the caudate. Hypointensity of the dentate nuclei (71%) and substantia nigra (52%) was also seen. Hyperintensity of the globus pallidus and putamen surrounded

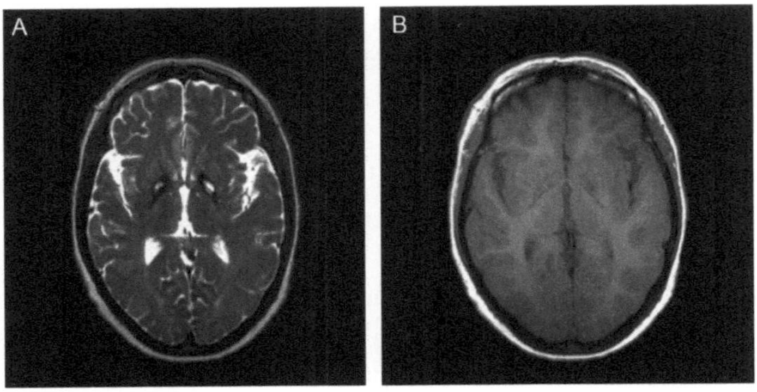

Figure 5.4 (A) A T_2-weighted MRI through the basal ganglia. Cavitation can be easily appreciated in the globus pallidus together with some hyperintensity extending into the caudate nucleus. (B) A T_1-weighted MRI of the same patient showing hypointensity within the same regions.

by a rim of hypointensity suggestive of cavitations and surrounding iron deposition was seen in 52% of patients. Again, the patients could be grouped into two groups based on their basal ganglial appearances with T_2-FSE; 48% showed small areas of hyperintensity within pallida, putamen, and caudates with hypointensity of substantia nigra and dentates, while the second group (52%) had confluent hyperintensity suggestive of probable cavitation of pallida and putamen with hypointensity of substantia nigra and dentate nuclei (McNeill et al., 2008). The authors subsequently compared the radiological features of neuroferritinopathy with other NBIA disorders and came to the conclusion that in cases in which basal ganglia iron deposition is seen, the features suggestive of neuroferritinopathy were (1) hypointensity of the globus pallidus, dentate, motor cortex, and substantia nigra, (2) hypointesity of at least one of the caudate, putamen, or thalamus, and (3) hyperintense regions in the caudate, pallidum, or putamen or confluent hyperintensity of the globus pallidus and putamen (McNeill et al., 2008; Fig. 5.4).

5.2. Clinical correlations

While cerebellar atrophy and generalized cerebral atrophy have been seen in association with cerebellar signs and cognitive impairment in neuroferritinopathy, there has been a paucity of information pertaining to clinical and radiological correlates in the disease. McNeil et al. have however recently described that T_2^* relaxation rate (R_2) in the thalamus does correlate with dystonia rating scale scores in patients carrying the c460*Ins*A mutation and may be a valid method of tracking disease progression in these patients (McNeill et al., 2012). Whether the same radiological markers are valid for the other genotypes of neuroferritinopathy remains to be determined.

6. OTHER CLINICAL INVESTIGATIONS

Routine hematological and biochemical blood tests are invariably normal with the exception of serum ferritin. A recent review of the ferritin levels of published cases of neuroferritinopathy showed that 64% of male and 84% of females had low serum ferritin when using a cutoff of 30 μg/l (Lehn et al., 2012). In addition to blood testing, a small number of patients have had CSF analysis, and this has shown normal constituents in the vast majority of cases (Chinnery et al., 2007; Wills et al., 2002), though a single case showed a mildly raised protein (Chinnery et al., 2007). In addition, nerve conduction studies have been reported to be invariably normal together with electroencephalograms that also are also generally normal, although may occasionally show some generalized but nonspecific slowing (Chinnery et al., 2007; Crompton et al., 2005; Mir, Edwards, Curtis, Bhatia, & Quinn, 2005; Ondo et al., 2010; Wills et al., 2002).

7. TREATMENT

There have been no formal prospective treatment trials of disease-modifying treatment in neuroferritinopathy, though attempts at inducing iron depletion have been attempted. Chinnery et al. described the results of three patients undergoing monthly venesections for 6 months. Two patients received desferrioxamine (4000 mg weekly for 14 months) and one patient took oral deferiprone (2 g three times a day for 2 months). All three treatments produced profound iron depletion, though only one patient showed any significant short-term clinical benefit that failed to last (Chinnery et al., 2007). In addition, Kubota et al. performed monthly venesections (400 ml/mo) for 2 months in a single patient, which failed to result in any clinical benefit (Kubota et al., 2009). Other attempts at disease modification have been tried with the use of coenzyme Q as an antioxidant therapy and have been tried in a small number of patients with no obvious benefits noted (author's unpublished observations).

Given that iron deposition can be seen in the basal ganglia decades before the onset of clinical symptoms (Keogh et al., 2012), iron chelation therapies may be best placed for patients in the presymptomatic or very early stages of disease though this also places significant ethical and financial burdens on such an approach.

7.1. Symptomatic management

The mainstay of treatment for patients with neuroferritinopathy remains symptomatic. Numerous agents have been tried for the control of dystonia, tremor, and chorea and are outlined in Table 5.4. The most promising agents appear to be botulinum for control of focal dystonia together with tetrabenazine for chorea (though three patients reported significant side effects) including dramatic improvements in some cases (Ondo et al., 2010), together with sulpiride and benzhexol for hyperkinetic movements and Parkinsonism, respectively. Benzodiazepines were also well tolerated and offered benefit in several patients, though muscle relaxants such as baclofen have not shown any efficacy. These data must of course be interpreted with caution given the low patient numbers (see Table 5.5), and in our experience, there is a marked heterogeneity of responses between patients and a multitude of agents are tried over time in order to achieve symptomatic improvement.

More invasive treatments have also been used in the care of patients. Firstly, deep brain stimulation was performed in a single case in our institution but provided no improvement and was later removed (author's unpublished observations). In addition, adjunctive treatments such as percutaneous enteral gastrostomies have been extremely effective in managing patients with severe dysphagia, with many patients surviving several years after the procedure. Given the complex needs of many of these patients that can involve speech and language, psychiatric, cognitive, and nutritional problems in addition to their movement disorder, we would advocate that patients are treated in a multidisciplinary environment.

8. FUTURE DIRECTIONS

Our understanding of neuroferritinopathy, as with many of the NBIA disorders, remains far from complete. The exact pathomechanistic basis of the disorder remains incompletely understood; in particular, the interplay between the role of iron deposition and oxidative stress on cell death needs to be clearly defined to understand the temporal and perhaps spatial progression of the disease. Such information is only likely to be forthcoming from tissue and animal models and this will set the direction for future therapeutic

Table 5.5 Table of symptomatic treatments utilized in patients with neuroferritinopathy

Group	Drug	Max dose/day	No with beneficial effect	No effect	Comments	Reference
Dopamine depleters	Tetrabenazine	125 mg	14	1	3 pts stopped due to depression. 1 pt—sedation plus parkinsonism	Chinnery et al. (2007), Lehn et al. (2012), Mir et al. (2005), and Ondo et al. (2010)
Dopamine blockers	Sulpiride	400 mg	3	1		Chinnery et al. (2007) and Kubota et al. (2009)
	Risperidone	–	–	1		Chinnery et al. (2007)
	Haloperidol	–	–	5	1 pt developed severe extrapyramidal side effects	Kubota et al. (2009), Lehn et al. (2012), and Vidal et al. (2004)
	Tiapride	–	–	3		Kubota et al. (2009)
	Olanzapine	–	–	1		Chinnery et al. (2007)
Anticholinergics	Benzhexol	30 mg	4	4		Chinnery et al. (2007), Kubota et al. (2009), Mancuso et al. (2005), and Mir et al. (2005)
	Amitriptyline	–	–	4		Chinnery et al. (2007), Mancuso et al. (2005), and Mir et al. (2005)
Dopaminergics	L-Dopa	375 mg	–	12		Chinnery et al. (2007), Devos et al. (2009), Lehn et al. (2012), and Mir et al. (2005)
	Apomorphine	–	–	2		Chinnery et al. (2007) and Mir et al. (2005)
	Amantadine	–	–	1		Chinnery et al. (2007)

Continued

Table 5.5 Table of symptomatic treatments utilized in patients with neuroferritinopathy—cont'd

Group	Drug	Max dose/day	No with beneficial effect	No effect	Comments	Reference
Cholinergics	Deanol	60 mg	–	2		Chinnery et al. (2007)
Muscle relaxants	Baclofen	60 mg	1	2		Chinnery et al. (2007) and Kubota et al. (2009)
	Dantrolene	400 mg	–	2		Chinnery et al. (2007)
Benzodiazepines	Diazepam	20 mg	3	2		Chinnery et al. (2007), Kubota et al. (2009), and Mir et al. (2005)
	Lorazepam	15 mg	2	–		Chinnery et al. (2007)
	Clonazepam	1 mg	2	–		Chinnery et al. (2007)
Antidepressants	Paroxetine	–	–	1		Chinnery et al. (2007)
	Mirtazapine	–	–	1		Chinnery et al. (2007)
Antiepileptics	Gabapentin	–	–	1		Chinnery et al. (2007)
	Sodium valproate	–	–	1		Chinnery et al. (2007)
	Carbamazepine	–	–	2		Chinnery et al. (2007) and Mir et al. (2005)
Neuromuscular agents	Botulinum toxin	–	8	–		Chinnery et al. (2007), Mancuso et al. (2005), and Mir et al. (2005)

Adapted from Lehn et al., 2012.

options. In addition, clinicians must strive to further define the phenotype of disease. With a greater understanding of the progression of the disorder and greater appreciation of its neurological, psychiatric, and cognitive phenotype, we may be able to better tailor care in the present and future. In addition, while neuroimaging techniques continue to advance and assist clinicians in making the diagnosis, further endeavors in determining imaging methods that can reflect disease progression are of paramount importance especially if disease-modifying treatment trials occur. Finally, and most importantly, it is critically important that a rational approach to a conducting a trial of a disease-modifying treatment is developed over the coming years, possibly using quantitative MRI as a trial end point.

REFERENCES

Anderson, J. C., Costantino, M. M., & Stratford, T. (2004). Basal ganglia: Anatomy, pathology, and imaging characteristics. *Current Problems in Diagnostic Radiology*, *33*(1), 28–41. http://dx.doi.org/10.1016/j.cpradiol.2003.09.004.

Arosio, P., Ingrassia, R., & Cavadini, P. (2009). Ferritins: A family of molecules for iron storage, antioxidation and more. *Biochimica et Biophysica Acta*, *1790*(7), 589–599. http://dx.doi.org/10.1016/j.bbagen.2008.09.004.

Arosio, P., & Levi, S. (2010). Cytosolic and mitochondrial ferritins in the regulation of cellular iron homeostasis and oxidative damage. *Biochimica et Biophysica Acta*, *1800*(8), 783–792. http://dx.doi.org/10.1016/j.bbagen.2010.02.005.

Arosio, P., Levi, S., Santambrogio, P., Cozzi, A., Luzzago, A., Cesareni, G., et al. (1991). Structural and functional studies of human ferritin H and L chains. *Current Studies in Hematology and Blood Transfusion*, (58), 127–131.

Barbeito, A. G., Garringer, H. J., Baraibar, M. A., Gao, X., Arredondo, M., Nunez, M. T., et al. (2009). Abnormal iron metabolism and oxidative stress in mice expressing a mutant form of the ferritin light polypeptide gene. *Journal of Neurochemistry*, *109*(4), 1067–1078. http://dx.doi.org/10.1111/j.1471-4159.2009.06028.x.

Bou-Abdallah, F., Biasiotto, G., Arosio, P., & Chasteen, N. D. (2004). The putative "nucleation site" in human H-chain ferritin is not required for mineralization of the iron core. *Biochemistry*, *43*(14), 4332–4337. http://dx.doi.org/10.1021/bi0498813.

Brass, S. D., Chen, N. K., Mulkern, R. V., & Bakshi, R. (2006). Magnetic resonance imaging of iron deposition in neurological disorders. *Topics in Magnetic Resonance Imaging*, *17*(1), 31–40. http://dx.doi.org/10.1097/01.rmr.0000245459.82782.e4.

Chasteen, N. D., & Harrison, P. M. (1999). Mineralization in ferritin: An efficient means of iron storage. *Journal of Structural Biology*, *126*(3), 182–194. http://dx.doi.org/10.1006/jsbi.1999.4118.

Chavhan, G. B., Babyn, P. S., Thomas, B., Shroff, M. M., & Haacke, E. M. (2009). Principles, techniques, and applications of T2*-based MR imaging and its special applications. *Radiographics*, *29*(5), 1433–1449. http://dx.doi.org/10.1148/rg.295095034.

Chinnery, P. F., Crompton, D. E., Birchall, D., Jackson, M. J., Coulthard, A., Lombes, A., et al. (2007). Clinical features and natural history of neuroferritinopathy caused by the FTL1 460InsA mutation. *Brain*, *130*(Pt 1), 110–119. http://dx.doi.org/10.1093/brain/awl319.

Cozzi, A., Corsi, B., Levi, S., Santambrogio, P., Albertini, A., & Arosio, P. (2000). Overexpression of wild type and mutated human ferritin H-chain in HeLa cells: In vivo role of ferritin ferroxidase activity. *Journal of Biological Chemistry*, *275*(33), 25122–25129. http://dx.doi.org/10.1074/jbc.M003797200.

Cozzi, A., Corsi, B., Levi, S., Santambrogio, P., Biasiotto, G., & Arosio, P. (2004). Analysis of the biologic functions of H- and L-ferritins in HeLa cells by transfection with siRNAs and cDNAs: Evidence for a proliferative role of L-ferritin. *Blood*, *103*(6), 2377–2383. http://dx.doi.org/10.1182/blood-2003-06-1842.

Cozzi, A., Levi, S., Corsi, B., Santambrogio, P., Campanella, A., Gerardi, G., et al. (2003). Role of iron and ferritin in TNF alpha-induced apoptosis in HeLa cells. *FEBS Letters*, *537*(1–3), 187–192.

Cozzi, A., Rovelli, E., Frizzale, G., Campanella, A., Amendola, M., Arosio, P., et al. (2010). Oxidative stress and cell death in cells expressing L-ferritin variants causing neuroferritinopathy. *Neurobiology of Disease*, *37*(1), 77–85. http://dx.doi.org/10.1016/j.nbd.2009.09.009.

Cozzi, A., Santambrogio, P., Corsi, B., Campanella, A., Arosio, P., & Levi, S. (2006). Characterization of the l-ferritin variant 460InsA responsible of a hereditary ferritinopathy disorder. *Neurobiology of Disease*, *23*(3), 644–652. http://dx.doi.org/10.1016/j.nbd.2006.05.004.

Cremonesi, L., Cozzi, A., Girelli, D., Ferrari, F., Fermo, I., Foglieni, B., et al. (2004). Case report: A subject with a mutation in the ATG start codon of L-ferritin has no haematological or neurological symptoms. *Journal of Medical Genetics*, *41*(6), e81.

Crompton, D. E., Chinnery, P. F., Bates, D., Walls, T. J., Jackson, M. J., Curtis, A. J., et al. (2005). Spectrum of movement disorders in neuroferritinopathy. *Movement Disorders*, *20*(1), 95–99. http://dx.doi.org/10.1002/mds.20284.

Crompton, D. E., Chinnery, P. F., Fey, C., Curtis, A. R., Morris, C. M., Kierstan, J., et al. (2002). Neuroferritinopathy: A window on the role of iron in neurodegeneration. *Blood Cells, Molecules & Diseases*, *29*(3), 522–531. http://dx.doi.org/S1079979602905899 [pii].

Curtis, A. R., Fey, C., Morris, C. M., Bindoff, L. A., Ince, P. G., Chinnery, P. F., et al. (2001). Mutation in the gene encoding ferritin light polypeptide causes dominant adult-onset basal ganglia disease. *Nature Genetics*, *28*(4), 350–354. http://dx.doi.org/10.1038/ng571ng571[pii].

Devos, D., Tchofo, P. J., Vuillaume, I., Destee, A., Batey, S., Burn, J., et al. (2009). Clinical features and natural history of neuroferritinopathy caused by the 458dupA FTL mutation. *Brain*, *132*(Pt 6), e109. http://dx.doi.org/10.1093/brain/awn274.

Dwork, A. J., Lawler, G., Zybert, P. A., Durkin, M., Osman, M., Willson, N., et al. (1990). An autoradiographic study of the uptake and distribution of iron by the brain of the young rat. *Brain Research*, *518*(1–2), 31–39.

Ferreira, C., Bucchini, D., Martin, M. E., Levi, S., Arosio, P., Grandchamp, B., et al. (2000). Early embryonic lethality of H ferritin gene deletion in mice. *Journal of Biological Chemistry*, *275*(5), 3021–3024.

Friedman, A., Arosio, P., Finazzi, D., Koziorowski, D., & Galazka-Friedman, J. (2011). Ferritin as an important player in neurodegeneration. *Parkinsonism & Related Disorders*, *17*(6), 423–430. http://dx.doi.org/10.1016/j.parkreldis.2011.03.016.

Gelman, N., Gorell, J. M., Barker, P. B., Savage, R. M., Spickler, E. M., Windham, J. P., et al. (1999). MR imaging of human brain at 3.0 T: Preliminary report on transverse relaxation rates and relation to estimated iron content. *Radiology*, *210*(3), 759–767.

Gregory, A., Polster, B. J., & Hayflick, S. J. (2009). Clinical and genetic delineation of neurodegeneration with brain iron accumulation. *Journal of Medical Genetics*, *46*(2), 73–80. http://dx.doi.org/10.1136/jmg.2008.061929.

Haacke, E. M., Cheng, N. Y., House, M. J., Liu, Q., Neelavalli, J., Ogg, R. J., et al. (2005). Imaging iron stores in the brain using magnetic resonance imaging. *Magnetic Resonance Imaging*, *23*(1), 1–25. http://dx.doi.org/10.1016/j.mri.2004.10.001.

Harrison, P. M., & Arosio, P. (1996). The ferritins: Molecular properties, iron storage function and cellular regulation. *Biochimica et Biophysica Acta, 1275*(3), 161–203.

Hayflick, S. J. (2006). Neurodegeneration with brain iron accumulation: From genes to pathogenesis. *Seminars in Pediatric Neurology, 13*(3), 182–185. http://dx.doi.org/10.1016/j.spen.2006.08.007.

Hikita, T., Abe, K., Sakoda, S., Tanaka, H., Murase, K., & Fujita, N. (2005). Determination of transverse relaxation rate for estimating iron deposits in central nervous system. *Neuroscience Research, 51*(1), 67–71. http://dx.doi.org/10.1016/j.neures.2004.09.006.

Ingrassia, R., Gerardi, G., Biasiotto, G., & Arosio, P. (2006). Mutations of ferritin H chain C-terminus produced by nucleotide insertions have altered stability and functional properties. *Journal of Biochemistry, 139*(5), 881–885. http://dx.doi.org/10.1093/jb/mvj101.

Jappelli, R., Luzzago, A., Tataseo, P., Pernice, I., & Cesareni, G. (1992). Loop mutations can cause a substantial conformational change in the carboxy terminus of the ferritin protein. *Journal of Molecular Biology, 227*(2), 532–543.

Jin, W., Takagi, H., Pancorbo, B., & Theil, E. C. (2001). "Opening" the ferritin pore for iron release by mutation of conserved amino acids at interhelix and loop sites. *Biochemistry, 40*(25), 7525–7532.

Keogh, M. J., Jonas, P., Coulthard, A., Chinnery, P. F., & Burn, J. (2012). Neuroferritinopathy: A new inborn error of iron metabolism. *Neurogenetics, 13*(1), 93–96. http://dx.doi.org/10.1007/s10048-011-0310-9.

Keogh, M. J., Singh, B., & Chinnery, P. F. (2013). Early neuropsychiatry features in neuroferritinopathy. *Movement Disorders, 28*(9), 1310–1313. http://dx.doi.org/10.1002/mds.25371.

Kidane, T. Z., Sauble, E., & Linder, M. C. (2006). Release of iron from ferritin requires lysosomal activity. *American Journal of Physiology. Cell Physiology, 291*(3), C445–C455. http://dx.doi.org/10.1152/ajpcell.00505.2005.

Koyama, M., & Yagishita, A. (2006). Pantothenate kinase-associated neurodegeneration with increased lentiform nuclei cerebral blood flow. *AJNR. American Journal of Neuroradiology, 27*(1), 212–213.

Kubota, A., Hida, A., Ichikawa, Y., Momose, Y., Goto, J., Igeta, Y., et al. (2009). A novel ferritin light chain gene mutation in a Japanese family with neuroferritinopathy: Description of clinical features and implications for genotype-phenotype correlations. *Movement Disorders, 24*(3), 441–445. http://dx.doi.org/10.1002/mds.22435.

Larson, J. A., Howie, H. L., & So, M. (2004). Neisseria meningitidis accelerates ferritin degradation in host epithelial cells to yield an essential iron source. *Molecular Microbiology, 53*(3), 807–820. http://dx.doi.org/10.1111/j.1365-2958.2004.04169.x.

Lehn, A., Boyle, R., Brown, H., Airey, C., & Mellick, G. (2012). Neuroferritinopathy. *Parkinsonism & Related Disorders, 18*(8), 909–915. http://dx.doi.org/10.1016/j.parkreldis.2012.06.021.

Lotfipour, A. K., Wharton, S., Schwarz, S. T., Gontu, V., Schafer, A., Peters, A. M., et al. (2012). High resolution magnetic susceptibility mapping of the substantia nigra in Parkinson's disease. *Journal of Magnetic Resonance Imaging, 35*(1), 48–55. http://dx.doi.org/10.1002/jmri.22752.

Maciel, P., Cruz, V. T., Constante, M., Iniesta, I., Costa, M. C., Gallati, S., et al. (2005). Neuroferritinopathy: Missense mutation in FTL causing early-onset bilateral pallidal involvement. *Neurology, 65*(4), 603–605. http://dx.doi.org/10.1212/01.wnl.0000178224.81169.c2.

Mancuso, M., Davidzon, G., Kurlan, R. M., Tawil, R., Bonilla, E., Di Mauro, S., et al. (2005). Hereditary ferritinopathy: A novel mutation, its cellular pathology, and pathogenetic insights. *Journal of Neuropathology and Experimental Neurology, 64*(4), 280–294.

McNeill, A., Birchall, D., Hayflick, S. J., Gregory, A., Schenk, J. F., Zimmerman, E. A., et al. (2008). T2* and FSE MRI distinguishes four subtypes of neurodegeneration with brain iron accumulation. *Neurology, 70*(18), 1614–1619. http://dx.doi.org/10.1212/01.wnl.0000310985.40011.d6.

McNeill, A., Gorman, G., Khan, A., Horvath, R., Blamire, A. M., & Chinnery, P. F. (2012). Progressive brain iron accumulation in neuroferritinopathy measured by the thalamic T2* relaxation rate. *American Journal of Neuroradiology,* http://dx.doi.org/10.3174/ajnr.A3036.

Mehlhase, J., Sandig, G., Pantopoulos, K., & Grune, T. (2005). Oxidation-induced ferritin turnover in microglial cells: Role of proteasome. *Free Radical Biology & Medicine, 38*(2), 276–285. http://dx.doi.org/10.1016/j.freeradbiomed.2004.10.025.

Milukits, M., Schranzhofer, M., Beug, H., & Mullner, E. (1999). Post-transcriptional control via iron-responsive elements: The impact of aberrations in hereditary disease. *Mutation Research, 437*, 219–230.

Mir, P., Edwards, M. J., Curtis, A. R., Bhatia, K. P., & Quinn, N. P. (2005). Adult-onset generalized dystonia due to a mutation in the neuroferritinopathy gene. *Movement Disorders, 20*(2), 243–245. http://dx.doi.org/10.1002/mds.20280.

Morris, C. M., Keith, A. B., Edwardson, J. A., & Pullen, R. G. (1992). Uptake and distribution of iron and transferrin in the adult rat brain. *Journal of Neurochemistry, 59*(1), 300–306.

Ohta, E., Nagasaka, T., Shindo, K., Toma, S., Nagasaka, K., Ohta, K., et al. (2008). Neuroferritinopathy in a Japanese family with a duplication in the ferritin light chain gene. *Neurology, 70*(16 Pt 2), 1493–1494. http://dx.doi.org/10.1212/01.wnl.0000310428.74624.95.

Ohta, E., & Takiyama, Y. (2012). MRI findings in neuroferritinopathy. *Neurology Research International, 2012*, 197438. http://dx.doi.org/10.1155/2012/197438.

Ondo, W. G., Adam, O. R., Jankovic, J., & Chinnery, P. F. (2010). Dramatic response of facial stereotype/tic to tetrabenazine in the first reported cases of neuroferritinopathy in the United States. *Movement Disorders, 25*(14), 2470–2472. http://dx.doi.org/10.1002/mds.23299.

Ory-Magne, F., Brefel-Courbon, C., Payoux, P., Debruxelles, S., Sibon, I., Goizet, C., et al. (2009). Clinical phenotype and neuroimaging findings in a French family with hereditary ferritinopathy (FTL498-499InsTC). *Movement Disorders, 24*(11), 1676–1683. http://dx.doi.org/10.1002/mds.22669.

Santambrogio, P., Levi, S., Cozzi, A., Rovida, E., Albertini, A., & Arosio, P. (1993). Production and characterization of recombinant heteropolymers of human ferritin H and L chains. *Journal of Biological Chemistry, 268*(17), 12744–12748.

Schroder, J. M. (2005). Ferritinopathy: Diagnosis by muscle or nerve biopsy, with a note on other nuclear inclusion body diseases. *Acta Neuropathologica, 109*(1), 109–114. http://dx.doi.org/10.1007/s00401-004-0949-5.

Strecker, K., Hesse, S., Wegner, F., Sabri, O., Schwarz, J., & Schneider, J. P. (2007). Eye of the tiger sign in multiple system atrophy. *European Journal of Neurology, 14*(11), e1–e2. http://dx.doi.org/10.1111/j.1468-1331.2007.01925.x.

Szumowski, J., Bas, E., Gaarder, K., Schwarz, E., Erdogmus, D., & Hayflick, S. (2010). Measurement of brain iron distribution in Hallervorden-Spatz syndrome. *Journal of Magnetic Resonance Imaging, 31*(2), 482–489. http://dx.doi.org/10.1002/jmri.22031.

Taylor, E. M., Crowe, A., & Morgan, E. H. (1991). Transferrin and iron uptake by the brain: Effects of altered iron status. *Journal of Neurochemistry, 57*(5), 1584–1592.

Thompson, K., Menzies, S., Muckenthaler, M., Torti, F. M., Wood, T., Torti, S. V., et al. (2003). Mouse brains deficient in H-ferritin have normal iron concentration but a protein profile of iron deficiency and increased evidence of oxidative stress. *Journal of Neuroscience Research, 71*(1), 46–63. http://dx.doi.org/10.1002/jnr.10463.

Vidal, R., Ghetti, B., Takao, M., Brefel-Courbon, C., Uro-Coste, E., Glazier, B. S., et al. (2004). Intracellular ferritin accumulation in neural and extraneural tissue characterizes a

neurodegenerative disease associated with a mutation in the ferritin light polypeptide gene. *Journal of Neuropathology and Experimental Neurology*, *63*(4), 363–380.
Vidal, R., Miravalle, L., Gao, X., Barbeito, A. G., Baraibar, M. A., Hekmatyar, S. K., et al. (2008). Expression of a mutant form of the ferritin light chain gene induces neurodegeneration and iron overload in transgenic mice. *Journal of Neuroscience*, *28*(1), 60–67. http://dx.doi.org/10.1523/JNEUROSCI.3962-07.2008.
Vymazal, J., Hajek, M., Patronas, N., Giedd, J. N., Bulte, J. W., Baumgarner, C., et al. (1995). The quantitative relation between T1-weighted and T2-weighted MRI of normal gray matter and iron concentration. *Journal of Magnetic Resonance Imaging*, *5*(5), 554–560.
Wills, A. J., Sawle, G. V., Guilbert, P. R., & Curtis, A. R. (2002). Palatal tremor and cognitive decline in neuroferritinopathy. *Journal of Neurology, Neurosurgery, and Psychiatry*, *73*(1), 91–92.
Zhao, G., Bou-Abdallah, F., Arosio, P., Levi, S., Janus-Chandler, C., & Chasteen, N. D. (2003). Multiple pathways for mineral core formation in mammalian apoferritin. The role of hydrogen peroxide. *Biochemistry*, *42*(10), 3142–3150. http://dx.doi.org/10.1021/bi027357v.

CHAPTER SIX

Aceruloplasminemia: An Update

Satoshi Kono[1]
First Department of Medicine, Hamamatsu University School of Medicine, Hamamatsu, Japan
[1]Corresponding author: e-mail address: satokono@hama-med.ac.jp

Contents

Abstract

Aceruloplasminemia is an inherited neurodegenerative disorder involving "neurodegeneration with brain iron accumulation," which is caused by genetic defects in the ceruloplasmin gene. Ceruloplasmin is a multicopper oxidase with ferroxidase activity that oxidizes ferrous iron following its transfer to extracellular transferrin. In the central nervous system, a glycosylphosphatidylinositol-linked ceruloplasmin bound to the cell membranes was found to be the major isoform of this protein. Aceruloplasminemia is characterized by diabetes, retinal degeneration, and progressive neurological symptoms, including extrapyramidal symptoms, ataxia, and dementia. Clinical and pathological studies and investigations of cell culture and murine models revealed that there is an iron-mediated cellular radical injury caused by a marked accumulation of iron in the affected parenchymal tissues. The aim of this chapter is to provide an overview of not only the clinical features, genetic and molecular pathogenesis, and treatment of aceruloplasminemia but also the biological and physiological features of iron metabolism.

International Review of Neurobiology, Volume 110
ISSN 0074-7742
http://dx.doi.org/10.1016/B978-0-12-410502-7.00007-7

1. INTRODUCTION

The recent molecular investigations of inherited neurodegenerative disorders have revealed the existence of genetic disorders termed "neurodegeneration with brain iron accumulation" (NBIA) due to genetic defects associated with iron metabolism. The characteristic syndrome of NBIA is a presentation with progressive extrapyramidal symptoms and excessive iron deposition in the brain, particularly affecting the basal ganglia. The main causes of the syndromes are mutations in neuroaxonal dystrophies pantothenate kinase-associated neurodegeneration (PKAN, formerly known as Hallervorden–Spatz disease, NBIA1) and PLA2G6-associated neurodegeneration (NBIA2). Intensive genetic approaches have identified additional genes that cause other NBIA syndromes, including Kufor–Rakeb disease (NBIA3, PARK9), fatty acid hydroxylase-associated neurodegeneration, neuroferritinopathy, mitochondrial membrane protein-associated neurodegeneration, Woodhouse–Sakati syndrome, beta-propeller protein-associated neurodegeneration, and aceruloplasminemia (Haack et al., 2012; Kalman et al., 2012). The molecular pathogenesis underlying the different forms of neurodegeneration has provided new insights into the pathways of brain iron metabolism and the role of iron accumulation in the brain pathology. This updated chapter of aceruloplasminemia provides an overview of the clinical features, pathological findings, therapeutic results, and the basis of the molecular genetics of the disease. Aceruloplasminemia is a model disease showing the link between iron accumulation and neurodegeneration. Understanding the role of ceruloplasmin within the central nervous system (CNS) is therefore fundamental to better understanding the iron-related pathology in the brain.

2. CERULOPLASMIN

Ceruloplasmin is a member of the multicopper oxidase family characterized by three types of spectroscopically distinct copper. This protein is a glycoprotein of the $\alpha 2$ globulin fraction of the serum and contains 95% of the copper in the plasma (Hellman & Gitlin, 2002). Ceruloplasmin consists of a single chain of 1046 amino acids forming six structural domains in which each sequential pair of these domains constitutes a large unit with about 340 amino acid residues (Yang et al., 1986). A crystallographic study of the spatial structure of ceruloplasmin at a 3.1 Å resolution revealed that there were precisely six copper ions present in the molecule and that there was important three-copper catalytic center, the so-called trinuclear cluster (Bento, Peixoto, Zaitsev, & Lindley, 2007). The copper ions present include three specific types; type I Cu^{2+} with strong absorption in the visible region with a maximum absorption at 610 nm, type 2 Cu^{2+} with no specific absorption, and type 3 Cu^{2+} with a maximum absorption at 330 nm. The type III Cu^{2+} and type II Cu^{2+} form the trinuclear cluster, which plays an important role in the oxidase reaction.

Despite the need for copper for the functions of ceruloplasmin, this protein plays no essential role in the transport or metabolism of the copper. The biological role of ceruloplasmin harboring a ferroxidase activity was demonstrated by Cartwright and colleagues in experiments using pigs with copper deficiency (Ragan, Nacht, Lee, Bishop, & Cartwright, 1969). The copper depletion resulted in a marked decrease in the circulating serum ceruloplasmin in association with iron deficiency anemia that could only be corrected by administration of copper and iron accumulation in parenchymal tissues, while the administration of exogenous ceruloplasmin resulted in the prompt release of tissue iron with subsequent incorporation into circulating transferrin. The essential function of ceruloplasmin is as a ferroxidase, utilizing the electron chemistry of bound copper ions to couple the oxidation of ferrous iron (Fe^{2+}) to the reduction of oxygen bound to the trinuclear cluster (Osaki, Johnson, & Frieden, 1966; Vachette et al., 2002).

In the reticuloendothelial system, the extracellular transport of iron by ferroportin, IREG 1, a membrane-bound protein, is regulated by the hepatic peptide, hepcidin, in response to inflammation and hepatic iron overload (Nemeth et al., 2004). Ferroportin transports intracellular ferrous iron (Fe^{2+}) to transferrin via the oxidization of ferrous iron (Fe^{2+}) to ferric iron (Fe^{3+}) via the ferroxidase of ceruloplasmin. Therefore, ceruloplasmin plays a role in the mobilization and oxidation of iron from the tissue stores associated with the subsequent incorporation of ferric iron into transferrin. On the other hand, in enterocytes of the duodenum, hephaestin, a membrane-bound copper protein homologous to ceruloplasmin, facilitates iron transport across the basolateral surface of the enterocytes by oxidizing ferrous iron to ferric iron to ensure incorporation of the latter into apo-transferrin (Vulpe et al., 1999). Other functional aspects of ceruloplasmin include its roles as an antioxidant and prooxidant. Ceruloplasmin is upregulated as an acute-phase protein by inflammation, trauma, pregnancy and diabetes, and after myocardial infarction, which is mostly attributed to its properties as an antioxidant (Samokyszyn, Miller, Reif, & Aust, 1989).

However, some studies have indicated that ceruloplasmin behaves as an oxidant by enhancing the oxidation of low-density lipoprotein in the vasculature (Fox, Mazumder, Ehrenwald, & Mukhopadhyay, 2000).

Ceruloplasmin is synthesized in the hepatocytes, where copper is incorporated into the apoprotein in the late secretory pathway prior to the secretion of the holoprotein containing six atoms of copper into the plasma (Sato & Gitlin, 1991). Copper enters cells via copper transport proteins, Ctrl and Ctr2, and then it binds to the copper chaperon, Atox1, which delivers copper to a copper-transporting ATPase, such as the Wilson protein encoded by the *ATP7B* gene (Bartnikas & Gitlin, 2001). The Wilson protein pumps the copper into the trans-Golgi network. Ceruloplasmin is initially synthesized as copper-free apoceruloplasmin in the endoplasmic reticulum (ER) and passes through the ER to the Golgi complex to obtain copper, resulting in change from apoceruloplasmin to the copper binding form, holoceruloplasmin. Holoceruloplasmin has a half-life of longer than 5 days in the plasma, while impairment in the incorporation of cooper results in the secretion of an apoceruloplasmin that lacks any ferroxidase activity and is rapidly degraded in the plasma (Hellman & Gitlin, 2002). The availability of copper within hepatocytes has no direct effect on the rate of ceruloplasmin synthesis or secretion.

The ceruloplasmin secreted into the plasma is considered to be involved in iron homeostasis. Although the liver is the predominant source of serum ceruloplasmin, the extrahepatic expression of ceruloplasmin has been shown in several tissues, including the CNS (Klomp, Farhangrazi, Dugan, & Gitlin, 1996). In the brain, ceruloplasmin is expressed in the astrocytes lining the brain microvasculature located in the basal ganglia, where a distinct form of ceruloplasmin is expressed as a glycosylphosphatidylinositol (GPI)-linked form by the alternative splicing of exons 19 and 20 (Patel & David, 1997; Patel, Dunn, & David, 2000). Earlier studies showed that the GPI-linked ceruloplasmin was located in leptomeningeal cells, the Müller glial cells in the retina, the Sertoli cells in the testes, and the Schwann cells in peripheral nerves (Mittal, Doroudchi, Jeong, Patel, & David, 2003; Salzer, Lovejoy, Linder, & Rosen, 1998); however, recent studies reported the expression of GPI-linked ceruloplasmin in various tissues (Kono et al., 2010; Mostad & Prohaska, 2011). Although the precise function of GPI-linked ceruloplasmin remains unknown, the GPI-linked ceruloplasmin likely plays an important role in the mobilization of iron and the antioxidant effects in the CNS (Jeong & David, 2003; Patel et al., 2002). GPI-linked ceruloplasmin may be associated with iron homeostasis and antioxidant defense by protecting the CNS from iron-mediated free-radical injury. The ferroxidase activity of GPI-linked ceruloplasmin is also essential for the stability of cell surface ferroportin (Fig. 6.1) (De Domenico et al., 2007; Kono et al., 2010). The requirement for a ferroxidase to maintain iron transport activity represents a novel mechanism of regulating cellular iron export.

The relationship between ceruloplasmin and neurodegenerative disease was revealed by the finding of a decrease in the ceruloplasmin concentration in the serum of patients with Wilson's disease, which is characteristic of hepatolenticular degeneration (Scheinberg & Gitlin, 1952). In Wilson's

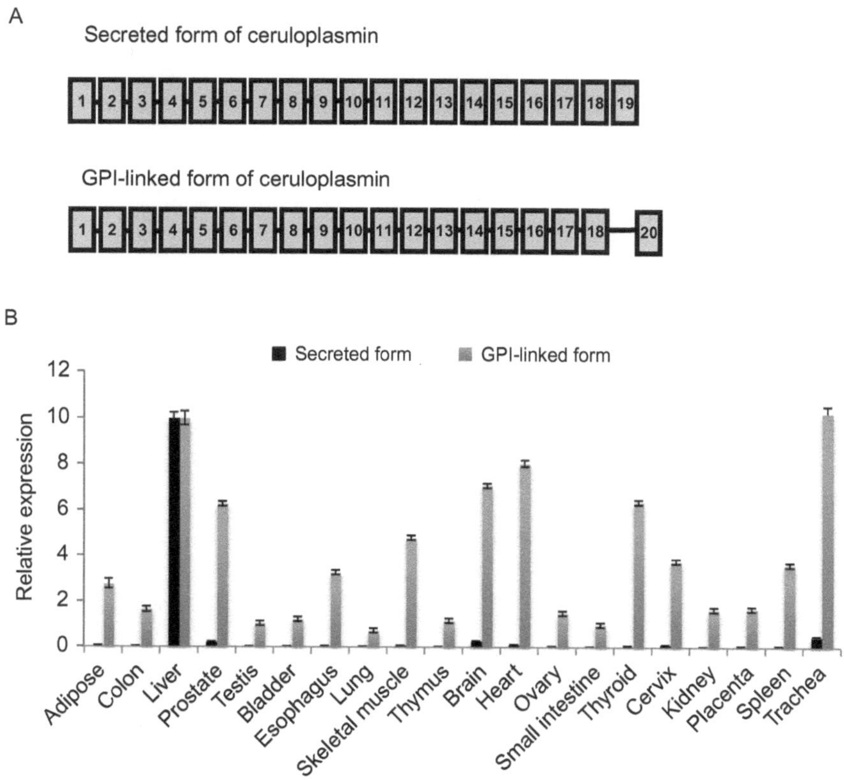

Figure 6.1 Schematic models of the secreted and GPI-linked forms of ceruloplasmin (A). The structure of the human ceruloplasmin gene consists of 20 exons. Alternative splicing at exon 18 allows for the secreted type or GPI-linked type of ceruloplasmin isoforms, as indicated. The RNA levels of the GPI-linked form and secreted form of ceruloplasmin in different human tissue specimens (B). A semiquantitative real-time PCR analysis was performed using human total RNA from several tissue specimens. Ceruloplasmin expression was calculated using a control mRNA plasmid standard curve and then was normalized to *GAPDH* expression. (For interpretation of the references to color in this figure legend, the reader is referred to the online version of this chapter.)

disease, the dysfunction of the Wilson protein caused by a mutation in the *ATP7B* gene disrupts the copper transport into the secretory pathway, resulting in a decrease in the serum ceruloplasmin level (Gitlin, 2003). The importance of ceruloplasmin in human biology is underscored by the autosomal recessive disorder aceruloplasminemia, which is an inherited disease of iron homeostasis.

3. CLINICAL FEATURES

In 1987, Miyajima et al. described the first case of aceruloplasminemia in a 52-year-old Japanese female suffering from blepharospasm, retinal degeneration, and diabetes mellitus (Miyajima et al., 1987). Subsequent evaluations revealed the presence of mild anemia, low plasma iron concentrations, elevated plasma ferritin levels, significant iron accumulation in the basal ganglia and liver on T2-weighted magnetic resonance images (MRI), and a complete absence of serum ceruloplasmin. Careful family studies in this original case showed that the lack of serum ceruloplasmin was inherited in an autosomal recessive fashion. A direct connection between iron accumulation in both the brain and liver and complete absence of serum ceruloplasmin was hypothesized. A genetic analysis of the ceruloplasmin gene revealed that this patient was homozygous, with a five-base insertion in exon 7, resulting in a frame shift mutation and a truncated open reading frame (Harris et al., 1995). The clinical findings and identification of a mutation in the ceruloplasmin gene confirmed that the disorder was a novel disorder of iron metabolism resulting from a lack of ceruloplasmin in the serum. The disorder was termed aceruloplasminemia (MIM 604290). An epidemiological study in Japan demonstrated that the prevalence of aceruloplasminemia was estimated to be approximately one per 2,000,000 in nonconsanguineous marriages, and subsequent studies have now identified more than 35 affected families from around the world (Miyajima, Kohno, Takahashi, Yonekawa, & Kanno, 1999).

Aceruloplasminemia patients present in the fourth or fifth decade of life with neurological symptoms (McNeill, Pandolfo, Kuhn, Shang, & Miyajima, 2008; Miyajima, 2003). These neurological features are usually progressive at the time of diagnosis and are associated with the iron accumulation in the basal ganglia and cerebellum as detected on T2-weighted MRI (Fig. 6.2A). Ophthalmologic examinations usually reveal evidence of peripheral retinal degeneration secondary to iron accumulation and photoreceptor cell loss. Although the neurological features dominate the clinical

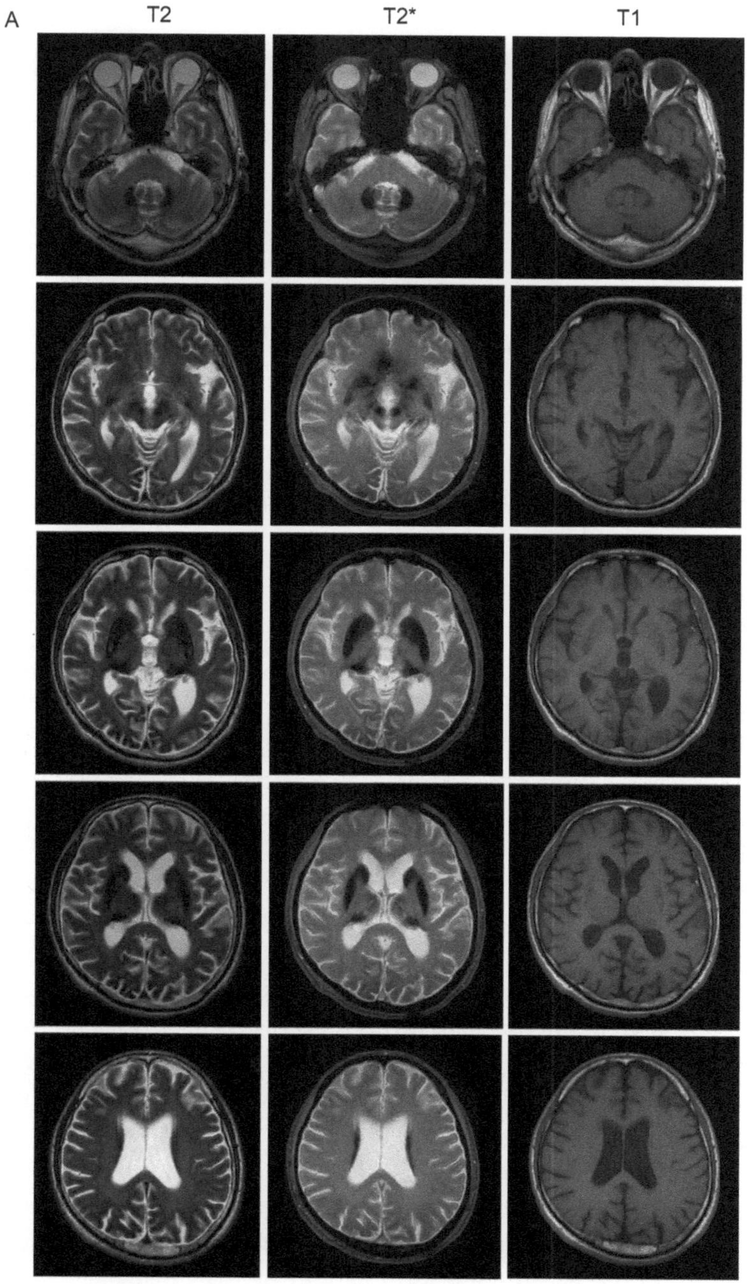

Figure 6.2 Magnetic resonance images (MRI) of an aceruloplasminemia patient. T1-, T2-, and T2*-weighted axial images of the brain showed signal attenuation of the dentate nucleus of the cerebellum, globus pallidum, putamen, caudate nucleus, and thalamus (A).

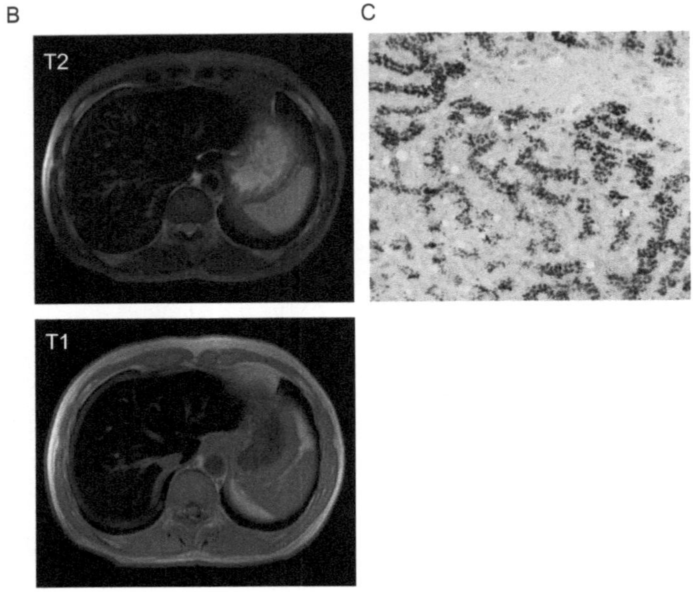

Figure 6.2—Cont'd T1- and T2-weighted axial images of the liver also showed signal attenuation (B). A liver biopsy specimen stained with Perl's stain showed iron in the hepatocytes (original magnification 200×) (C). (For interpretation of the references to color in this figure legend, the reader is referred to the online version of this chapter.)

features in most patients, all individuals have evidence of systemic iron accumulation at the time of diagnosis. The laboratory findings demonstrated microcytic anemia, decreased serum iron content, and an increased serum ferritin concentration, usually greater than 1000 ng/ml. T2-weighted MRI of the liver shows low-intensity signals associated with the iron accumulation (Fig. 6.2B). Liver biopsy samples reveal normal hepatic architecture and histology without cirrhosis or fibrosis; however, they do demonstrate excess iron accumulation (>1200 μg/g dry weight) within hepatocytes and reticuloendothelial cells (Fig. 6.2C).

Aceruloplasminemia patients also present with diabetes or evidence of abnormal glucose tolerance. Autopsy studies have revealed significant iron accumulation within the endocrine portion of the pancreas, with marked diminution in the β-cell population within the islets of Langerhans (Kato et al., 1997; Kawanami et al., 1996; Morita et al., 1995). Thus, the diagnosis of aceruloplasminemia in a symptomatic individual relies upon the demonstration of the complete absence of serum ceruloplasmin and abnormal laboratory findings, as well as MRI findings suggesting iron overload in both the liver and brain. The neuroimaging studies in aceruloplasminemia patients are strongly supported by the characteristic MRI findings of abnormal low intensities reflecting iron accumulation in the liver and brain, including the basal ganglia, thalamus, and dentate nucleus on both T1- and T2-weighted images. A T2* MRI study showed that hypointensity in both the basal ganglia, including the caudate nucleus, putamen, and pallidum, and the thalamus was typical in aceruloplasminemia patients and could be used to distinguish these patients from those with other NBIA, including PKAN, neuroferritinopathy, and infantile neuroaxonal dystrophy (McNeill, Birchall, et al., 2008). Functional neuroimaging studies using fluorodeoxyglucose-positron emission tomography demonstrated hypometabolism in the basal ganglia (Haemers, Kono, Goldman, Gitlin, & Pandolfo, 2004; Miyajima et al., 2005; Miyajima, Takahashi, et al., 2005, 2002). The clinical characteristics of the patients are summarized in Table 6.1. In general, aceruloplasminemia patients present neurological symptoms, including extrapyramidal signs, in the fourth or fifth decade of life. Although the neurological findings dominate the clinical features in most patients, some patients have been recognized prior to the onset of neurological symptoms due to biochemical abnormalities indicating changes in iron metabolism, the presence of diabetes or evidence of abnormal glucose tolerance, and abnormal MRI findings of the liver and the brain (Hatanaka, Okano, Oda, Yamamoto, & Yoshida, 2003; Hellman et al., 2000; Mariani et al., 2004; Ogimoto et al., 2011). The clinical diagnosis should be made based on the measurement of the serum ceruloplasmin level and the MRI findings in the brain and abdomen.

4. GENETIC AND MOLECULAR FEATURES

Ceruloplasmin is a single-copy gene on chromosome 3 in the human genome; however, the presence of a processed pseudogene on chromosome 8 encoding the carboxyl-terminal 563 amino acids of this protein must be taken into account when performing molecular diagnostic testing (Yang et al., 1986). The human ceruloplasmin gene contains 20 exons with total length of about 65 kb (Patel et al., 2000). The genetic analyses of aceruloplasminemia patients have identified more than 40 distinct mutations in the ceruloplasmin gene (Fig. 6.3) (Bethlehem, van Harten, & Hoogendoorn, 2010; Bosio et al., 2002; Daimon et al., 1995, 2000; Fasano, Bentivoglio, & Colosimo, 2007; Fasano et al., 2008; Harris, Migas, Hughes, Logan, & Gitlin, 1996; Hatanaka et al., 2003; Hellman et al., 2000; Hellman, Kono, Mancini, et al., 2002; Hellman, Kono,

Table 6.1 The clinical characteristics of patients with aceruloplasminemia

Clinical manifestations in 71 patients with aceruloplasminemia

- Anemia (80%)
- Retinal degeneration (76%)
- Diabetes mellitus (70%)
- Neurological symptoms (68%)

1. Ataxia (71%): dysarthria > gait ataxia > limb ataxia
2. Involuntary movement (64%): dystonia (blepharospasm, grimacing, neck dystonia) > chorea > tremors
3. Parkinsonism (20%): rigidity > akinesia
4. Cognitive dysfunction (60%): apathy > forgetfulness

Onset of clinical manifestations

- Diabetes mellitus: under 30 years old, 18%; 30–39 years old, 35%; 40–49 years old, 31%; over 50 years old, 16%
- Neurological symptoms: under 40 years old, 7%; 40–49 years old, 38%; 50–59 years old, 42%; over 60 years old, 13%

Laboratory findings

- Undetectable serum ceruloplasmin
- Elevated serum ferritin
- Decreased serum iron, iron-refractory microcytic anemia
- Low serum copper and normal urinary copper levels

MRI (magnetic resonance imaging) findings

- Low intensity on both T1- and T2-weighted MRI in the liver and the basal ganglia, including the caudate nucleus, putamen and pallidum, and the thalamus

Liver biopsy results

- Excess iron accumulation (>1000 μg/g dry weight) within hepatocytes and reticuloendothelial cells
- Normal hepatic architecture and histology without cirrhosis or fibrosis
- Normal copper accumulation

Miyajima, & Gitlin, 2002; Kohno, Miyajima, Takahashi, & Inoue, 2000; Kono et al., 2007; Kono, Suzuki, Oda, et al., 2006; Kono, Suzuki, Takahashi, et al., 2006; Kuhn et al., 2005; Loreal et al., 2002; Mariani et al., 2004; Muroi et al., 2006; Okamoto et al., 1996; Perez-Aguilar, Burguera, Benlloch, Berenguer, & Rayon, 2005; Shang, Jiang, Burgunder, Chen, & Zhou, 2006; Takahashi et al., 1996; Takeuchi et al., 2002; Yazaki et al., 1998; Yoshida et al., 1995). Most of mutations detected are unique to specific families, where there is often a history of

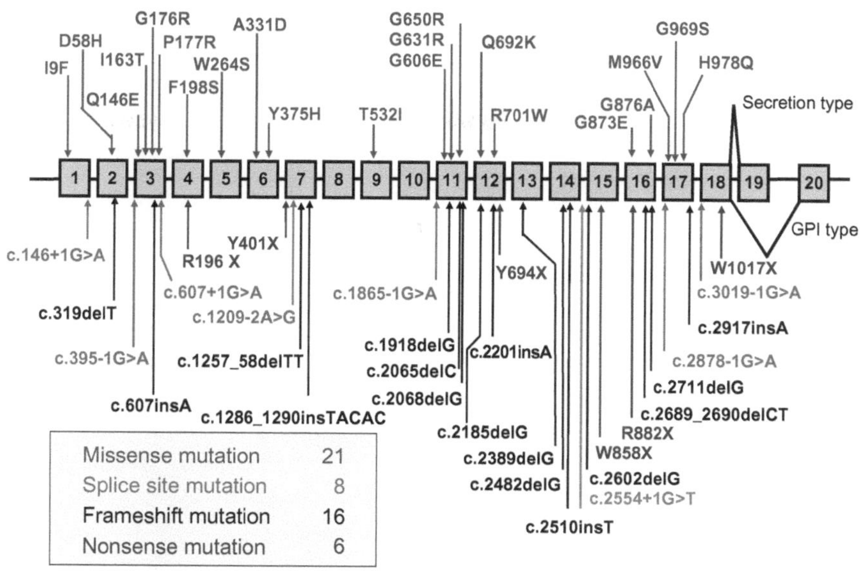

Figure 6.3 Genetic mutations characterized in patients with aceruloplasminemia and their family members. The indicated mutations are referenced in the text and some mutations are included in our unpublished data. (For interpretation of the references to color in this figure legend, the reader is referred to the online version of this chapter.)

consanguinity. The majority of mutations are truncated mutations, leading to the formation of a premature stop codon. The ferroxidase activity of ceruloplasmin is dependent upon the trinuclear copper cluster, the ligands for which are encoded by exon 18 (Vachette et al., 2002). The truncated mutations identified are predicted to result in the formation of a protein lacking the copper cluster sites presumed to be critical for enzymatic function. The symptoms, onset, and prognosis in single cases have demonstrated that there is no genotype–phenotype association (McNeill, Pandolfo, et al., 2008). This finding suggests that unknown genetic or environmental factors may regulate iron accumulation in the brain.

Biosynthesis studies of mutant ceruloplasmin in mammalian cell culture systems without endogenous ceruloplasmin were performed to investigate the molecular pathogenesis of aceruloplasminemia (di Patti et al., 2009; Hellman, Kono, Mancini, et al., 2002; Hellman, Kono, Miyajima, et al., 2002; Kono, Suzuki, Oda, et al., 2006; Kono et al., 2007, 2010). The biosynthesis studies of missense mutants revealed three distinct pathological mechanisms (Fig. 6.4). The first group comprising the I9F, G176R, P177R, D58H, F198S, W264S, A331D, G606E, and G873E mutants was retained in the ER (Hellman, Kono, Miyajima, et al., 2002; Kono,

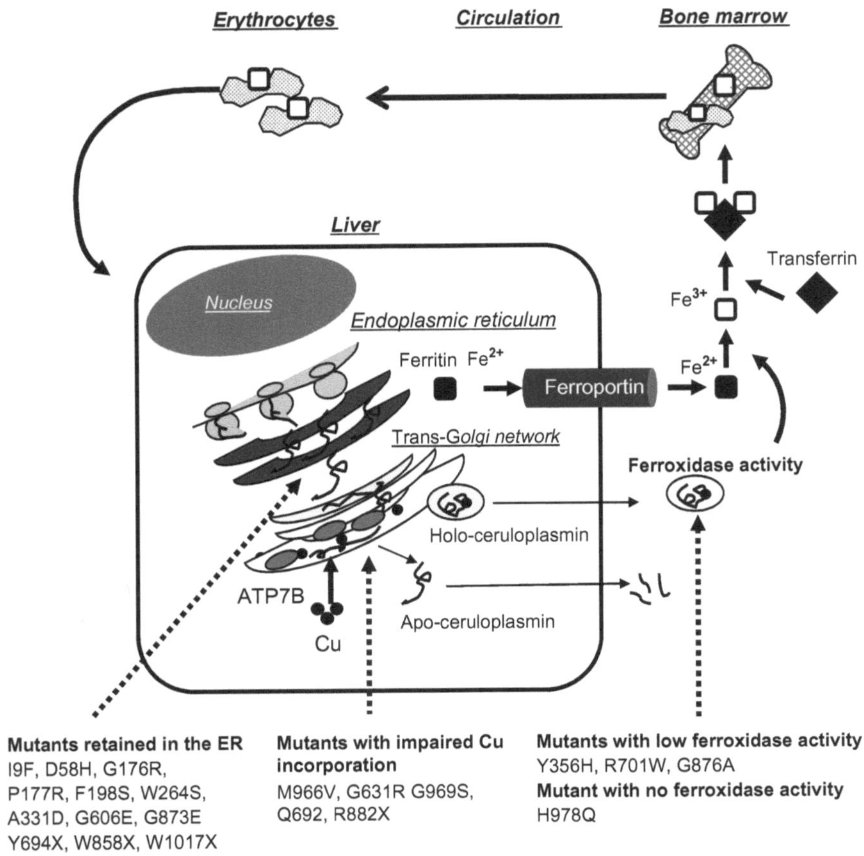

Figure 6.4 A model for the interaction between iron and copper homeostasis in normal subjects and in the aceruloplasminemia patients. In normal subjects, iron is continuously recycled between the bone marrow and hepatocytes, with serum transferrin acting as a shuttle to deliver iron from hepatocytes to the bone marrow. The role of ceruloplasmin, formed as holoceruloplasmin, is as a ferroxidase mediating ferrous iron oxidation and subsequent transfer to transferrin. In normal subjects, copper enters the cell and binds the copper chaperons, which deliver the copper to ATP-7B. The ATP-7B pumps the copper into the trans-Golgi network. Ceruloplasmin is initially synthesized as apoceruloplasmin and incorporates the copper into the apoprotein in the Golgi body, resulting in the formation of holoceruloplasmin prior to extracellular secretion. Mutant ceruloplasmin biosyntheses were investigated using a cell culture system without endogenous ceruloplasmin. The mutant proteins revealed three distinct pathological mechanisms, including mutants associated with mistrafficking, resulting in the retention of the protein in the ER; mutants altering the intrinsic protein structure, resulting in abrogation of copper incorporation into apoceruloplasmin; and mutants with impaired ferroxidase activity.

Suzuki, Oda, et al., 2006; Kono et al., 2010). The mutants presumably result from the misfolding of ceruloplasmin in the ER. The amino acid sequence of a G(FLI) (LI)GP repeat motif is believed to affect folding during the early secretory pathway (Hellman, Kono, Miyajima, et al., 2002). The G873E, G176R, and P177R mutants affect the conserved repeated G (FLI) (LI) GP motif, which is consistent with this hypothesis, whereas the G876A mutant was not retained in the ER (Kono et al., 2010). The mutants located beside this motif are speculated to have other molecular mechanisms involved in the cellular trafficking of ceruloplasmin. The second group of mutants, including the G631R, Q692K, M966V, and G969S mutants, was synthesized and was found to be secreted with normal kinetics but failed to incorporate copper during the late secretory pathway, resulting in apoceruloplasmin (Hellman, Kono, Mancini, et al., 2002; Kono, Suzuki, Takahashi, et al., 2006; Kono et al., 2010). The G631R and G969S mutations are located in the nearby type I copper binding His637 and His975 sites, respectively. The Q692K and M966V mutations are also located near the type I copper binding sites of M690 and His975, respectively. A site-directed mutagenesis analysis of the type I copper binding site indicated that these mutants failed to incorporate copper into the apoceruloplasmin (Hellman, Kono, Mancini, et al., 2002). These biochemical studies demonstrated that the type I copper binding site did not affect either the protein folding for intracellular trafficking from the ER to the Golgi body or the subsequent protein secretion from the cell. However, the copper binding site may play an essential role in the protein structure for copper incorporation into the apoceruloplasmin. The third group containing the Y356H, R701W, and G876A mutants reconstituted both the apo- and holoproteins and was secreted extracellularly (Kono et al., 2010). However, the mutants had impaired ferroxidase activity, which is required for ferroportin stability. These mutants may have altered iron binding sites or changes in the trinuclear copper cluster, which are essential for the oxidase activity of the protein. It will be necessary to analyze the crystal structure of the mutant ceruloplasmin proteins in order to obtain insight into the mechanism of ferroxidase activity.

A biogenesis study of nonsense mutations, including Y694X, W858X, and R882X, demonstrated that the Y694X and W858X mutants were retained in the ER, while the R882X mutant was secreted (Kono et al., 2007). Subsequent site-directed mutagenesis analyses revealed that the truncated mutant containing the cysteine residue at amino acid 881(Cys-881) was able to pass through the ER and was secreted, while the truncated mutant protein without Cys-881 appeared to accumulate in the ER, leading to ER stress and eventually resulting in cell death. Thus, Cys-881 is necessary for the secretion of almost all of the truncated ceruloplasmin, although a recent biogenesis study of the W1017X mutant with Cys-881 showed that the mutant was exclusively retained in the ER (Hida et al., 2010).

The clinical phenotype in most patients shows little variation, regardless of the specific mutation (McNeill, Pandolfo, et al., 2008). While almost all patients have a complete absence of serum ceruloplasmin, H978Q and G969S mutations were reported in patients who had detectable levels of serum ceruloplasmin but who presented with the clinical features of aceruloplasminemia. The H978Q mutation, which is located at one of the type I copper binding sites and constitutes holoceruloplasmin, was speculated to be devoid of ferroxidase activity (Takeuchi et al., 2002). The G969S mutation that led to apoceruloplasmin in the serum was suggested to be less fragile than the wild-type apoceruloplasmin, allowing it to remain in circulation for a longer period of time (Kono, Suzuki, Takahashi, et al., 2006). These findings have important diagnostic implications, indicating that the presence of ceruloplasmin in the serum of patients with the typical clinical features of aceruloplasminemia requires ceruloplasmin gene analysis before the diagnosis of aceruloplasminemia can be conclusively ruled out.

Aceruloplasminemia is an autosomal recessive inherited disease, and heterozygous individuals with a partial ceruloplasmin deficiency may have normal iron metabolism and no clinical symptoms. However, symptomatic heterozygous patients have recently been reported. The first case report showed three Japanese patients from two families with half the normal ceruloplasmin levels in the serum and who developed cerebellar ataxia from the fourth decade of life (Miyajima, Kono, et al., 2001). They were all heterozygous for a W858X mutation, and their serum iron concentrations and transferrin saturation levels were normal. At autopsy, pathological and biochemical examinations showed marked loss of Purkinje cells and a large amount of iron deposition in the cerebellum and small deposits in the basal ganglia, thalamus, and liver. The W858X mutation is frequently detected in Japanese patients; however, most of the carriers heterozygous for the W858X mutation are asymptomatic (Daimon et al., 1995; Takahashi et al., 1996). The second case report was of a young patient who presented with subacute progressive extrapyramidal movement disorders (Kuhn et al., 2005). Although her brain MRI showed no iron accumulation, the iron content in a liver biopsy specimen exceeded the normal expected range. A genetic analysis of her ceruloplasmin gene revealed that she was heterozygous for a R701W mutation. However, her father, who was also heterozygous for the R701W mutation, was asymptomatic and the pathological effects of the mutation on the neurological symptoms are unclear. These observations indicate that the clinical features in heterozygous patients show variation regarding the specific mutations and suggest that there may be a dominant-negative effect of the mutations overriding the influence of ceruloplasmin function in iron metabolism. *In vitro* biogenesis studies showed a potential function of the dominant-negative effect of mutant ceruloplasmin occurring via silencing of the wild-type ceruloplasmin function. The W858X mutant accumulated in the ER, leading to the ER stress, which resulted in cell death (Kono, Suzuki, Oda, et al., 2006). The R701W mutant induced the subcellular relocalization of the copper-transporting ATPase, ATP7B, in the Golgi complex and fragmentation of the Golgi complex, resulting in the failure of copper loading in wild-type ceruloplasmin (di Patti et al., 2009). The Arg701 site is located in one of the repeat CX(R/K) motifs consisting of large exposed loops connecting domains. A mutagenesis study of the motifs revealed that the external loops play an important role in copper incorporation.

5. MURINE MODELS OF ACERULOPLASMINEMIA

The generation of murine models of aceruloplasminemia provided a critical clue to study the role of ceruloplasmin in iron homeostasis. Three distinct research groups generated ceruloplasmin knockout mice, which develop hepatic and reticuloendothelial iron overload. The first knockout mice generated on a C57BL/6 J genetic background were reported by Harris, Durley, Man, and Gitlin (1999). The mice have an increased iron content, with lipid peroxidation in the brain (Texel, Xu, & Harris, 2008). However, there is no evidence of neurological symptoms in these mice. Double-knockout mice lacking both ceruloplasmin and hephaestin were generated by crossing the ceruloplasmin knockout mice with *sla* mice, which are hephaestin knockout, sex-linked anemic mice. Hephaestin is a ceruloplasmin homolog and is also a multicopper oxidase with ferroxidase activity, which is abundantly expressed in the neurons in the murine brain, as well as in the enterocytes in the duodenum. The knockout mice lacking both ceruloplasmin and hephaestin exhibited a neurodegenerative phenotype and retinal degeneration consistent with the aceruloplasminemia patients (Hahn et al., 2004; Schulz, Vulpe, Harris, & David, 2011; Texel et al., 2008). In the mice, hephaestin expression may play a more important role to maintain the cellular redox environment than it does in the CNS in the humans.

The iron homeostasis associated with ceruloplasmin was investigated using the ceruloplasmin knockout mice. When the mice were injected with damaged red blood cells in order to induce an increased reticuloendothelial iron overload or when the mice received a phlebotomy to accelerate reticuloendothelial iron transport to the bone marrow, the mice failed to show an increase in their serum iron levels. Only upon injection of holoceruloplasmin were the mice able to release iron from a storage compartment for delivery to a synthetic compartment, the bone marrow, thus confirming the essential role for ceruloplasmin in regulating efficient iron efflux. Of interest, the copper metabolism is normal in ceruloplasmin knockout mice, and none of the clinical pathology associated with aceruloplasminemia is secondary to copper deficiency or toxicity.

The second group of knockout mice generated on a C57BL/6 J genetic background was reported by Patel et al. (2002). The mice developed a neurodegenerative phenotype and showed increased iron deposition in several regions of the CNS, such as the cerebellum and brainstem. Increased lipid peroxidation due to iron-mediated cellular radical injury was also seen in some regions. Cerebellar neuronal cells from neonatal mice were also more susceptible to oxidative stress *in vitro*. These mice showed deficits in motor coordination that were associated with a loss of brainstem dopaminergic neurons. These results indicate that ceruloplasmin plays an important role in maintaining iron homeostasis in the brain and in protecting the brain from iron-mediated free-radical injury.

The third group of knockout mice, this time generated on a C57BL/10 and BALB/c genetic background, was reported by Yamamoto et al. (2002). Although the mice showed hepatic iron overload, there was no evidence of iron accumulation in the brain, even after treatment with rotenone, a mitochondrial complex 1 inhibitor that enhances oxidative stress (Hineno, Kaneko, Yoshida, & Ikeda, 2011).

6. PATHOGENESIS

The abnormalities of iron homeostasis observed in patients with aceruloplasminemia can be understood by considering the cellular physiology of systemic iron metabolism. A small amount of total iron is delivered from absorption in enterocytes, while a large amount of the iron arising from recycling of the heme iron from aging red blood cells is turned over within the reticuloendothelial system. The recycled iron is released from endothelial cells in the liver and the spleen and binds to the transferrin in the plasma, resulting in its return to the bone marrow for erythropoiesis. Ceruloplasmin functions as an important factor in the iron cycle by performing iron oxidation, which is required to sustain the iron release and uptake by transferrin (Fig. 6.4). The lack of plasma ferroxidase activity in ceruloplasmin results in increased extracellular ferrous iron, which is rapidly taken up into cells. Pathological studies in patients with aceruloplasminemia showed that iron accumulates within hepatocytes, pancreatic endocrine cells, and astrocytes (Kaneko, Yoshida, et al., 2002; Kawanami et al., 1996; Morita et al., 1995). Although the mechanism underlying the neurodegeneration in aceruloplasminemia has not been clarified, its pathogenesis is presumably secondary to consistent accumulation of iron within neurons and astrocytes. The electronic properties of iron enable the metal to take part in chemical reactions because the Fenton catalysis of iron plays an important role in cellular redox chemistry by reducing H_2O_2 to the highly cytotoxic hydroxyl (OH•) radical, which may be injurious to neural and other cellular substrates. The antioxidant activity of ceruloplasmin can be mainly ascribed to its ferroxidase activity, which effectively inhibits ferrous ion-stimulated lipid peroxidation and ferrous ion-dependent formation of hydroxyl radicals in the Fenton reaction. A direct role for iron in the oxidant-mediated neuronal injury is supported by findings of increased lipid peroxidation and subsequent mitochondrial dysfunction in the brain tissues, cerebral spinal fluid, and erythrocytes of the aceruloplasminemia patients (Kohno, Miyajima, Takahashi, Suzuki, & Hishida, 2000; Miyajima, Adachi, et al., 2001; Miyajima, Fujimoto, Kohno, Kaneko, & Gitlin, 1998; Miyajima, Kono, Takahashi, & Sugimoto, 2002; Miyajima, Takahashi, Serizawa, Kaneko, & Gitlin, 1996; Yoshida et al., 2000).

The pathological findings in the brain showed severe iron deposition in both the astrocytes and neurons and neuronal loss in the same regions associated with the highest iron accumulation and necrosis (Fig. 6.5) (Kawanami et al., 1996; Morita et al., 1995). The neurodegenerative changes were observed in the cerebral cortex, as well as in the basal ganglia, dentate nuclei, and cerebellar cortices. The distribution in order of the iron level is the globus pallidus > putamen > the cerebellar cortex and cerebral cortex (Miyajima, 2003). The characteristic histopathologic findings of the patients were deformed astrocytes and globular structures, which were observed more frequently in the striatum than in the cerebral cortex, which occurred in parallel with significant iron deposition and neuronal loss (Kaneko et al.,

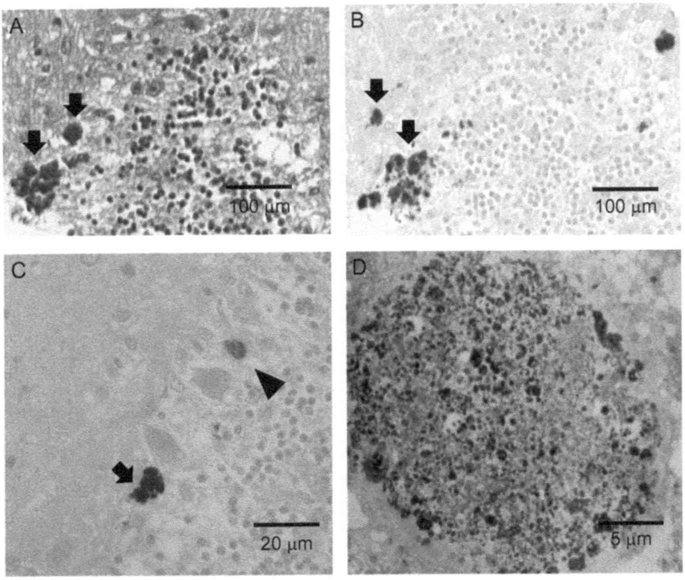

Figure 6.5 The histopathologic findings of cerebellar cortices in the brains of patients with aceruloplasminemia. Globular structures, indicated by arrows, were seen in the Purkinje cell layer. Many of the globular structures contained brown materials (A: H&E stain) and revealed siderous features (B: Prussian blue stain). At the cellular layer, iron deposits in Purkinje cells, indicated by an arrowhead, with a decreased number of cells (C: Prussian blue stain) were seen. The electron microscopic findings of the globular structures indicated that they contain many electron-dense bodies (D). (For interpretation of the references to color in this figure legend, the reader is referred to the online version of this chapter.)

2012; Kaneko, Yoshida, et al., 2002). The globular structures were immunologically reactive for a glial marker protein, suggesting that they were ballooned foot processes of astrocytes. The deformed astrocytes were more frequently observed in the basal ganglia in which marked iron deposition was observed. GFAP is one of the proteins most severely modified by oxidative stress in the brains of aceruloplasminemia patients (Kaneko, Nakamura, et al., 2002). Intense ferrous iron deposition was demonstrated in the terminal astrocytic processes and the globular structures (Oide, Yoshida, Kaneko, Ohta, & Arima, 2006). The morphological changes of astrocytes may be related to iron-induced tissue damage.

The iron accumulation was observed in neurons and astrocytes. This finding indicates that the neurons take up significant amounts of iron due to alternative sources of nontransferrin-bound iron complexed to molecules, such as citrate and ascorbate, because astrocytes without any expression of ceruloplasmin are not able to transport iron to transferrin that binds to transferrin receptor 1 on neurons. A recent pathological study of a murine model of aceruloplasminemia showed that the neuronal cell loss may result from iron deficiency in regions where the iron in astrocytes is not able to be mobilized for uptake into neurons, and the excess iron accumulation in astrocytes could also result in oxidative damage to these cells, with subsequent loss of the glial-derived growth factors critical for neurons (Jeong & David, 2006).

7. IRON HOMEOSTASIS

Ceruloplasmin plays an essential role in cellular iron efflux by oxidizing the ferrous iron exported from ferroportin. Ferroportin is post-translationally regulated through internalization triggered by hepcidin binding (Nemeth et al., 2004). Previous studies showed that the ferroxidase activity of GPI-linked ceruloplasmin was essential for the stability of cell surface ferroportin in rat glioma cells lines (De Domenico et al., 2007). The *in vitro* biological analyses suggested that the ceruloplasmin mutants had impaired ferroportin stability on the cell surface, resulting in exacerbated iron accumulation (di Patti et al., 2009; Kono et al., 2010). The hepatic expression of ferroportin proteins and the mRNA levels were analyzed to evaluate the involvement of ferroportin in the pathogenesis of aceruloplasminemia (Kono et al., 2010). The hepatic ferroportin protein levels decrease despite the presence of high ferroportin mRNA levels in two aceruloplasminemia patients. Decreased ferroportin protein levels in the liver may be due to degradation due to the absence of ceruloplasmin rather than due to a decreased synthesis of ferroportin at the transcriptional level. Clinical analyses of the hepcidin level in patients with aceruloplasminemia revealed that the serum hepcidin levels and hepatic hepcidin mRNA levels are lower than in control subjects (Kaneko et al., 2010; Kono et al., 2010). Analyses of ceruloplasmin knockout mice showed a decreased hepatic hepcidin mRNA levels in comparison to those of wild-type and heterozygous mice (Guo et al., 2009). The low serum hepcidin levels may induce increased iron absorption in the intestine, where the ceruloplasmin homolog hephaestin retains ferroxidase activity that is involved in basolateral intestinal iron transport. Therefore, the low hepcidin level in the serum and the loss of cell surface ferroportin due to mutant ceruloplasmin may enhance the cellular iron accumulation, contributing to the pathology of aceruloplasminemia.

8. TREATMENT OF ACERULOPLASMINEMIA

Aceruloplasminemia is a fatal disease, and its early diagnosis and early treatment of patients are issues of paramount importance. Iron-mediated lipid peroxidation and oxidative stress are considered to be the main cause of the neuronal degeneration in aceruloplasminemia patients. To reduce the iron accumulation, systemic iron chelation therapy has been introduced in some patients. Desferrioxamine (deferoxamine) is a high-affinity iron chelator that combines with ferric iron. It has been shown to cross the blood–brain barrier and to promote the excretion of excess iron in patients with inherited and acquired forms of iron overload (Summers, Jacobs, Tudway, Perera, & Ricketts, 1979). The administration of desferrioxamine was effective for reducing the hepatic iron overload and leading to a partial improvement of the neurological symptoms and brain iron accumulation, as reported in a single case report (Miyajima et al., 1997). However, subsequent studies showed little effect of desferrioxamine on the central nervous symptoms, despite normalization of the serum ferritin and hepatic iron concentrations and improvement in the insulin requirement and the regional brain iron levels in T2*-weighted MRI (Loreal et al., 2002; Mariani et al., 2004; Pan, Tang, Chen, Song, & Shang, 2011). The desferrioxamine therapy was often discontinued because of a concomitant decrease in hemoglobin, and the serum iron level was observed after several months of the therapy, suggesting that desferrioxamine sequestered the iron available for erythropoiesis. Combination therapy with fresh frozen plasma for 6 weeks to replenish the blood ceruloplasmin levels and, thereafter, administration of deferoxamine for an additional 6 weeks to deplete ferric iron stores showed unprecedented improvement in neurological symptoms (Yonekawa, Okabe, Asamoto, & Ohta, 1999). Deferiprone, which has a lower molecular weight and more lipophilic properties, had no beneficial effects in a patient (Mariani et al., 2004). Deferasirox, an oral iron-chelating agent, did not lead to any improvement in the neurological symptoms or brain iron accumulation quantified by MRI (Finkenstedt et al., 2010; Roberti Mdo, Borges Filho, Goncalves, & Lima, 2011), while deferasirox therapy has been reported to lead to mild improvement in neurological symptoms including cognitive performance, gait, and balance in an aceruloplasminemia patient who had no response to both desferrioxamine and fresh frozen plasma therapy (Skidmore et al., 2008). Short-term iron chelation therapy is therefore effective for reducing the hepatic iron overload and improving the diabetic mellitus but is ineffective for the treatment of neurological symptoms due to brain iron accumulation. In many reports of single cases, the side effects of the iron chelation therapy prohibited the long-term treatment that may be required to mobilize iron from the brain. However, it seems rational to suggest that the therapy should be initiated early in the course of aceruloplasminemia in order to remove the iron before it induces neurodegeneration.

In comparison with iron chelation therapy, oral zinc sulfate therapy (administered for 1.5 years) led to dramatic neurological improvement in a patient with extrapyramidal and cerebellar-mediated movement disorder caused by a heterozygous mutation in the ceruloplasmin gene (Kuhn et al., 2007). Although the patient was bedridden before the zinc treatment, she was able to stand for a short time and walk a few steps after undergoing this treatment. The antioxidant properties of zinc, as well as its effects on iron absorption, are well established (Donangelo, Woodhouse, King, Viteri, & King, 2002; Powell, 2000). While the mechanisms of antioxidation are not fully understood, the induction of metallothionein synthesis is considered to be one relevant aspect. The zinc therapy could be used as an alternative treatment when iron chelation therapy is discontinued due to side effects or progression of the symptoms, because the zinc therapy shows no side effects and may ameliorate the neurological symptoms in aceruloplasminemia patients.

9. CONCLUSIONS

About 30 years have passed since the first identification of patients lacking circulating serum ceruloplasmin who presented with neurodegeneration associated with brain iron accumulation. Our understanding how an absence of ceruloplasmin results in neurodegeneration has grown considerably. A number of investigations of the clinical features of aceruloplasminemia have revealed an essential role for ceruloplasmin in iron homeostasis. They have also shown that ceruloplasmin regulates the efficiency of iron efflux by its function as a ferroxidase mediating the oxidation of Fe^{2+} to Fe^{3+}. GPI-linked ceruloplasmin is the predominant form expressed in astrocytes in the brain. Ceruloplasmin also stabilizes the expression of ferroportin, an iron efflux transporter expressed on the cell membrane. An iron-mediated oxidative stress mechanism plays an important role in the degeneration observed in aceruloplasminemia. An insufficiency of ceruloplasmin synthesis or its decreased activity in the cells of the CNS is currently regarded as one of the mechanisms underlying the development of a number of neurodegenerative disorders, such as Alzheimer's and Parkinson's disease (Ke & Ming Qian, 2003; Zecca, Youdim, Riederer, Connor, & Crichton, 2004). It seems unlikely that all of the functions of ceruloplasmin in the nervous system have already been discovered, but what is already known leads us to believe that the protein plays an important role in the metabolism and development of nervous tissue. A careful analysis of the genetically based pathogenesis of NBIA, including aceruloplasminemia, will likely lead to the discovery of novel proteins and will provide new insights into a variety of aspects of iron metabolism.

ACKNOWLEDGMENTS

The author expresses their gratitude to Prof. Hiroaki Miyajima (Hamamatsu University School of Medicine) for the critical reading of the article and discussions and Makiko Sakao (Hamamatsu University School of Medicine) for performing the pathological examinations.

REFERENCES

Bartnikas, T. B., & Gitlin, J. D. (2001). How to make a metalloprotein. *Nature Structural Biology*, *8*(9), 733–734.

Bento, I., Peixoto, C., Zaitsev, V. N., & Lindley, P. F. (2007). Ceruloplasmin revisited: Structural and functional roles of various metal cation-binding sites [research support, Non-U.S. Gov't]. *Acta Crystallographica Section D: Biological Crystallography*, *63*(Pt 2), 240–248.

Bethlehem, C., van Harten, B., & Hoogendoorn, M. (2010). Central nervous system involvement in a rare genetic iron overload disorder [case reports]. *The Netherlands Journal of Medicine*, *68*(10), 316–318.

Bosio, S., De Gobbi, M., Roetto, A., Zecchina, G., Leonardo, E., Rizzetto, M., et al. (2002). Anemia and iron overload due to compound heterozygosity for novel ceruloplasmin mutations. *Blood*, *100*(6), 2246–2248.

Daimon, M., Kato, T., Kawanami, T., Tominaga, M., Igarashi, M., Yamatani, K., et al. (1995). A nonsense mutation of the ceruloplasmin gene in hereditary ceruloplasmin deficiency with diabetes mellitus [case reports]. *Biochemical and Biophysical Research Communications*, *217*(1), 89–95.

Daimon, M., Susa, S., Ohizumi, T., Moriai, S., Kawanami, T., Hirata, A., et al. (2000). A novel mutation of the ceruloplasmin gene in a patient with heteroallelic ceruloplasmin gene mutation (HypoCPGM). *The Tohoku Journal of Experimental Medicine*, *191*(3), 119–125.

De Domenico, I., Ward, D. M., di Patti, M. C., Jeong, S. Y., David, S., Musci, G., et al. (2007). Ferroxidase activity is required for the stability of cell surface ferroportin in cells expressing GPI-ceruloplasmin. *The EMBO Journal*, *26*(12), 2823–2831.

di Patti, M. C., Maio, N., Rizzo, G., De Francesco, G., Persichini, T., Colasanti, M., et al. (2009). Dominant mutants of ceruloplasmin impair the copper loading machinery in aceruloplasminemia. *The Journal of Biological Chemistry*, *284*(7), 4545–4554.

Donangelo, C. M., Woodhouse, L. R., King, S. M., Viteri, F. E., & King, J. C. (2002). Supplemental zinc lowers measures of iron status in young women with low iron reserves. *Journal of Nutrition, 132*(7), 1860–1864.

Fasano, A., Bentivoglio, A. R., & Colosimo, C. (2007). Movement disorder due to aceruloplasminemia and incorrect diagnosis of hereditary hemochromatosis. *Journal of Neurology, 254*(1), 113–114.

Fasano, A., Colosimo, C., Miyajima, H., Tonali, P. A., Re, T. J., & Bentivoglio, A. R. (2008). Aceruloplasminemia: A novel mutation in a family with marked phenotypic variability. *Movement Disorders, 23*(5), 751–755.

Finkenstedt, A., Wolf, E., Hofner, E., Gasser, B. I., Bosch, S., Bakry, R., et al. (2010). Hepatic but not brain iron is rapidly chelated by deferasirox in aceruloplasminemia due to a novel gene mutation. *Journal of Hepatology, 53*(6), 1101–1107.

Fox, P. L., Mazumder, B., Ehrenwald, E., & Mukhopadhyay, C. K. (2000). Ceruloplasmin and cardiovascular disease. *Free Radical Biology & Medicine, 28*(12), 1735–1744.

Gitlin, J. D. (2003). Wilson disease. *Gastroenterology, 125*(6), 1868–1877.

Guo, P., Cui, R., Chang, Y. Z., Wu, W. S., Qian, Z. M., Yoshida, K., et al. (2009). Hepcidin, an antimicrobial peptide is downregulated in ceruloplasmin-deficient mice. *Peptides, 30*(2), 262–266.

Haack, T. B., Hogarth, P., Kruer, M. C., Gregory, A., Wieland, T., Schwarzmayr, T., et al. (2012). Exome sequencing reveals de novo WDR45 mutations causing a phenotypically distinct, X-linked dominant form of NBIA. *The American Journal of Human Genetics, 91*(6), 1144–1149.

Haemers, I., Kono, S., Goldman, S., Gitlin, J. D., & Pandolfo, M. (2004). Clinical, molecular, and PET study of a case of aceruloplasminaemia presenting with focal cranial dyskinesia. *Journal of Neurology, Neurosurgery & Psychiatry, 75*(2), 334–337.

Hahn, P., Qian, Y., Dentchev, T., Chen, L., Beard, J., Harris, Z. L., et al. (2004). Disruption of ceruloplasmin and hephaestin in mice causes retinal iron overload and retinal degeneration with features of age-related macular degeneration. *Proceedings of the National Academy of Sciences of the United States of America, 101*(38), 13850–13855.

Harris, Z. L., Durley, A. P., Man, T. K., & Gitlin, J. D. (1999). Targeted gene disruption reveals an essential role for ceruloplasmin in cellular iron efflux. *Proceedings of the National Academy of Sciences of the United States of America, 96*(19), 10812–10817.

Harris, Z. L., Migas, M. C., Hughes, A. E., Logan, J. I., & Gitlin, J. D. (1996). Familial dementia due to a frameshift mutation in the caeruloplasmin gene. *QJM, 89*, 355–359.

Harris, Z. L., Takahashi, Y., Miyajima, H., Serizawa, M., MacGillivray, R. T., & Gitlin, J. D. (1995). Aceruloplasminemia: Molecular characterization of this disorder of iron metabolism. *Proceedings of the National Academy of Sciences of the United States of America, 92*(7), 2539–2543.

Hatanaka, Y., Okano, T., Oda, K., Yamamoto, K., & Yoshida, K. (2003). Aceruloplasminemia with juvenile-onset diabetes mellitus caused by exon skipping in the ceruloplasmin gene. *Internal Medicine, 42*(7), 599–604.

Hellman, N. E., & Gitlin, J. D. (2002). Ceruloplasmin metabolism and function. *Annual Review of Nutrition, 22*, 439–458.

Hellman, N. E., Kono, S., Mancini, G. M., Hoogeboom, A. J., De Jong, G. J., & Gitlin, J. D. (2002). Mechanisms of copper incorporation into human ceruloplasmin. *The Journal of Biological Chemistry, 277*(48), 46632–46638.

Hellman, N. E., Kono, S., Miyajima, H., & Gitlin, J. D. (2002). Biochemical analysis of a missense mutation in aceruloplasminemia. *The Journal of Biological Chemistry, 277*(2), 1375–1380.

Hellman, N. E., Schaefer, M., Gehrke, S., Stegen, P., Hoffman, W. J., Gitlin, J. D., et al. (2000). Hepatic iron overload in aceruloplasminaemia. *Gut, 47*(6), 858–860.

Hida, A., Kowa, H., Iwata, A., Tanaka, M., Kwak, S., & Tsuji, S. (2010). Aceruloplasminemia in a Japanese woman with a novel mutation of CP gene: Clinical presentations and analysis of genetic and molecular pathogenesis. *Journal of Neurological Sciences*, *298*(1–2), 136–139.

Hineno, A., Kaneko, K., Yoshida, K., & Ikeda, S. (2011). Ceruloplasmin protects against rotenone-induced oxidative stress and neurotoxicity. *Neurochemical Research*, *36*(11), 2127–2135.

Jeong, S. Y., & David, S. (2003). Glycosylphosphatidylinositol-anchored ceruloplasmin is required for iron efflux from cells in the central nervous system. *The Journal of Biological Chemistry*, *278*(29), 27144–27148.

Jeong, S. Y., & David, S. (2006). Age-related changes in iron homeostasis and cell death in the cerebellum of ceruloplasmin-deficient mice. *Journal of Neuroscience*, *26*(38), 9810–9819.

Kalman, B., Lautenschlaeger, R., Kohlmayer, F., Buchner, B., Kmiec, T., Klopstock, T., et al. (2012). An international registry for neurodegeneration with brain iron accumulation. *Orphanet Journal of Rare Diseases*, *7*, 66.

Kaneko, K., Hineno, A., Yoshida, K., Ohara, S., Morita, H., & Ikeda, S. (2012). Extensive brain pathology in a patient with aceruloplasminemia with a prolonged duration of illness. *Human Pathology*, *43*(3), 451–456.

Kaneko, Y., Miyajima, H., Piperno, A., Tomosugi, N., Hayashi, H., Morotomi, N., et al. (2010). Measurement of serum hepcidin-25 levels as a potential test for diagnosing hemochromatosis and related disorders. *Journal of Gastroenterology*, *45*(11), 1163–1171.

Kaneko, K., Nakamura, A., Yoshida, K., Kametani, F., Higuchi, K., & Ikeda, S. (2002). Glial fibrillary acidic protein is greatly modified by oxidative stress in aceruloplasminemia brain. *Free Radical Research*, *36*(3), 303–306.

Kaneko, K., Yoshida, K., Arima, K., Ohara, S., Miyajima, H., Kato, T., et al. (2002). Astrocytic deformity and globular structures are characteristic of the brains of patients with aceruloplasminemia. *Journal of Neuropathology and Experimental Neurology*, *61*(12), 1069–1077.

Kato, T., Daimon, M., Kawanami, T., Ikezawa, Y., Sasaki, H., & Maeda, K. (1997). Islet changes in hereditary ceruloplasmin deficiency. *Human Pathology*, *28*(4), 499–502.

Kawanami, T., Kato, T., Daimon, M., Tominaga, M., Sasaki, H., Maeda, K., et al. (1996). Hereditary caeruloplasmin deficiency: Clinicopathological study of a patient. *Journal of Neurology, Neurosurgery & Psychiatry*, *61*(5), 506–509.

Ke, Y., & Ming Qian, Z. (2003). Iron misregulation in the brain: A primary cause of neurodegenerative disorders. *Lancet Neurology*, *2*(4), 246–253.

Klomp, L. W., Farhangrazi, Z. S., Dugan, L. L., & Gitlin, J. D. (1996). Ceruloplasmin gene expression in the murine central nervous system. *Journal of Clinical Investigation*, *98*(1), 207–215.

Kohno, S., Miyajima, H., Takahashi, Y., & Inoue, Y. (2000). Aceruloplasminemia with a novel mutation associated with parkinsonism. *Neurogenetics*, *2*(4), 237–238.

Kohno, S., Miyajima, H., Takahashi, Y., Suzuki, H., & Hishida, A. (2000). Defective electron transfer in complexes I and IV in patients with aceruloplasminemia. *Journal of Neurological Sciences*, *182*(1), 57–60.

Kono, S., Suzuki, H., Oda, T., Miyajima, H., Takahashi, Y., Shirakawa, K., et al. (2006). Biochemical features of ceruloplasmin gene mutations linked to aceruloplasminemia. *Neuromolecular Medicine*, *8*(3), 361–374.

Kono, S., Suzuki, H., Oda, T., Shirakawa, K., Takahashi, Y., Kitagawa, M., et al. (2007). Cys-881 is essential for the trafficking and secretion of truncated mutant ceruloplasmin in aceruloplasminemia. *Journal of Hepatology*, *47*(6), 844–850.

Kono, S., Suzuki, H., Takahashi, K., Takahashi, Y., Shirakawa, K., Murakawa, Y., et al. (2006). Hepatic iron overload associated with a decreased serum ceruloplasmin level in a novel clinical type of aceruloplasminemia. *Gastroenterology*, *131*(1), 240–245.

Kono, S., Yoshida, K., Tomosugi, N., Terada, T., Hamaya, Y., Kanaoka, S., et al. (2010). Biological effects of mutant ceruloplasmin on hepcidin-mediated internalization of ferroportin. *Biochimica et Biophysica Acta, 1802*(11), 968–975.
Kuhn, J., Bewermeyer, H., Miyajima, H., Takahashi, Y., Kuhn, K. F., & Hoogenraad, T. U. (2007). Treatment of symptomatic heterozygous aceruloplasminemia with oral zinc sulphate. *Brain & Development, 29*(7), 450–453.
Kuhn, J., Miyajima, H., Takahashi, Y., Kunath, B., Hartmann-Klosterkoetter, U., Cooper-Mahkorn, D., et al. (2005). Extrapyramidal and cerebellar movement disorder in association with heterozygous ceruloplasmin gene mutation. *Journal of Neurology, 252*(1), 111–113.
Loreal, O., Turlin, B., Pigeon, C., Moisan, A., Ropert, M., Morice, P., et al. (2002). Aceruloplasminemia: New clinical, pathophysiological and therapeutic insights. *Journal of Hepatology, 36*(6), 851–856.
Mariani, R., Arosio, C., Pelucchi, S., Grisoli, M., Piga, A., Trombini, P., et al. (2004). Iron chelation therapy in aceruloplasminaemia: Study of a patient with a novel missense mutation. *Gut, 53*(5), 756–758.
McNeill, A., Birchall, D., Hayflick, S. J., Gregory, A., Schenk, J. F., Zimmerman, E. A., et al. (2008). T2* and FSE MRI distinguishes four subtypes of neurodegeneration with brain iron accumulation. *Neurology, 70*(18), 1614–1619.
McNeill, A., Pandolfo, M., Kuhn, J., Shang, H., & Miyajima, H. (2008). The neurological presentation of ceruloplasmin gene mutations. *European Neurology, 60*(4), 200–205.
Mittal, B., Doroudchi, M. M., Jeong, S. Y., Patel, B. N., & David, S. (2003). Expression of a membrane-bound form of the ferroxidase ceruloplasmin by leptomeningeal cells. *Glia, 41*(4), 337–346.
Miyajima, H. (2003). Aceruloplasminemia, an iron metabolic disorder. *Neuropathology, 23*(4), 345–350.
Miyajima, H., Adachi, J., Kohno, S., Takahashi, Y., Ueno, Y., & Naito, T. (2001). Increased oxysterols associated with iron accumulation in the brains and visceral organs of acaeruloplasminaemia patients. *QJM, 94*(8), 417–422.
Miyajima, H., Fujimoto, M., Kohno, S., Kaneko, E., & Gitlin, J. D. (1998). CSF abnormalities in patients with aceruloplasminemia. *Neurology, 51*(4), 1188–1190.
Miyajima, H., Kohno, S., Takahashi, Y., Yonekawa, O., & Kanno, T. (1999). Estimation of the gene frequency of aceruloplasminemia in Japan. *Neurology, 53*(3), 617–619.
Miyajima, H., Kono, S., Takahashi, Y., & Sugimoto, M. (2002). Increased lipid peroxidation and mitochondrial dysfunction in aceruloplasminemia brains. *Blood Cells, Molecules & Diseases, 29*(3), 433–438.
Miyajima, H., Kono, S., Takahashi, Y., Sugimoto, M., Sakamoto, M., & Sakai, N. (2001). Cerebellar ataxia associated with heteroallelic ceruloplasmin gene mutation. *Neurology, 57*(12), 2205–2210.
Miyajima, H., Nishimura, Y., Mizoguchi, K., Sakamoto, M., Shimizu, T., & Honda, N. (1987). Familial apoceruloplasmin deficiency associated with blepharospasm and retinal degeneration. *Neurology, 37*(5), 761–767.
Miyajima, H., Takahashi, Y., Kamata, T., Shimizu, H., Sakai, N., & Gitlin, J. D. (1997). Use of desferrioxamine in the treatment of aceruloplasminemia. *Annals of Neurology, 41*(3), 404–407.
Miyajima, H., Takahashi, Y., Kono, S., Hishida, A., Ishikawa, K., & Sakamoto, M. (2005). Frontal lobe dysfunction associated with glucose hypometabolism in aceruloplasminemia. *Journal of Neurology, 252*(8), 996–997.
Miyajima, H., Takahashi, Y., Kono, S., Sugimoto, M., Suzuki, Y., Hishida, A., et al. (2002). Glucose and oxygen hypometabolism in aceruloplasminemia brains. *Internal Medicine, 41*(3), 186–190.
Miyajima, H., Takahashi, Y., Serizawa, M., Kaneko, E., & Gitlin, J. D. (1996). Increased plasma lipid peroxidation in patients with aceruloplasminemia. *Free Radical Biology & Medicine, 20*(5), 757–760.

Morita, H., Ikeda, S., Yamamoto, K., Morita, S., Yoshida, K., Nomoto, S., et al. (1995). Hereditary ceruloplasmin deficiency with hemosiderosis: A clinicopathological study of a Japanese family. *Annals of Neurology*, *37*(5), 646–656.

Mostad, E. J., & Prohaska, J. R. (2011). Glycosylphosphatidylinositol-linked ceruloplasmin is expressed in multiple rodent organs and is lower following dietary copper deficiency. *Experimental Biology and Medicine (Maywood, NJ)*, *236*(3), 298–308.

Muroi, R., Yagyu, H., Kobayashi, H., Nagata, M., Sato, N., Ideno, J., et al. (2006). Early onset insulin-dependent diabetes mellitus as an initial manifestation of aceruloplasminaemia. *Diabetic Medicine*, *23*(10), 1136–1139.

Nemeth, E., Tuttle, M. S., Powelson, J., Vaughn, M. B., Donovan, A., Ward, D. M., et al. (2004). Hepcidin regulates cellular iron efflux by binding to ferroportin and inducing its internalization. *Science*, *306*(5704), 2090–2093.

Ogimoto, M., Anzai, K., Takenoshita, H., Kogawa, K., Akehi, Y., Yoshida, R., et al. (2011). Criteria for early identification of aceruloplasminemia. *Internal Medicine*, *50*(13), 1415–1418.

Oide, T., Yoshida, K., Kaneko, K., Ohta, M., & Arima, K. (2006). Iron overload and antioxidative role of perivascular astrocytes in aceruloplasminemia. *Neuropathology and Applied Neurobiology*, *32*(2), 170–176.

Okamoto, N., Wada, S., Oga, T., Kawabata, Y., Baba, Y., Habu, D., et al. (1996). Hereditary ceruloplasmin deficiency with hemosiderosis. *Human Genetics*, *97*(6), 755–758.

Osaki, S., Johnson, D. A., & Frieden, E. (1966). The possible significance of the ferrous oxidase activity of ceruloplasmin in normal human serum. *The Journal of Biological Chemistry*, *241*(12), 2746–2751.

Pan, P. L., Tang, H. H., Chen, Q., Song, W., & Shang, H. F. (2011). Desferrioxamine treatment of aceruloplasminemia: Long-term follow-up. *Movement Disorders*, *26*(11), 2142–2144.

Patel, B. N., & David, S. (1997). A novel glycosylphosphatidylinositol-anchored form of ceruloplasmin is expressed by mammalian astrocytes. *The Journal of Biological Chemistry*, *272*(32), 20185–20190.

Patel, B. N., Dunn, R. J., & David, S. (2000). Alternative RNA splicing generates a glycosylphosphatidylinositol-anchored form of ceruloplasmin in mammalian brain. *The Journal of Biological Chemistry*, *275*(6), 4305–4310.

Patel, B. N., Dunn, R. J., Jeong, S. Y., Zhu, Q., Julien, J. P., & David, S. (2002). Ceruloplasmin regulates iron levels in the CNS and prevents free radical injury. *Journal of Neuroscience*, *22*(15), 6578–6586.

Perez-Aguilar, F., Burguera, J. A., Benlloch, S., Berenguer, M., & Rayon, J. M. (2005). Aceruloplasminemia in an asymptomatic patient with a new mutation. Diagnosis and family genetic analysis. *Journal of Hepatology*, *42*(6), 947–949.

Powell, S. R. (2000). The antioxidant properties of zinc. *Journal of Nutrition*, *130*(5S Suppl.), 1447S–1454S.

Ragan, H. A., Nacht, S., Lee, G. R., Bishop, C. R., & Cartwright, G. E. (1969). Effect of ceruloplasmin on plasma iron in copper-deficient swine. *American Journal of Physiology*, *217*(5), 1320–1323.

Roberti Mdo, R., Borges Filho, H. M., Goncalves, C. H., & Lima, F. L. (2011). Aceruloplasminemia: A rare disease—diagnosis and treatment of two cases. *Revista Brasileira de Hematologia e Hemoterapia*, *33*(5), 389–392.

Salzer, J. L., Lovejoy, L., Linder, M. C., & Rosen, C. (1998). Ran-2, a glial lineage marker, is a GPI-anchored form of ceruloplasmin. *Journal of Neuroscience Research*, *54*(2), 147–157.

Samokyszyn, V. M., Miller, D. M., Reif, D. W., & Aust, S. D. (1989). Inhibition of superoxide and ferritin-dependent lipid peroxidation by ceruloplasmin. *The Journal of Biological Chemistry*, *264*(1), 21–26.

Sato, M., & Gitlin, J. D. (1991). Mechanisms of copper incorporation during the biosynthesis of human ceruloplasmin. *The Journal of Biological Chemistry*, *266*(8), 5128–5134.

Scheinberg, I. H., & Gitlin, D. (1952). Deficiency of ceruloplasmin in patients with hepatolenticular degeneration (Wilson's disease). *Science*, *116*(3018), 484–485.

Schulz, K., Vulpe, C. D., Harris, L. Z., & David, S. (2011). Iron efflux from oligodendrocytes is differentially regulated in gray and white matter. *Journal of Neuroscience*, *31*(37), 13301–13311.

Shang, H. F., Jiang, X. F., Burgunder, J. M., Chen, Q., & Zhou, D. (2006). Novel mutation in the ceruloplasmin gene causing a cognitive and movement disorder with diabetes mellitus. *Movement Disorders*, *21*(12), 2217–2220.

Skidmore, F. M., Drago, V., Foster, P., Schmalfuss, I. M., Heilman, K. M., & Streiff, R. R. (2008). Aceruloplasminaemia with progressive atrophy without brain iron overload: Treatment with oral chelation. *Journal of Neurology, Neurosurgery & Psychiatry*, *79*(4), 467–470.

Summers, M. R., Jacobs, A., Tudway, D., Perera, P., & Ricketts, C. (1979). Studies in desferrioxamine and ferrioxamine metabolism in normal and iron-loaded subjects. *British Journal of Haematology*, *42*(4), 547–555.

Takahashi, Y., Miyajima, H., Shirabe, S., Nagataki, S., Suenaga, A., & Gitlin, J. D. (1996). Characterization of a nonsense mutation in the ceruloplasmin gene resulting in diabetes and neurodegenerative disease. *Human Molecular Genetics*, *5*(1), 81–84.

Takeuchi, Y., Yoshikawa, M., Tsujino, T., Kohno, S., Tsukamoto, N., Shiroi, A., et al. (2002). A case of aceruloplasminaemia: Abnormal serum ceruloplasmin protein without ferroxidase activity. *Journal of Neurology, Neurosurgery & Psychiatry*, *72*(4), 543–545.

Texel, S. J., Xu, X., & Harris, Z. L. (2008). Ceruloplasmin in neurodegenerative diseases. *Biochemical Society Transactions*, *36*(Pt 6), 1277–1281.

Vachette, P., Dainese, E., Vasyliev, V. B., Di Muro, P., Beltramini, M., Svergun, D. I., et al. (2002). A key structural role for active site type 3 copper ions in human ceruloplasmin. *The Journal of Biological Chemistry*, *277*(43), 40823–40831.

Vulpe, C. D., Kuo, Y. M., Murphy, T. L., Cowley, L., Askwith, C., Libina, N., et al. (1999). Hephaestin, a ceruloplasmin homologue implicated in intestinal iron transport, is defective in the sla mouse. *Nature Genetics*, *21*(2), 195–199.

Yamamoto, K., Yoshida, K., Miyagoe, Y., Ishikawa, A., Hanaoka, K., Nomoto, S., et al. (2002). Quantitative evaluation of expression of iron-metabolism genes in ceruloplasmin-deficient mice. *Biochimica et Biophysica Acta*, *1588*(3), 195–202.

Yang, F., Naylor, S. L., Lum, J. B., Cutshaw, S., McCombs, J. L., Naberhaus, K. H., et al. (1986). Characterization, mapping, and expression of the human ceruloplasmin gene. *Proceedings of the National Academy of Sciences of the United States of America*, *83*(10), 3257–3261.

Yazaki, M., Yoshida, K., Nakamura, A., Furihata, K., Yonekawa, M., Okabe, T., et al. (1998). A novel splicing mutation in the ceruloplasmin gene responsible for hereditary ceruloplasmin deficiency with hemosiderosis. *Journal of Neurological Sciences*, *156*(1), 30–34.

Yonekawa, M., Okabe, T., Asamoto, Y., & Ohta, M. (1999). A case of hereditary ceruloplasmin deficiency with iron deposition in the brain associated with chorea, dementia, diabetes mellitus and retinal pigmentation: Administration of fresh-frozen human plasma. *European Neurology*, *42*(3), 157–162.

Yoshida, K., Furihata, K., Takeda, S., Nakamura, A., Yamamoto, K., Morita, H., et al. (1995). A mutation in the ceruloplasmin gene is associated with systemic hemosiderosis in humans. *Nature Genetics*, *9*(3), 267–272.

Yoshida, K., Kaneko, K., Miyajima, H., Tokuda, T., Nakamura, A., Kato, M., et al. (2000). Increased lipid peroxidation in the brains of aceruloplasminemia patients. *Journal of Neurological Sciences*, *175*(2), 91–95.

Zecca, L., Youdim, M. B., Riederer, P., Connor, J. R., & Crichton, R. R. (2004). Iron, brain ageing and neurodegenerative disorders. *Nature Reviews Neuroscience*, *5*(11), 863–873.

CHAPTER SEVEN

Therapeutic Advances in Neurodegeneration with Brain Iron Accumulation

Giovanna Zorzi*, Nardo Nardocci*,[1]
*Department of Child Neurology, Fondazione IRCCS Istituto Neurologico "Carlo Besta", Milan, Italy
[1]Corresponding author: e-mail address: nnardocci@istituto-besta.it

Contents

Abstract

Neurodegeneration with brain iron accumulation (NBIA) includes a heterogeneous group of genetically defined progressive disorders with iron accumulation in the basal ganglia. Current conventional medical or surgical therapeutic options for these disorders remain unsatisfactory and do not prevent the disease to progress to a severe and most disabling condition for the patients. In the last decade, research has been focused on the role of iron in the pathophysiological process of these disorders. The availability of chelating agents with specific properties that have been demonstrated to be effective in other disorders with regional iron accumulation and MR techniques that allow a quantitative assessment of iron have very recently brought an increasing interest in the possible efficacy of chelating agents in NBIA, and preliminary results of pilot trials are now available.

This chapter aims to provide an overview of the results of conventional medical and surgical treatments as well as of more innovative therapy in NBIA.

International Review of Neurobiology, Volume 110
ISSN 0074-7742
http://dx.doi.org/10.1016/B978-0-12-410502-7.00008-9

1. INTRODUCTION

Neurodegeneration with brain iron accumulation (NBIA) constitutes a group of genetically determined disorders with brain iron accumulation (Gregory & Hyflick, 2011). Phenotypically, they are characterized by prominent dystonia and parkinsonism associated with cognitive and psychiatric disturbances; onset can be from early childhood to adulthood, and the course of the disease, although invariably progressive, can be very variable, even within the same disease type. The clinical and the genetic spectrum of these conditions is continuously expanding, but still in a considerable proportion of patients no genetic alteration is identified (idiopathic NBIA) (Schneider, Hardy, & Bhatia, 2012). NBIA due to pantothenate kinase II (PANK2) degeneration, also named pantothenate kinase-associated neurodegeneration (PKAN) or NBIA type I, is the most common form, accounting for approximately half of all patients with NBIA.

Treating NBIA disorders is a challenging issue that requires expertise and a comprehensive approach to the patient. Conventional pharmacotherapy and surgical treatment are primarily symptomatic, aimed to reduce abnormal movements and spasticity that represent the most disabling symptoms. The advances in pathophysiological mechanism have led to the availability of innovative and more rational treatments. Here, we will review the main therapeutic approach in NBIA, from conventional strategies to new therapies under investigation.

2. PHARMACOTHERAPY

Pharmacotherapy is used for the symptomatic treatment of the movement disorder, spasticity, and psychiatric disturbances in patients with NBIA.

Several drugs may be efficacious for *dystonia* including anticholinergics, baclofen, typical and atypical neuroleptics, benzodiazepines, and L-dopa. Anticholinergic drugs, such as trihexyphenidyl, are the first line of the treatment of generalized and segmental dystonia (Hoon et al., 2001). Some patients need up to 60–80 mg per day of trihexyphenidyl but may experience dose-related blurred vision, drowsiness, confusion, memory difficulty, and hallucinations. Children usually tolerate higher doses than adults. The starting dose has to be low and increased very slowly so the accurate evaluation of efficacy requires some months.

Although antidopaminergic drugs may be beneficial in the treatment of dystonia, the potential clinical benefit is usually limited by the development of side effects, especially sedation. Dopamine-depleting drugs, however, such as tetrabenazine, have proven useful in some patients with dystonia. Tetrabenazine, an inhibitor of vesicular monoamine transporter 2, has the advantage over other antidopaminergic drugs in that it does not cause tardive dyskinesia (Kenney & Jankovic, 2006). Recently, two patients from one family with neuroferritinopathy were reported to have a dramatic improvement of orofacial dyskinesias with low doses of tetrabenazine (Ondo, Adam, Jankovic, & Chinnery, 2010).

Modest improvements with levodopa have been reported in patients with secondary and degenerative dystonias (Jankovic, 2004) and in those patients with associated parkinsonism. As an exception to this, some patients with late-onset PLAN, MPAN, and with the very recently identified form of NBIA named beta-propeller-associated neurodegeneration (Saitsu et al., 2013), previously known as SENDA, have been reported to respond well to L-dopa treatment at short- and long-term follow-up (Gregory, Polster, & Hayflick, 2009; Schneider et al., 2012). However, they may be prone to developing L-dopa-induced dyskinesia.

Benzodiazepines (diazepam, lorazepam, and clonazepam) can provide additional benefit for patients whose response to anticholinergic drugs is unsatisfactory.

Spasticity is common in most patients with NBIA and may vary in severity from a subtle neurological sign to severe spasticity causing pain and orthopedic deformities. The most used antispasticity drugs are those acting on a GABAergic system such as baclofen, gabapentin, and benzodiazepines and on alpha-2 adrenergic system such as tizanidine and those that block calcium release into the muscles such as dantrolene. It is important to note that there are no evidence-based guidelines for the choice and titration of these drugs but are based on clinical experiences (Shakespeare, Boggild, & Young, 2003).

Psychiatric symptoms such as aggressive behavior, depression, nervousness, and irritability have been described in PKAN patients, especially in atypical cases (Hayflick et al., 2003), and may require specific intervention; drugs such as benzodiazepine, selective serotonin receptor inhibitor, and antidepressant may be useful in children and adult patients.

Few cases of NBIA presenting with clear psychotic symptoms (visual and auditory hallucination, ideation, and psychomotor agitation) have been described (Morphy, Feldman, & Kilburn, 1989; Oner, Oner, Deda, &

Icagasioglu, 2003; Panas, Spngos, et al., 2007; Sunwoo et al., 2009). In the reported cases, symptoms resolved with olanzapine, clozapine, aripiprazole, and risperidone, and in these patients, the use of atypical neuroleptic is recommended (del Valle-López, Pérez-García, Sanguino-Andrés, & González-Pablos, 2011).

Obsessive–compulsive disorder and complex tics may be prominent symptoms especially in patients with late-onset PKAN, resembling Tourette syndrome (Nardocci et al., 1994; Pellecchia et al., 2005; Scarano, Pellecchia, Filla, & Barone, 2002).

3. BOTULINUM-TOXIN INJECTIONS

Botulinum-toxin injections are the treatment of choice in adult-onset focal and segmental dystonia (Jankovic, 2004). In the context of NBIA, this local treatment may be helpful in reducing the most disabling symptoms, such as the oral mandibular or cranial involvement that can be frequently seen in NBIA, particular in patients with PKAN.

4. SURGICAL TREATMENT OPTIONS IN NBIA

4.1. Intrathecal baclofen infusion

Intrathecal baclofen has been reported as effective in patients with dystonia associated with spasticity; among them, it has also been reported to be effective in several patients with NBIA (Albright, Barry, Fasick, Barron, & Shultz, 1996; Hou, Ondo, & Jankovic, 2001).

In addition, a case series recently described the use of intraventricular baclofen in patients with treatment-refractory dystonia, including one child with PKAN. This patient's dystonia improved significantly with intraventricular baclofen administration. Intraventricular delivery of baclofen is of interest because delivery at this site may better treat upper-body and facial dystonia, such as blepharospasm and oromandibular dystonia that may be particularly disabling in NBIA patients (Albright & Ferson, 2009).

4.2. Stereotactic surgery

Surgical interventions with stereotactic procedures in NBIA are currently seen as a promising treatment option in NBIA patients.

Since early 1990s, *thalamotomy and pallidotomy* have been performed to treat some cases of PKAN presenting with status dystonicus and the ablative procedure resolved the life-threatening dystonia (Balas, Kovacs, & Hollody,

2006; Justesen, Penn, Kroin, & Egel, 1999; Kyriagis et al., 2004; Tsukamoto et al., 1992).

The advent of the technique of *deep brain stimulation* (DBS) has surely increased the number of patients with NBIA undergoing stereotactic surgery, the majority of them affected by classic PKAN.

After the initial observation by Umemura, Jaggi, Dolinskas, Stern, and Baltuch (2004), several other patients have been reported demonstrating the efficacy of pallidal DBS on dystonia in PKAN, with varying follow-up from 1 to 4 years, in single patients or small case series (Castelnau et al., 2005; Grandas, Fernandez-Carballal, Guzman-de-Villoria, & Ampuero, 2011; Isaac, Wright, Bhattacharyya, Baxter, & Rowe, 2007; Krause et al., 2006; Lim et al., 2012; Mikati, Yehya, Darwish, Karam, & Comair, 2009). In some patients, the benefit was sustained at long-term follow-up (Adamovicova, Jech, Urgosik, Spackova, & Krepelova, 2011; Krause et al., 2006); in other cases, there was a loss of benefit due to disease progression (Timmermann et al., 2010).

A recent multinational study reported outcome of 23 patients with NBIA treated with pallidal DBS. The primary outcome measure was dystonia severity using the Burke–Fahn and Marsden dystonia rating scale (BFMDRS) and the Barry–Albright dystonia scale. A mean improvement of 28.5% in dystonia severity was seen at 2–6 months and 25.7% improvement was seen at 9–15 months. Two-thirds of the patients treated experienced an improvement of their dystonia of 20% or more. Although overall improvements were modest, global quality-of-life ratings showed a median improvement of 83.3% at 9–15 months, suggesting that even this relatively small effect led to clinically meaningful improvements (Timmermann et al., 2010). There was amelioration of cognitive performance in children with PKAN under pallidal DBS (Mahoney, Selway, & Lin, 2011).

A sustained improvement (3 years of follow-up) of dystonia in PKAN was also seen in one patient with DBS of the subthalamic nucleus (Ge et al., 2011).

The reports of all patients with NBIA treated with DBS have been reviewed and summarized in a recent paper (Schneider et al., 2013).

5. INNOVATIVE THERAPEUTIC STRATEGIES

The results of the aforementioned therapeutic options, being purely symptomatic, remain overall unsatisfactory and do not prevent the disease to progress to a severe and most disabling condition. In the last decade, research has been focused on the role of iron in the pathophysiological process of NBIA. It is not clear yet whether iron accumulation is the effective cause or whether it is simply an epiphenomenon (Schneider, Hardy, & Bhatia, 2009). Nevertheless, iron remains a clinical relevant clue to these disorders.

The identification of new chelating agents with specific characteristics (Kakhlon, Breuer, Munnich, & Cabantchik, 2010) and the advances in radiological techniques that have allowed a quantitative assessment of iron by magnetic resonance (Aquino et al., 2009; Stankiewicz et al., 2007), together with the preliminary experience of successful treatment with chelators in patients with other conditions characterized by regional iron accumulation (Boddaert et al., 2007), has very recently brought an increasing interest about the possible efficacy of chelating agents in NBIA.

5.1. Chelating treatment

The active deferiprone (3-hydroxy-1,2-dimethylpyridin-4-one, DFP) is a chelating agent with properties of iron relocator and is therefore most suitable for treating conditions characterized by regional iron accumulation (Kakhlon et al., 2010). It has the ability to donate iron to physiological acceptor, which means that it can transfer iron from cellular pools to circulating transferrin. It has high permeability across cell membranes, gains access to mitochondria, and reduces the formation of iron free radicals. Moreover, it has been shown to cross the blood–brain barrier with obvious therapeutic implication for neurodegenerative disorders (Boddaert et al., 2007).

In the pioneer study of Chinnery et al. (2007), three adult patients with neuroferritinopathy were put under chelating treatment (two with intravenous desferrioxamine, 4 g weekly for 14 months and one with oral deferiprone 2 g tid for 2 months), and the response to therapy was documented by clinical observation and videotaping. In one case, dystonia remarkably worsened, while the remaining two patients did not experienced any benefit.

In 2008, Forni reported the result of iron chelation therapy in a 61-year-old woman affected by adult-onset idiopathic NBIA. She was treated with deferiprone15 mg/kg per day for 6 months; there was a marked improvement after 6 months of therapy without side effects. MR images at 7 months demonstrated a reduction of iron deposit in the basal ganglia. Both clinical status and MRI changes were assessed subjectively (Forni et al., 2008).

Another recent study described the long-term improvement under deferiprone (30 mg/kg per day) in a case of idiopathic NBIA with the onset at 47 years of age presenting with cerebellar ataxia and parkinsonism. The clinical status was assessed by standardized scales (motor UPDRS, ICARS, and SARA); after 6 months, there was 30% improvement in ataxia, 35% in dysarthria, and 45% in orofacial dystonia. After the first year of treatment, there was a slight progression of symptoms, but at the end of the follow-up (32 months), the clinical status was overall better than the pretreatment phase. Quantitative analysis of T2* MRI changes revealed a marked but transient decrease of iron concentration in the dentate nuclei, sustained decrease in the substantia nigra of a lesser extent, and no changes in the red nuclei (Kwiatkowski et al., 2012).

More recently, a young patient with classic PKAN treated with oral deferiprone (20 mg/kg/die) was reported with a sustained improvement of dystonia after 1 year of chelating therapy without remarkable side effects (Pratini, Sweeters, Vichinsky, & Neufeld, 2013).

Besides these reports of single cases, two studies on larger series of patients have been published. Our group reported the results of the first phase II pilot open trial in PKAN, assessing the clinical and radiological effects of deferiprone at a dose of 25 mg/kg per day over a 6-month period. Among the nine patients who completed the study, six had classic disease and three had atypical disease. At enrollment, median age was 26 years (range 7–39 years) and median disease duration was 11 years. Deferiprone was well tolerated overall; side effects included nausea and gastralgia (44%), but no serious adverse event occurred. The authors observed a significant (median, 30%) reduction in globus pallidus iron content, ranging from 15% to 61%. However, there was no clinical benefit, as rated on the BFMDRS and SF-36 scale. We suggest that the discrepancy between radiological and clinical data may have multiple different explanations: the relatively short treatment duration, long disease duration, or neuronal damage too advanced to allow for a rescue of function (Zorzi et al., 2011).

A similar study was conducted for a longer period on a series of six patients with different forms of NBIA (four PKAN and two with idiopathic NBIA) treated with deferiprone 15 mg/kg per day and assessed at 6 and 12 months. Mean age at enrollment was 36.5 and mean disease duration was 7.5 (range 3–13) years. Clinical rating scales (UPDRS/III, ICARS, and UDRS) and blinded video rating documented a slight improvement in three cases (two PKAN and one NBIA) and no change in the remaining three cases. The improvement was observed after 6 months of treatment and persisted at 12 months. Quantitative analysis of brain iron through T2* relaxometry was possible only in three patients and demonstrated a reduction of iron content in the globus pallidus (Abbruzzese et al., 2011).

A large multicenter placebo-control study in PKAN patients treated with deferiprone is currently ongoing and will hopefully clarify the potential efficacy of chelating treatment in NBIA.

5.2. Other innovative therapeutic strategies under investigation

Since the gene identification of the gene defects in some NBIA, there have been significant insights into NBIA disease biology that hopefully will allow to identify more rationale and causative therapeutic strategies. Most of the data refer to PANK2 deficiency, which is the most common form of NBIA; little is known about the functioning of the other known genes associated to NBIA.

5.2.1 Pantothenate

Pantothenate is the substrate of the defective enzyme PANK2. The existence of residual enzyme activity in some individuals with PKAN, mainly patients with late-onset PKAN, has raised the possibility of treatment using high-dose pantothenate. Pantothenate has no known toxicity in humans. However, the efficacy of pantothenate supplementation in ameliorating symptoms has not been systematically proven; some individuals with an atypical disease course have anecdotally reported improvement in their symptoms under pantothenate treatment (Gregory et al., 2009).

5.2.2 Pantethine

Pantethine, a vitamin B5 analog, has the ability to potentially circumvent the enzymatic defect in PKAN and restore the ability to synthesize adequate quantities of coenzyme A. Pantethine was administered to drosophila fumble mutants, which feature mutation of dPANK, the drosophila homolog of human PANK enzymes, resulting in rescue of the phenotype (Rana et al., 2010). Although this therapy has significant potential as a rational therapy, several limitations exist. This includes incomplete data regarding the bioavailability, pharmacokinetics, and ability of this compound to cross the blood–brain barrier in humans. Furthermore, there are important differences between drosophila and human physiology, including a single PANK isoform in drosophila but four distinct isoforms in humans, with *PANK2* being specifically targeted to mitochondria. Despite these limitations, the identification of pantethine as a potential therapeutic agent has generated much interest within the NBIA community.

5.2.3 Lipid metabolism

Pantothenate kinase is a regulator in the synthesis of lipid metabolism and bile acids, pathways that both require coenzyme A. In a recent paper, using a global metabolic profiling of plasma from patients with PANK2 deficiency, reduced lipid and cholesterol biosynthesis and impaired bile acid metabolism were found. The authors suggested that these findings may raise a novel therapeutic hypothesis with dietary fats and bile acid supplementation (Leoni et al., 2012).

6. CONCLUSIONS

There is an increasing interest in NBIA and its treatment, aimed to better understand their pathogenesis and to find more effective and rationale treatments. Among innovative therapy, the use of iron chelators is emerging, but other strategies are under investigation that will possibly modify the clinical course of these severe conditions.

REFERENCES

Abbruzzese, G., Cossu, G., Balocco, M., Marchese, R., Murgia, D., Melis, M., et al. (2011). A pilot trial of deferiprone for neurodegeneration with brain iron accumulation. *Haematologica*, *96*(11), 1708–17011.

Adamovicova, M., Jech, R., Urgosik, D., Spackova, N., & Krepelova, A. (2011). Pallidal stimulation in siblings with pantothenate kinase-associated neurodegeneration: Four-year follow-up. *Movement Disorders*, *26*, 184–187.

Albright, A. L., Barry, M. J., Fasick, P., Barron, W., & Shultz, B. (1996). Continuous intrathecal baclofen infusion for symptomatic generalized dystonia. *Neurosurgery*, *38*, 934–939.

Albright, A. L., & Ferson, S. S. (2009). Intraventricular baclofen for dystonia: Techniques and outcomes. Clinical article. *Journal of Neurosurgery-Pediatrics*, *3*(1), 11–14.

Aquino, D., Bizzi, A., Grisoli, M., Garavaglia, B., Bruzzone, M. G., Nardocci, N., et al. (2009). Age-related iron deposition in the basal ganglia: Quantitative analysis in healthy subjects. *Radiology*, *252*, 165–172.

Balas, I., Kovacs, N., & Hollody, K. (2006). Staged bilateral stereotactic pallidothalamotomy for life-threatening dystonia in a child with Hallervorden-Spatz disease. *Movement Disorders*, *21*, 82–85.

Boddaert, N., Le Quan Sang, K. H., Rötig, A., Leroy-Willig, A., Gallet, S., et al. (2007). Selective iron chelation in Friedreich ataxia: Biologic and clinical implications. *Blood*, *110*, 401–408.

Castelnau, P., Cif, L., Valente, E. M., Vayssiere, N., Hemm, S., Gannau, A., et al. (2005). Pallidal stimulation improves pantothenate kinase-associated neurodegeneration. *Annals of Neurology*, *57*, 738–741.

Chinnery, P. F., Crompton, D. E., Birchall, D., Jackson, M. J., Coulthard, A., Lombes, A., et al. (2007). Clinical features and natural history of neuroferritinopathy caused by the FTL1 460InsA mutation. *Brain, 130,* 110–119.

del Valle-López, P., Pérez-García, R., Sanguino-Andrés, R., & González-Pablos, E. (2011). Adult onset Hallervorden-Spatz disease with psychotic symptoms. *Actas Españolas de Psiquiatría, 39*(4), 260–262.

Forni, G. L., Balocco, M., Cremonesi, L., Abruzzese, G., Parodi, R. C., & Marchese, R. (2008). Regression of symptoms after selective iron chelation therapy in a case of neurodegeneration with brain iron accumulation. *Movement Disorders, 23*(6), 904–907.

Ge, M., Zhang, K., Ma, Y., Meng, F. G., Hu, W. H., Yang, A. C., et al. (2011). Bilateral subthalamic nucleus stimulation in the treatment of neurodegeneration with brain iron accumulation type 1. *Stereotactic and Functional Neurosurgery, 89*(3), 162–166.

Grandas, F., Fernandez-Carballal, C., Guzman-de-Villoria, J., & Ampuero, I. (2011). Treatment of a dystonic storm with pallidal stimulation in a patient with PANK2 mutation. *Movement Disorders, 26*(5), 921–922.

Gregory, A., & Hyflick, S. J. (2011). Genetics of neurodegeneration with brain iron accumulation. *Current Neurology and Neuroscience Reports, 11*(3), 254–261.

Gregory, A., Polster, B. J., & Hayflick, S. J. (2009). Clinical and genetic delineation of neurodegeneration with brain iron accumulation. *Journal of Medical Genetics, 46,* 73–80.

Hayflick, S. J., Westaway, S. K., Levinson, B., Zhou, B., Johnson, M. A., Ching, K. H., et al. (2003). Genetic, clinical, and radiographic delineation of Hallervorden-Spatz syndrome. *New England Journal of Medicine, 348*(1), 33–40.

Hoon, A. H., Jr., Freese, P. O., Reinhardt, E. M., Wilson, M. A., Lawrie, W. T., Jr., Harryman, S. E., et al. (2001). Age-dependent effects of trihexyphenidyl in extrapyramidal cerebral palsy. *Pediatric Neurology, 25,* 55–58.

Hou, J. G., Ondo, W., & Jankovic, J. (2001). Intrathecal baclofen for dystonia. *Movement Disorders, 16,* 1201–1202.

Isaac, C., Wright, I., Bhattacharyya, D., Baxter, P., & Rowe, J. (2007). Pallidal stimulation for pantothenate kinase-associated neurodegeneration dystonia. *Archives of Disease in Childhood, 93,* 239–240.

Jankovic, J. (2004). Dystonia: Medical therapy and botulinum toxin. In S. Fahn, M. Hallett, & D. R. DeLong (Eds.), *Advances in neurology: 94. Dystonia 4* (pp. 275–286). Philadelphia: Lippincott, Williams & Wilkins.

Justesen, C. R., Penn, R. D., Kroin, J. S., & Egel, R. T. (1999). Stereotactic pallidotomy in a child with Hallervorden-Spatz disease. Case report. *Journal of Neurosurgery, 90,* 551–554.

Kakhlon, O., Breuer, W., Munnich, A., & Cabantchik, Z. I. (2010). Iron redistribution as a therapeutic strategies for treating diseases of localized iron accumulation. *Canadian Journal of Physiology and Pharmacology, 88*(3), 187–196.

Kenney, C., & Jankovic, J. (2006). Tetrabenazine in the treatment of hyperkinetic movement disorders. *Expert Review of Neurotherapeutics, 6,* 7–17.

Krause, M., Fogel, W., Tronnier, V., Pohle, S., Hörtnagel, K., Thyen, U., et al. (2006). Long-term benefit to pallidal deep brain stimulation in a case of dystonia secondary to pantothenate kinase-associated neurodegeneration. *Movement Disorders, 21,* 2255–2257.

Kwiatkowski, A., Ryckewaert, G., Jissendi Tchofo, P., Moreau, C., Vuillaume, I., Chinnery, P. F., et al. (2012). Long term improvement under deferiprone in a case of neurodegeneration with brain iron accumulation. *Parkinsonism & Related Disorders, 18,* 110–112.

Kyriagis, M., Grattan-Smith, P., Scheinberg, A., Teo, C., Nakaji, N., & Waugh, M. (2004). Status dystonicus and Hallervorden-Spatz disease: Treatment with intrathecal baclofen and pallidotomy. *Journal of Paediatrics and Child Health, 40,* 322–325.

Leoni, V., Strittmatter, L., Zorzi, G., Zibordi, F., Dusi, S., Garavaglia, B., et al. (2012). Metabolic consequences of mitochondrial coenzyme A deficiency in patients with PANK2 mutations. *Molecular Genetics and Metabolism, 105*(3), 463–471.

Lim, B. C., Ki, C. S., Cho, A., Hwang, H., Kim, K. J., Hwang, Y. S., et al. (2012). Pantothenate kinase-associated neurodegeneration in Korea: Recurrent R440P mutation in PANK2 and outcome of deep brain stimulation. *European Journal of Neurology, 19*, 556–561.

Mahoney, R., Selway, R., & Lin, J. P. (2011). Cognitive functioning in children with pantothenate-kinase neurodegeneration undergoing deep brain stimulation. *Developmental Medicine and Child Neurology, 53*(3), 275–279.

Mikati, M. A., Yehya, A., Darwish, H., Karam, P., & Comair, Y. (2009). Deep brain stimulation as a mode of treatment of early onset pantothenate kinase-associated neurodegeneration. *European Journal of Paediatric Neurology, 13*, 61–64.

Morphy, M. A., Feldman, J. A., & Kilburn, G. (1989). Hallervorden-Spatz disease in a psychiatric setting. *Journal of Clinical Psychiatry, 50*, 66–68.

Nardocci, N., Rumi, V., Combi, M. L., Angelini, L., Mirabile, D., & Bruzzone, M. G. (1994). Complex tics, stereotypies, and compulsive behavior as clinical presentation of a juvenile progressive dystonia suggestive of Hallervorden-Spatz disease. *Movement Disorders, 9*, 369–371.

Ondo, W. G., Adam, O. R., Jankovic, J., & Chinnery, P. F. (2010). Dramatic response of facial stereotype/tic to tetrabenazine in the first reported cases of neuroferritinopathy in the United States. *Movement Disorders, 25*(14), 2470–2472.

Oner, O., Oner, P., Deda, G., & Icagasioglu, D. (2003). Psychotic disorder in a case with Hallervorden–Spatz disease. *Acta Psychiatrica Scandinavica, 108*, 394–398.

Panas, M., Spngos, K., Koutsis, G., Tsivgoulis, G., Sfagos, K., Kalfakis, N., et al. (2007). Psychosis as presenting symptoms in adult onset Hallervorden-Spatz syndrome. *Acta Neuropsychiatrica, 19*, 122–124.

Pellecchia, M. T., Valente, E. M., Cif, L., Salvi, S., Albanese, A., & Scarano, V. (2005). The diverse phenotype and genotype of pantothenate kinase-associated neurodegeneration. *Neurology, 64*(10), 1810–1812.

Pratini, N. R., Sweeters, N., Vichinsky, E., & Neufeld, J. A. (2013). Treatment of classic pantothenate kinase-associated neurodegeneration with deferiprone and intrathecal baclofen. *American Journal of Physical Medicine & Rehabilitation, 92*, 728–733.

Rana, A., Seinen, E., Siudeja, K., Muntendam, R., Srinivasan, B., van der Want, J. J., et al. (2010). Pantethine rescues a Drosophila model for pantothenate kinase-associated neurodegeneration. *Proceedings of the National Academy of Sciences of the United States of America, 107*, 6988–6993.

Saitsu, H., Nishimura, T., Muramatsu, K., Kodera, H., Kumada, S., Sugai, K., et al. (2013). De novo mutations in the autophagy gene WDR45 cause static encephalopathy of childhood with neurodegeneration in adulthood. *Nature Genetics, 45*, 445–449.

Scarano, V., Pellecchia, M. T., Filla, A., & Barone, P. (2002). Hallervorden-Spatz syndrome resembling a typical Tourette syndrome. *Movement Disorders, 17*, 618–620.

Schneider, S. A., Dusek, P., Hardy, J., Westenberger, A., Jankovic, J., & Bhatia, K. P. (2013). Genetics and pathophysiology of neurodegeneration with brain iron accumulation (NBIA). *Current Neuropharmacology, 11*, 59–79.

Schneider, S. A., Hardy, J., & Bhatia, K. P. (2009). Iron accumulation in syndromes of neurodegeneration with brain accumulation—Causative or consequential? *Journal of Neurology, Neurosurgery, and Psychiatry, 80*, 589–590.

Schneider, S. A., Hardy, J., & Bhatia, K. P. (2012). Syndromes of neurodegeneration with brain iron accumulation (NBIA): An update on clinical presentations, histological and genetic underpinnings, and treatment considerations. *Movement Disorders, 27*, 42–53.

Shakespeare, D., Boggild, M., & Young, C. A. (2003). Antispasticity agents for multiple sclerosis: A systematic review. *Cochrane Database of Systematic Reviews, 1*, 7–8.

Stankiewicz, J., Panter, S. S., Neema, M., Arora, A., Batt, C. E., & Bakshi, R. (2007). Iron in chronic brain disorders: Imaging and neurotherapeutic implications. *Neurotherapeutics, 4*, 371–386.

Sunwoo, Y. K., Lee, J. S., Kim, W. H., Shin, Y. B., Lee, M. J., Cho, I. H., et al. (2009). Psychiatric disorder in two siblings with Hallervorden-Spatz disease. *Psychiatry Investigation, 6*, 226–229.

Timmermann, L., Pauls, K. A., Wieland, K., Jech, R., Kurlemann, G., Sharma, N., et al. (2010). Dystonia in neurodegeneration with brain iron accumulation: Outcome of bilateral pallidal stimulation. *Brain, 133*, 701–712.

Tsukamoto, H., Inui, K., Taniike, M., Nishimoto, J., Midorikawa, M., Yoshimine, T., et al. (1992). A case of Hallervorden-Spatz disease: Progressive and intractable dystonia controlled by bilateral thalamotomy. *Brain & Development, 14*, 269–272.

Umemura, A., Jaggi, J. L., Dolinskas, C. A., Stern, M. B., & Baltuch, G. H. (2004). Pallidal deep brain stimulation for longstanding severe generalized dystonia in Hallervorden-Spatz syndrome: Case report. *Journal of Neurosurgery, 100*, 706–709.

Zorzi, G., Zibordi, F., Chiapparini, L., Bertini, E., Russo, L., Piga, A., et al. (2011). Iron-related MRI images in patients with pantothenate kinase-associated neurodegeneration (PKAN) treated with deferiprone: Results of a phase II pilot trial. *Movement Disorders, 26*(9), 1756–1759.

CHAPTER EIGHT

The Neuropathology of Neurodegeneration with Brain Iron Accumulation

Michael C. Kruer[1]

Sanford Children's Health Research Center, Sanford Children's Hospital, Sioux Falls, South Dakota, USA

[1]Corresponding author: e-mail address: e-mail address: michael.kruer@sanfordhealth.org

Contents

International Review of Neurobiology, Volume 110
ISSN 0074-7742
http://dx.doi.org/10.1016/B978-0-12-410502-7.00009-0

Abstract

Neuropathology plays a key role in characterizing the pathogenesis of neurodegenerative diseases including forms of neurodegeneration with brain iron accumulation (NBIA). Despite important differences, several genetically diverse forms of NBIA nevertheless share common features in addition to iron deposition, such as the presence of neuroaxonal spheroids. Multiple forms of NBIA also demonstrate tau or synuclein pathology, suggesting parallels with both Alzheimer and Parkinson diseases. This chapter summarizes what has been learned from the study of human patient tissues. Gross and microscopic findings are delineated, and similarities and differences between forms of NBIA are presented. Neuropathologic findings often help characterize fundamental features of disease and provide a springboard for more focused hypothesis-driven studies. Lessons learned from neuropathology thus contribute much to the characterization of the molecular mechanisms of disease.

1. INTRODUCTION

Neuropathology has played a central role in defining the biology of neurodegenerative diseases. Postmortem studies of affected patients often provide important clues to disease pathogenesis. These findings can then be followed up with more specific biochemical assays or studies using *in vitro* or animal models. In fact, the lack of human autopsy material can be a limiting factor in disease research, particularly in the case of rare diseases such as neurodegeneration with brain iron accumulation (NBIA).

The recognition of the brain iron deposition syndromes began with the work of Julius Hallervorden and Hugo Spatz, German neuropathologists who noted a gross discoloration of the globus pallidus in postmortem brains from patients with extrapyramidal features. It was their work that led to the recognition of the brain iron accumulation disorders, and the two were credited for their discovery via the initial eponym for the disease. However, given their association or at the very least, tacit acquiescence with the mass euthanasia programs of the Third Reich (Shevell, 1992), the disorder was renamed once the causative gene was identified (Zhou et al., 2001).

Since the first recognition of the NBIA disorders, many papers delineating the neuropathology of NBIA have appeared in the literature. Early reports, published before the recognition of distinct subtypes of the disorder and the molecular genetic characterization of causative genes, relied on clinical findings and the unifying feature of brain iron to classify a disorder as "NBIA." Although these important early manuscripts characterized a rich diversity of phenotypic findings, interpretation is complicated by the fact that these reports likely comprise a heterogenous group of disorders, only some of which would be classified as NBIA today. Furthermore, important differences exist among subtypes of NBIA, particularly as some forms of the disease can be grouped based on shared pathologic features that are not universal among forms of NBIA. In addition, although animal models are a valuable contributor to efforts to dissect mechanisms of disease, such models do not always faithfully recapitulate "core" features of the human condition (Kuo et al., 2005). For example, existing mouse models for subtypes of NBIA conspicuously lack iron accumulation (Kuo, Hayflick, & Gitschier, 2007; Malik et al., 2008; Potter et al., 2011; Schultheis et al., 2013; Wada et al., 2009; Zöller et al., 2008). Animal models thus require careful corroboration with human disease histopathology to identify important similarities and differences between species. For these reasons, this chapter will focus on molecularly characterized cases of human brain pathology.

Other neurodegenerative diseases, including Parkinson disease, Huntington disease, and multiple system atrophy, are known to feature brain iron accumulation. The overlap of the NBIA disorders with these more common forms of neurodegeneration has thus generated significant interest and suggests an element of shared pathophysiology (Schneider & Bhatia, 2013). However, significant differences also exist, such as the predominant sites of iron deposition (pallidal in most forms of NBIA and nigral in other neurodegenerative diseases) and in clinical features of disease. Furthermore, whether iron accumulation occurs as a result of converging molecular mechanisms or simply as an epiphenomenon is not known.

Pathologic hallmarks of NBIA include the disease-defining deposition of iron, the presence of neuroaxonal spheroids, and the variable accumulation of α-synuclein-positive Lewy bodies and/or tau pathology. Microscopic visualization of iron deposits has largely been performed using Perls' stain (also referred to as the Prussian or Berlin blue stain), which detects Fe^{3+}; Turnbull blue, which reacts more vigorously with Fe^{2+}; and/or ferritin immunostaining (typically using antibodies raised against ferritin light chain). Ferritin-associated iron largely occurs as Fe^{3+}. Surprisingly, little is known about the subcellular location of iron deposits or the chemical form iron deposition occurs in. Correspondingly, little is also known about the origin and composition of spheroids, although ultrastructural characterization of spheroids in putative pantothenate kinase-associated neurodegeneration (PKAN) and infantile neuroaxonal dystrophy has been performed (Malandrini, Bonuccelli, et al., 1995; Malandrini, Cavallaro, et al., 1995). Reports have suggested that spheroids begin within the axon and spread retrogradely to the cell body, perhaps indicated an impairment of axonal transport (Malandrini, Bonuccelli, et al., 1995; Malandrini, Cavallaro, et al., 1995).

Recognizing the importance of studies of human histopathology, substantial challenges still exist in interpreting findings from this work. In particular, separating primary pathology from secondary effects, particularly in patients with end-stage disease, can be difficult. Downstream mechanisms can complicate interpretation, begging the question as to whether a response is compensatory or maladaptive. Furthermore, given the paucity of autopsy material from patients with NBIA, findings are derived from a small number of cases, sometimes challenging efforts to distinguish incidental findings or age-related pathology from cardinal features of the disease. Despite these inherent difficulties, analysis of autopsy and biopsy material has determined that subtypes of NBIA have distinct pathologic features that have provided important clues to the biological origins of disease (Table 8.1). These

Table 8.1 Neuropathologic features of subtypes of NBIA

Subtype	Associated gene	Iron deposition	Pathologic features
PKAN	*PANK2*	GP, SN	NFTs, spheroids
PLAN	*PLA2G6*	GP, SN	Lewy bodies, NFTs, spheroids
MPAN	*c19orf12*	GP, SN	Lewy bodies, spheroids, tau
ACP	*CP*	P, CN, GP, Th, DN	GFSB; bizarre astrocytes
NFT	*FTL*	P, CN, GP, Th, DN	Distorted, iron-laden nuclei
FAHN	*FA2H*	GP, SN	???
KRS	*ATP13A2*	CN, P	???
BPAN	*WDR45*	SN, GP	Lewy bodies, NFTs, spheroids

GP, globus pallidus; SN, substantia nigra; CN, caudate nucleus; P, putamen; Th, thalamus; DN, dentate nucleus; NFTs, neurofibrillary tangles; GFSB, grumose or foamy spheroid bodies.

findings have clarified the nosological relationship between forms of NBIA and highlighted relationships to other neurodegenerative diseases.

2. PANTOTHENATE KINASE-ASSOCIATED NEURODEGENERATION (PKAN)

The most common form of NBIA, PKAN, was the first form of the disease to have the causative gene identified. Affected patients develop progressive dystonia and parkinsonism associated with a variable degree of intellectual decline. Neuroimaging demonstrates the characteristic "eye-of-the-tiger" appearance of the globus pallidus, with T2 hypointensity indicative of iron deposition surrounding a central region of hyperintensity. The eye-of-the-tiger sign is considered pathognomonic for this form of NBIA (Hayflick, Hartman, Coryell, Gitschier, & Rowley, 2006), although similar but "atypical" findings can occur in other subtypes of the disease. The pathology of PKAN has been largely described by Kruer et al. (2011) and Li et al. (2012).

2.1. Gross brain pathology

The brains of patients with PKAN do not show significant gross atrophy, consistent with neuroimaging findings. Sectioning of the brain reveals a rusty discoloration largely confined to the globus pallidus. The remainder of the brain is grossly unremarkable.

2.2. Histologic findings

Histologically, pathology is most apparent within the globus pallidus but diffuse cortical and subcortical involvement occurs. Although inflammation is not a prominent feature, CD163+ microglial infiltration of the globus pallidus is seen. Loss of both myelin (assessed by myelin basic protein staining) and axons (as determined by phosphorylated neurofilament heavy-chain immunohistochemistry) is prominent in the globus pallidus and adjacent white matter tracts. Neuronal degeneration is the predominant pathology, and reactive fibrillary astrogliosis occurs in affected brain regions. The cerebellum is largely spared. In addition, midbrain pathology is relatively mild, with preservation of the pigmented neurons of the substantia nigra. Rarefaction of the neuropil is seen most prominently in the globus pallidus. A demarcated region of profound rarefaction occurs within the central globus pallidus corresponding to the site of the "eye of the tiger" seen on neuroimaging.

2.3. Iron deposition

Perls' staining of PKAN brain tissue shows a widespread perivascular deposition of iron largely confined to the globus pallidus but not strictly respecting the borders of this structure. Iron deposition occurs to a lesser extent in the substantia nigra. A coarse granular cytoplasmic appearance of both neurons and astrocytes is appreciable by Perls' stain when compared to controls. Degenerating neurons show loss of this granularity. In addition, a fine "iron dusting" of the neuropil occurs. No significant iron deposition has been appreciated in oligodendrocytes. Occasional microglia contain coarse granular iron. Ferritin staining of both neurons and astrocytes is prominent in PKAN and in general mirrors the pattern seen by Perls' staining. However, this correlation is incomplete, suggesting that iron not only occurs as ferritin deposits but also occurs in other forms.

2.4. Spheroids

A pathologic hallmark of NBIA, eosinophilic spheroids are seen in PKAN brains in the globus pallidus and to a lesser extent in the pars reticulata of the substantia nigra. Such spheroids range from 20 to 70 μm in diameter and may take on a variety of morphologies. In many if not all cases, these bodies are clearly neuronally derived. The larger spheroids have been alternately characterized as degenerating neurons or ovoid bodies, while the smaller structures are felt to represent classical "neuroaxonal spheroids," sometimes surrounded by a myelin rim. It is likely that given the neuronal degeneration that occurs, these findings exist along a continuum. Spheroids are inconsistently tau- and amyloid precursor protein (APP)-positive and inconsistently iron-positive while usually not appearing heavily iron-laden.

Electron microscopic analysis of spheroids has been performed in autopsy samples with clinical and neuroimaging features of PKAN (Malandrini, Bonuccelli, et al., 1995). Although not molecularly confirmed, these cases can be classified as *probable PKAN*. Findings from EM demonstrated dense osmiophilic bodies associated with vesicles and amorphous material.

2.5. Tau and synuclein pathology

Gallyas silver stain and tau-positive neurofibrillary tangles are a feature of PKAN but affected patients vary widely in their burden of tau pathology, suggesting this is a secondary feature of the disease. Although PKAN was once purported to represent a synucleinopathy, more recent studies of molecularly confirmed cases have shown that these early reports likely represent another form of NBIA as Lewy bodies and synuclein deposition are conspicuously absent in PKAN. Although ubiquitin immunostaining is widespread in PKAN and shows overlap with spheroids, markers of other neurodegenerative disease, including TDP-43 and FUS, are absent.

2.6. Other affected organ systems

Although pigmentary retinopathy is seen in PKAN, studies reporting ocular pathology have not yet been published. Unlike other forms of NBIA, peripheral neuropathy is not a feature of PKAN. The cause of death is PKAN that is often unknown, although largely ascribed to cardiopulmonary failure. Prominent autonomic signs are not seen in PKAN, but it is not known if cardiac pathology exists. A single report described myopathy as a feature of PKAN (Malandrini, Cavallaro, et al., 1995), but this was not in a molecularly confirmed case.

Important aspects of PKAN neuropathology are portrayed in Figs. 8.1–8.3.

3. PHOSPHOLIPASE-ASSOCIATED NEURODEGENERATION (PLAN)

In contrast to PKAN, phospholipase-associated neurodegeneration (PLAN) is related to mutations in *PLA2G6* and widely considered a synucleinopathy, cross-listed as PARK14 (OMIM #612953). Iron deposition is not a universal feature of PLAN, present in roughly half of childhood-onset cases and cerebellar atrophy usually precedes iron accumulation on patient MRI scans. Patients presenting with adult-onset dystonia-parkinsonism do not typically show iron deposition (Paisan-Ruiz et al., 2009). The pathology of PLAN has been described primarily by Gregory et al. (2008) and Paisán-Ruiz et al. (2010).

3.1. Gross pathology

Gross brain pathology shows diffuse cortical and cerebellar atrophy, with cerebellar atrophy predominating. A rust-colored appearance of the globus pallidus may be appreciable, while pallor of the substantia nigra may also occur.

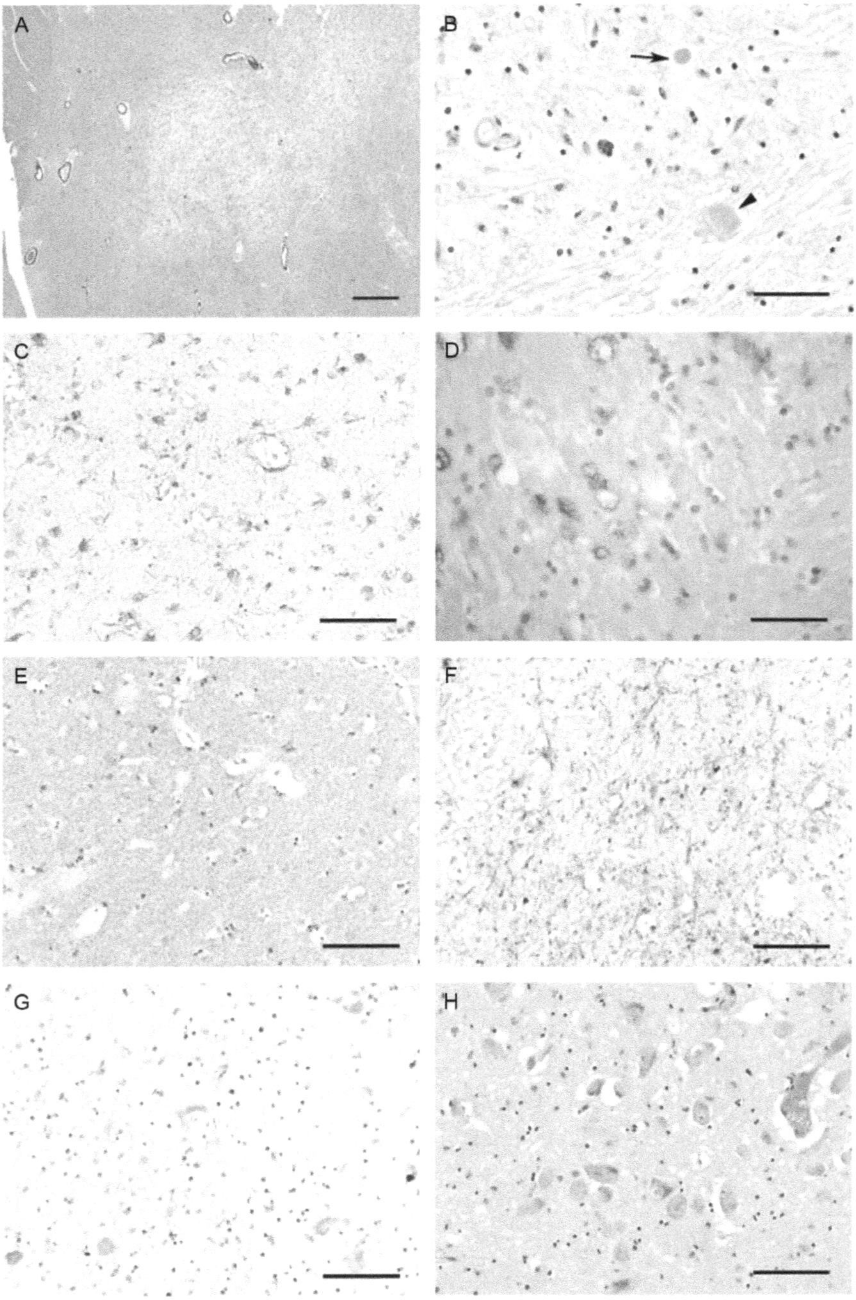

Figure 8.1 Histopathologic features of affected tissues in PKAN. (A) Low-magnification view of hematoxylin and eosin stain of globus pallidus, with periodic acid Schiff's stain added to highlight vascular structures, showing the rarefied area that corresponds to

3.2. Histologic findings

Histologically, widespread neuronal loss occurs with corresponding astrogliosis. Purkinje cell degeneration leads to torpedo morphology and eventual dropout. Cerebellar granule cell loss may be prominent with associated astrogliosis. Atrophy of the molecular layer also occurs.

3.3. Iron deposition

Perls' stain shows perivascular and intracellular iron accumulation largely confined to the globus pallidus, although the substantia nigra variably shows iron deposition. Affected neurons accumulate iron leading to a coarse intracellular appearance that recedes in degenerating cells.

3.4. Spheroids

Neuroaxonal spheroids are a prominent feature of PLAN, occurring throughout the brain and spinal cord. Spheroids can also be appreciated in peripheral nerve. Spheroids are eosinophilic, neuronally derived structures 30–100 μm in diameter. They may stain positively for neurofilament, ubiquitin, and α-synuclein. They are highlighted by the Bielschowsky silver stain method and have been observed to be surrounded with a rim of residual myelin. Electron microscopy of spheroids from a *probable PLAN* case has shown accumulation of membrano-tubular or granulo-vesicular material within neurons (Malandrini, Bonuccelli, et al., 1995; Malandrini, Cavallaro, et al., 1995).

the "eye of the tiger" observed radiographically (scale bar: 600 μm). (B) Both large degenerating neurons (arrowhead) and smaller neuroaxonal spheroids (arrow) were present in the globus pallidus in PKAN. Note also hemosiderin in the background (scale bar: 50 μm). (C) Glial fibrillary acidic protein immunohistochemistry of the globus pallidus, demonstrating widespread gliosis. Larger residual cells are astrocytes (scale bar: 100 μm). (D) Focal macrophage infiltrates in the globus pallidus as assessed by CD163 immunohistochemistry (scale bar: 50 μm). (E) Normal preservation of synaptic content in putamen as assessed by synaptophysin immunohistochemistry; for comparison, (E)–(G) were acquired at the same magnification and exposure under same light intensity (scale bar: 100 μm). (F) Relative synapse preservation more peripherally in the globus pallidus externa adjacent to the internal capsule as assessed by synaptophysin immunohistochemistry (scale bar: 100 μm). (G) Marked synapse loss in central region of globus pallidus as assessed by synaptophysin immunohistochemistry (scale bar: 100 μm). (H) Hematoxylin and eosin stain of the substantia nigra, showing good preservation of midbrain pigmented neurons (scale bar: 200 μm). *(Modified from Kruer et al. (2011) with permission).* (For color version of this figure, the reader is referred to the online version of this chapter.)

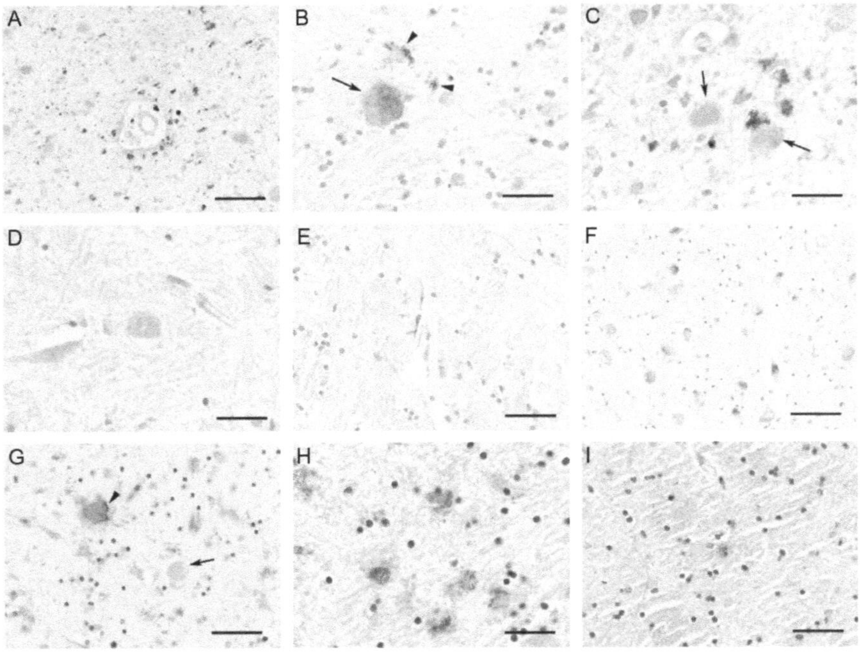

Figure 8.2 Iron deposits in PKAN. (A) Low-magnification view of the globus pallidus in PKAN, stained with Perls' stain for iron. Some perivascular accentuation of iron deposits is present, with focal collections of hemosiderin-laden macrophages as may be encountered with normal aging (scale bar: 200 μm). (B) Perls' stain for iron, demonstrating cellular localization of iron in the globus pallidus in PKAN. Degenerating neurons with relative preservation of cytoplasm demonstrate increased cytoplasmic iron staining (arrow) compared with neurons in control globus pallidus (D), as well as astrocytes with more intense, dense granular accumulation of cytoplasmic iron (arrowheads). In contrast, oligodendroglia and microglia displayed no increase in iron (scale bar: 50 μm). (C) Perls' iron stain of the globus pallidus in PKAN, showing progressive decrease in iron content as neurons degenerate. Iron-positive astrocytes are more conspicuous and greatly outnumber those present in normal globus pallidus (refer to E for comparison). Note also increased iron diffusely present in the neuropil, manifest as a pale blue background tinge (scale bar: 50 μm). (D) Perls' stain of control globus pallidus for comparison, showing very rare neurons with detectable iron accumulation (scale bar: 100 μm). (E) Perls' stain of control globus pallidus. Rare astrocytes, predominantly in a juxtavascular distribution, were found to contain increased iron. The numbers of these were markedly less than encountered in PKAN (scale bar: 100 μm). (F) Perls' stain of the midbrain in PKAN; no detectable iron was present in the substantia nigra (scale bar: 200 μm). (G) Ferritin immunohistochemistry of the globus pallidus in PKAN, showing ferritin association of iron in some neurons, sometimes in a peripheral distribution (arrowhead). As with iron stains, ferritin staining tended not to be present in extensively degenerated neurons (arrow) (scale bar: 50 μm). (H) Ferritin immunohistochemistry of astrocytes in the globus pallidus in PKAN. As with iron stains, astrocytes demonstrated the most intense cytoplasmic ferritin (scale bar: 50 μm). (I) Ferritin immunohistochemistry of astrocytes in control globus pallidus. Very rare astrocytes display detectable labeling (scale bar: 100 μm). *(Modified from Kruer et al. (2011) with permission).* (For color version of this figure, the reader is referred to the online version of this chapter.)

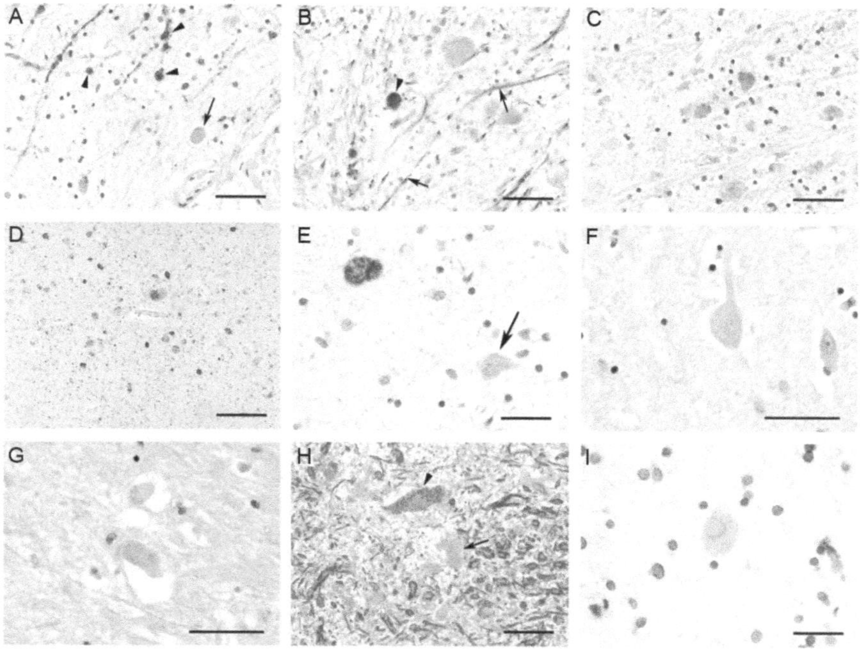

Figure 8.3 Features of neuroaxonal spheroids and degenerating neurons. (A) Immunohistochemistry for amyloid precursor protein demonstrates intense staining of neuroaxonal spheroids (arrowhead) and a lower amount of staining, or no staining, of degenerating neurons (arrow) in the globus pallidus in PKAN (scale bar: 50 μm). (B) Immunohistochemistry for high-molecular-weight neurofilament protein reveals limited staining, mostly of smaller neuroaxonal spheroids (arrowheads), which are larger than residual axons (arrows), but significantly smaller than negatively staining degenerating neurons in the background (scale bar: 50 μm). (C) Tau-2 immunohistochemistry reveals variable expression of tau in degenerating neurons; an area of relatively intense staining is depicted (scale bar: 50 μm). (D) The spectrum of degenerating neurons and neuroaxonal spheroids in the globus pallidus is uniformly positive by anti-ubiquitin immunohistochemistry (scale bar: 200 μm). (E) Higher magnification of ubiquitin immunohistochemistry of globus pallidus, showing strong granular staining of degenerating neurons (upper left) as well as finely granular cytoplasmic staining in rare intact neurons (arrow) (scale bar: 50 μm). (F) The neuron in the center has developed eosinophilic cytoplasmic granularity and an indistinct nuclear contour compared with the intact neuron on the right (scale bar: 50 μm). (G) The lower neuron contains peripheral cytoplasmic lipofuscin pigment as well as paranuclear eosinophilic cytoplasm; the nucleus displays smudging. The upper neuron displays similar cytoplasmic changes, but the nucleus has degenerated and is no longer apparent (scale bar: 50 μm). (H) Toluidine blue-stained plastic-embedded section of relatively intact (upper) and degenerated (lower) neurons in the globus pallidus. Neurons with advanced degeneration did not have morphologically recognizable nuclei or distinct cytoplasmic boundaries; however, small axons tended to be relatively preserved (scale bar: 50 μm). (I) Lamin A/C immunohistochemistry of structures showing advanced degeneration in the globus pallidus reveals a residual nuclear outline, indicating their cellular origin. Nuclear size indicates neuronal origin; note pallor of staining compared with surrounding viable glial nuclei (scale bar: 50 μm). *(Modified from Kruer et al. (2011) with permission).* (For color version of this figure, the reader is referred to the online version of this chapter.)

3.5. Tau and synuclein pathology

Abundant tau and synuclein pathology can be seen in PLAN brain, although tau pathology is not an invariant feature. Widespread phosphorylated tau, Gallyas silver, and AT100-positive neurofibrillary tangles and neuropil threads can occur in PLAN brain and can be found in hippocampus and frontal and temporal cortex. Similarly, Lewy burden can be mild to severe, with both Lewy bodies and Lewy neurites observed. Lewy bodies can be confined to the substantia nigra or found diffusely throughout the mesial temporal structures and neocortex. These bodies stain positively for phosphorylated α-synuclein (Ser129).

3.6. Other affected organ systems

Spheroids can be observed via biopsy of the skin (cutaneous nerves), sural nerve, rectum, or conjunctiva. Peripheral nerve pathology includes thinning of myelin and a dystrophic appearance and massive enlargement of axons. Although optic atrophy is a typical feature of PLAN, optic nerve pathology has not yet been reported.

Depictions of PLAN pathology can be found in Figs. 8.4 and 8.5.

4. MITOCHONDRIAL MEMBRANE PROTEIN-ASSOCIATED NEURODEGENERATION (MPAN)

Mitochondrial membrane protein-associated neurodegeneration (MPAN) was first identified among a founder population in Poland with a *PANK2*- and *PLA2G6*-negative form of NBIA. MPAN is caused by mutations in *c19orf12*, a mitochondrial membrane protein of cryptic function. Intermediate in phenotype, MPAN shares features of both PKAN and PLAN. Intriguingly, one of the early cases of MPAN was identified among a cohort of patients with idiopathic Parkinson disease (Hartig et al., 2011). The pathology of MPAN has been described by Hartig et al. (2011) and Hogarth et al. (2013).

4.1. Histologic findings

Similar to PKAN, the globus pallidus is selectively affected, showing rarefaction of the neuropil and widespread loss of neurons with associated astrogliosis. Large-scale neuronal loss within the substantia nigra also occurs. The cerebellum is largely spared with the exception of rare Lewy bodies and Lewy neurites within the dentate nucleus.

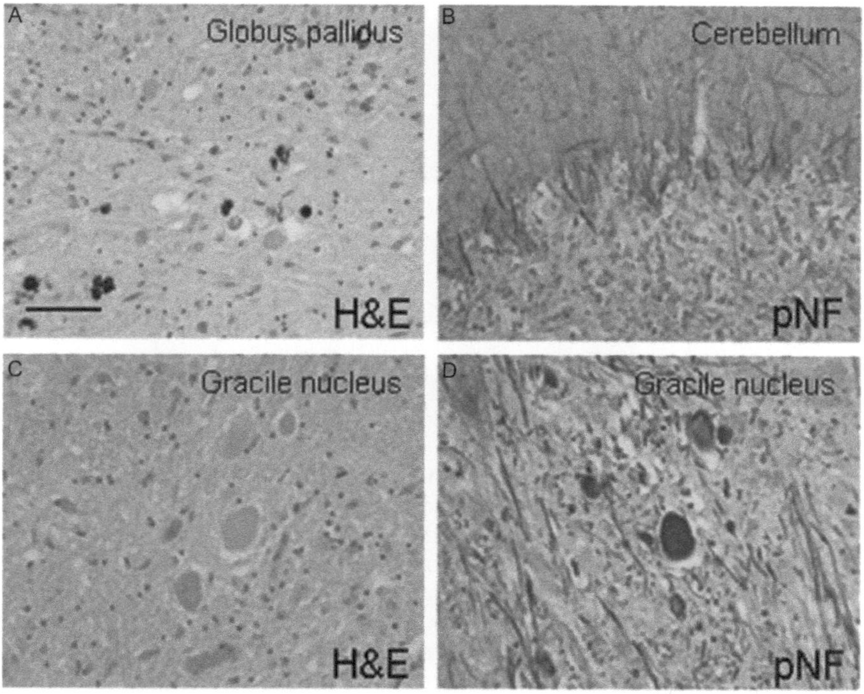

Figure 8.4 Histologic findings in PLAN. (A) Significant degree of pigment deposition in the globus pallidus. (B) Empty baskets highlighting significant Purkinje cell loss in the cerebellar cortex. (C) Large neuroaxonal swellings in the gracile nucleus, which were often immunoreactive for neurofilament (D). (A and C) Hematoxylin and eosin (H&E); (B and D) phospho-neurofilament immunohistochemistry (pNF) (RT97 antibody). The bar on (A) represents 40 μm. *(Modified from Paisán-Ruiz et al. (2012) with permission).* (For color version of this figure, the reader is referred to the online version of this chapter.)

4.2. Iron deposition

Iron deposition is greater in the globus pallidus as compared to the substantia nigra. Perls' stain shows iron in a perivascular distribution, within microglia, within neuronal concretions, and more diffusely within astrocytes.

4.3. Spheroids

Spheroids in MPAN occur within the neocortex, basal ganglia, cerebellum, brainstem, and spinal cord. Spheroids stain positively for ubiquitin, lightly positive for APP, and inconsistently positive for tau. Spheroids do not show immunoreactivity for TDP-43 or α-synuclein.

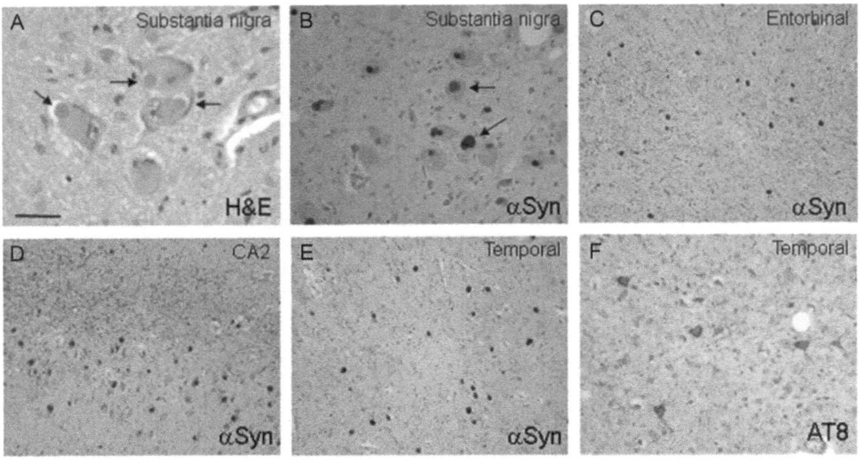

Figure 8.5 Tau and synuclein pathology in PLAN. (A and B) Frequent Lewy bodies in substantia nigra neurons. Severe Lewy pathology is demonstrated in the entorhinal cortex (C), CA2 hippocampal subregion (D), and temporal neocortex (E). The tau pathology was extensive in some cases and is demonstrated here in the temporal cortex. (A) Hematoxylin and eosin (H&E); (B–E) α-synuclein immunohistochemistry (αSyn); (F) tau immunohistochemistry (AT8 antibody). The bar on (A) represents 80 μm. *(Modified from Paisán-Ruiz et al. (2012) with permission).* (For color version of this figure, the reader is referred to the online version of this chapter.)

4.4. Tau and synuclein pathology

Widespread Lewy bodies and neurites are found within the globus pallidus, corpus striatum, neocortex, and substantia nigra. Relative Lewy body burdens in MPAN, PLAN, and Parkinson disease with dementia were recently compared (Hogarth et al., 2013). MPAN brain contained a Lewy body content more than 5 times that of PLAN and ~30 times greater than seen in individuals with Parkinson disease with dementia.

Hyperphosphorylated tau inclusions within neurons can be observed within the hippocampus. Perinuclear tau accumulation occurs in occasional pyramidal neurons, reminiscent of some forms of tauopathy. Only a few tau-positive bundles stain by silver staining methods, and fibrils and neurofibrillary tangles are not typical. Finally, apparently extracellular deposits of tau and synuclein have been noted in affected regions.

4.5. Other affected organ systems

Peripheral nerve biopsies from patients with MPAN have shown axonal spheroids.

An overview of MPAN pathology can be found in Figs. 8.6 and 8.7.

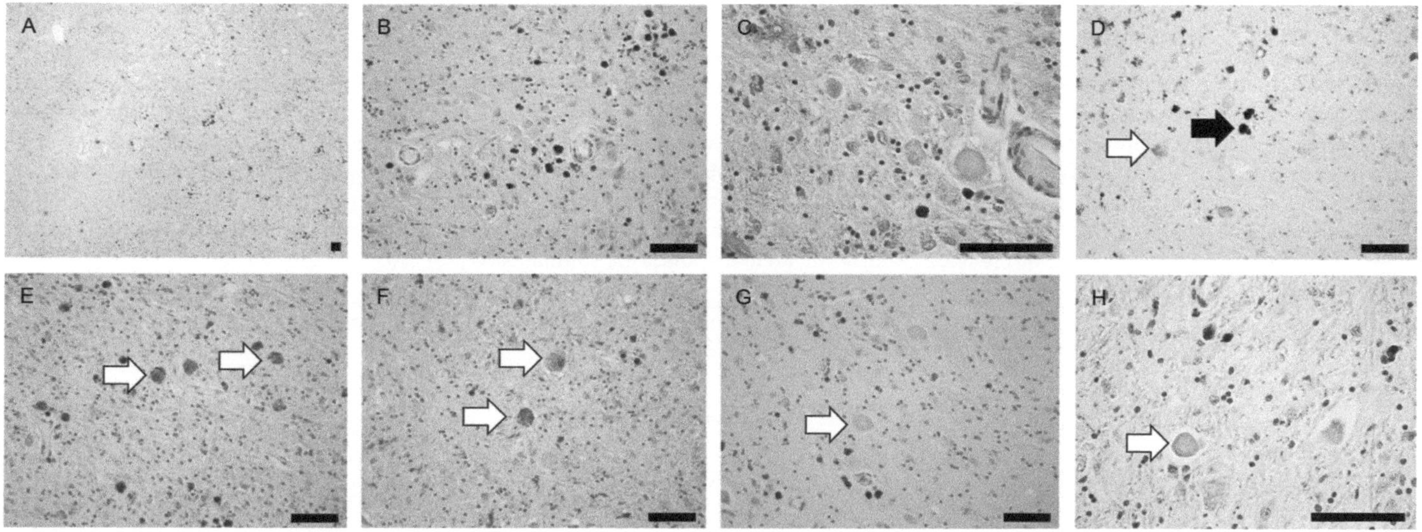

Figure 8.6 Histopathologic features of mitochondrial membrane protein-associated neurodegeneration. (A) Low-magnification (4×) view of hematoxylin and eosin stain of the globus pallidus, showing central pallor and widespread dark hemosiderin deposits. (B) At higher magnification (20×), the predominantly perivascular distribution of hemosiderin and the near complete loss of neurons in the globus pallidus are apparent; residual small nuclei are glial-derived. (C) Eosinophilic spheroidal structures as described in pantothenate kinase-associated neurodegeneration (PKAN) are readily apparent; some of these harbor residual nuclear outlines and lipofuscin pigment, indicating an origin in degenerating neurons (40×). (D) Perls' stain (blue) for iron highlights densely stained hemosiderin deposits (white arrowhead) as well as more diffuse iron in the globus pallidus and increased iron associated with eosinophilic spheroids (black arrowhead, 20×). (E) Ubiquitin immunohistochemistry shows uniform, strong staining of spheroids (white arrowheads) as in PKAN (20×). (F) An immunostain for tau labels a subpopulation of spheroids (white arrowheads); occasional tau-positive neurites are in the background (20×). (G) TDP-43 immunostaining does not label any structures in the globus pallidus (white arrowhead); the brown signal is derived from hemosiderin deposits (20×). (H) Spheroids in the globus pallidus are not labeled with an antibody to α-synuclein (white arrowhead, 40×). Scale bars: 100 μm. *(Modified from Hogarth et al. (2013) with permission)*. (For color version of this figure, the reader is referred to the online version of this chapter.)

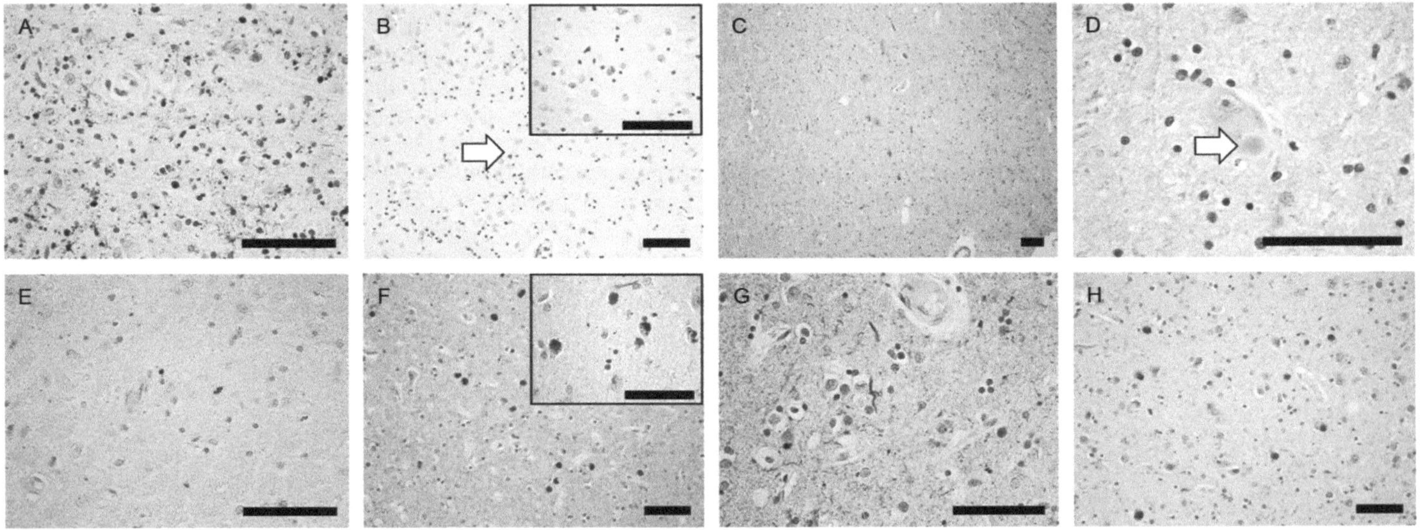

Figure 8.7 Additional features of MPAN. (A) Lewy neurites are abundant in relatively intact areas at the periphery of the globus pallidus (40 ×). (B) α-Synuclein immunohistochemistry of the putamen developed with Vector Red identifies Lewy bodies (white arrowhead) and neurites (20 ×), inset at higher magnification (60 ×). (C) Hematoxylin and eosin stain of the midbrain demonstrates sclerosis of the substantia nigra with only isolated residual neurons (10 ×); (D) many of these contain Lewy bodies (white arrowhead, 60 ×). (E) Perls' stain (blue) of the midbrain shows only occasional areas of minute parenchymal iron deposits, predominantly in a perivascular distribution (40 ×). (F) α-Synuclein immunohistochemistry revealed abundant Lewy bodies in the neocortex (20 ×), inset at higher magnification (60 ×); (G) α-synuclein immunohistochemistry also highlighted Lewy neurites in a widespread distribution, including the cerebral cortex (depicted, 40 ×), basal ganglia, pons, and midbrain. (H) Neocortical Lewy bodies were ubiquitin positive, as in the case of sporadic Lewy body disease (20 ×). Scale bars: 100 μm. *(Modified from Hogarth et al. (2013) with permission).* (For color version of this figure, the reader is referred to the online version of this chapter.)

5. KUFOR RAKEB SYNDROME (KRS)

Mutations in *ATP13A2* were first identified in patients with a pallidopyramidal syndrome (KRS) and have since been identified in patients with juvenile parkinsonism (Crosiers et al., 2011; Paisán-Ruiz et al., 2010; Santoro et al., 2011) and neuronal ceroid lipofuscinosis (Bras, Verloes, Schneider, Mole, & Guerreiro, 2012). Several individuals have T2 hypointensity of the basal ganglia suggestive of brain iron deposition by MRI (Brüggemann et al., 2010; Schneider et al., 2010). For this reason, patients with *ATP13A2* mutations have been characterized as having a form of NBIA, although like patients with PLAN, brain iron deposits appear to represent an inconsistent feature. A peripheral nerve biopsy in KRS showed reduced myelin fiber density, CD68+ endoneurial macrophage infiltration, and degenerating axons with axonal loss and perineurial and endoneurial edema (Paisán-Ruiz et al., 2010). Cytoplasmic inclusions were seen in vascular smooth muscle cells, endoneurium, and perineurium. Electron microscopic analysis of this nerve biopsy as well as skin biopsy from KRS patients (Malandrini, Rubegni, Battisti, Berti, & Federico, 2013) has shown electron-dense lamellated deposits ~1 μm in size. No postmortem studies of affected patients have yet been reported.

6. BETA PROPELLER PROTEIN-ASSOCIATED NEURODEGENERATION (BPAN)

A recently identified X-linked dominant form of NBIA, beta propeller protein-associated neurodegeneration (BPAN), features a course atypical from other forms of NBIA. Intellectual disability is noted in affected patients in early childhood, inconsistently accompanied by mild neuromotor impairment. BPAN patients then develop dystonia-parkinsonism in adulthood, sometimes over a short period of time.

Although no brain pathology from patients with mutations in *WDR45* has yet been reported, a report almost three decades ago by Eidelberg et al. (1987) is highly suggestive of BPAN. Affected patients demonstrated early intellectual disability, followed by a rapidly progressive form of dystonia-parkinsonism in young adulthood. Although tissue is not available to confirm that these patients harbored a WDR45 mutation (Eidelberg, personal communication), they can be considered *probable BPAN*.

6.1. Gross pathology

In the three patients reported, findings were remarkably similar. Gross cortical atrophy was seen with relative sparing of the cerebellum. Rust-colored staining of the globus pallidus was evident *ex vivo*.

6.2. Histologic findings

H&E staining demonstrated widespread cortical neuronal loss, with corresponding loss of horizontal lamination. The globus pallidus featured substantial neuronal loss, as did the substantia nigra, with depigmentation of remaining neurons. Diffuse fibrillary astrocytosis occurred within the globus pallidus and substantia nigra. Corticospinal tracts, Clarke's column, and spinocerebellar tracts all exhibited myelin loss with co-occurring fibrillary astrocytosis.

6.3. Iron deposition

Iron deposition as measured by Perls' stain was prominent within the globus pallidus and substantia nigra, particularly perivascularly and throughout the neuropil. Intracellular iron was largely observed within neurons in both the globus pallidus and substantia nigra.

6.4. Spheroids

Numerous eosinophilic axonal spheroids, 5–50 μm in diameter, were seen in the globus pallidus, substantia nigra, and gracile and cuneate nuclei. Axonal swellings occurred within the brainstem nuclei and cerebellum.

6.5. Tau and synuclein pathology

Tau and synuclein pathology was quite prominent in the cases examined. Approximately one-third of surviving neurons featured neurofibrillary tangle accumulation, with relative sparing of primary sensory and motor cortex, while Lewy bodies or Lewy neurites were seen in about one-fourth of remaining cells. Neurofibrillary tangles were widespread within pyramidal neurons and observed in the hippocampus, cortex, basal forebrain, subthalamic nucleus, and reticular formation. Tangles consisted of paired helical filaments 20–25 nm in diameter co-occurring alongside accumulated straight filaments. No neuritic plaques or amyloid deposition was appreciated.

6.6. Other pathology

Rare Hirano bodies and granulovacuolar degeneration were seen in the hippocampus. Anterior horn cell degeneration (chromatolysis) with axonal swelling was observed in the only case for whom the spinal cord was able to be examined. Manganese accumulation was detected within tangle-bearing hippocampal neurons by laser microprobe mass analysis.

Putative BPAN pathology is delineated in Figs. 8.8 and 8.9.

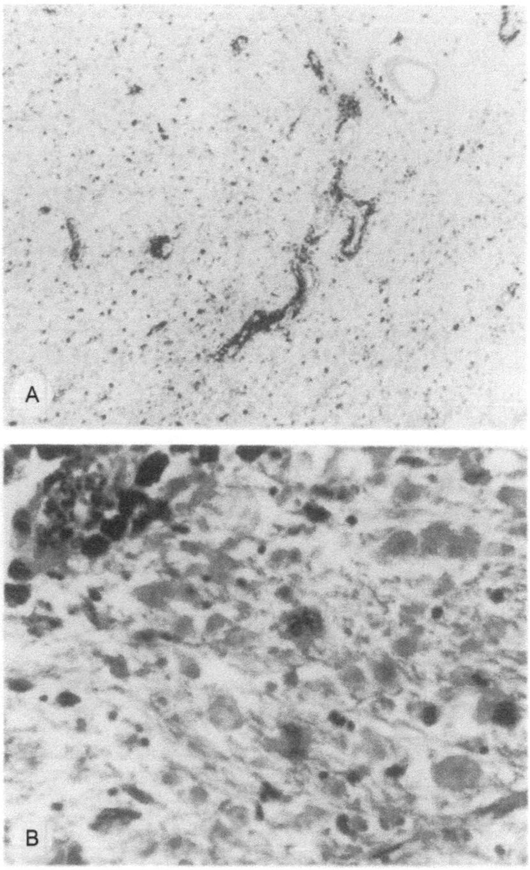

Figure 8.8 Iron deposition and neuroaxonal spheroids in putative BPAN. (A) Perivascular and parenchymal deposits iron deposits within the pars reticularis of substantia nigra demonstrated by Perls' stain. (B) Higher magnification showing axonal spheroids and perivascular pigment accumulation. *(Modified from Eidelberg et al. (1987) with permission).*

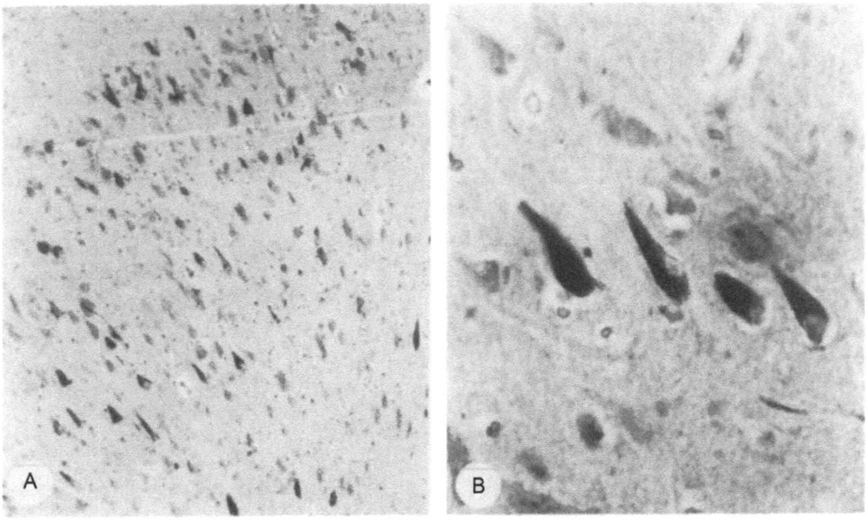

Figure 8.9 Argyrophilic lesions in putative BPAN. (A) Most Bodian-positive elements within the hippocampus stain positively to PHF. (B) Higher magnification. *(Modified from Eidelberg et al. (1987) with permission).*

7. NEUROFERRITINOPATHY (NFT)

Neuroferritinopathy (NFT) is an autosomal dominant form of adult-onset NBIA that leads to a dystonia–chorea syndrome with dementia. It is caused by mutations in *FTL*, the ferritin light chain, which is predicted to lead to the formation of insoluble ferritin precipitates (Baraibar, Muhoberac, Garringer, Hurley, & Vidal, 2010). NFT pathology has been delineated by Mancuso et al. (2005) and Hautot et al. (2007).

7.1. Gross pathology

Grossly, the brains of patients with NFT exhibit diffuse atrophy of the cerebellum along with cavitary lesions of the putamen and rusty discoloration of the putamen, globus pallidus, and substantia nigra.

7.2. Histologic findings

Microscopic findings include a diffuse loss of neurons and glia with profound rarefaction of the putamen and globus pallidus. There is large-scale loss of oligodendrocytes in the paradentate white matter. There is a disproportionate loss of myelin compared to axons. Affected neurons exhibit pathognomonic distorted, enlarged, and vacuolated nuclei, most prominently in the putamen, while affected glia are predominantly carbonic anhydrase II-positive oligodendrocytes within white matter tracts. Extracellular hyaline deposits positive for both iron and ferritin can be seen. No substantial inflammatory infiltrate is observed.

7.3. Iron deposition

Intracellular and extracellular (perivascular) iron deposits are appreciable by both Turnbull and Perls' stain and ferritin immunohistochemistry. Distorted nuclei react prominently for both Fe^{2+} and Fe^{3+}. Birefringent crystals occur in neurons within the subthalamic nucleus, dentate nucleus, and putamen. These have been hypothesized to represent nonerythrocytic iron.

7.4. Spheroids

Neuroaxonal spheroids that were immunoreactive to ubiquitin, tau, and neurofilaments were observed in the putamen, although these were not prominent.

7.5. Tau and synuclein pathology

Tau pathology is not prominent in NFT, although spheroids can be tau-reactive. Synuclein pathology has not been described in NFT.

7.6. Other findings

A superconducting quantum interference device magnetometry analysis showed that the primary iron compounds in NFT are biogenic magnetite (Fe_3O_4) and maghemite (Fe_2O_3) (Hautot et al., 2007). Widespread oxidative stress, as evidenced by 4-HNE-, HO-1-, and MnSOD-positive cells and caspase-3-positive neurons, has been observed.

Figures 8.10–8.14 summarize the neuropathologic features of NFT.

8. ACERULOPLASMINEMIA (ACP)

Aceruloplasminemia (ACP) occurs as a result of mutations in CP, which encodes ceruloplasmin. Ceruloplasmin is a plasma membrane glycophosphatidylinositol-anchored ferroxidase found largely on the surface of astrocytes in the brain. Ceruloplasmin interconverts Fe^{2+} and Fe^{3+} as part of the process of iron efflux and delivery to other cell types. Affected patients exhibit dystonia, chorea, diabetes mellitus, and pigmentary retinopathy.

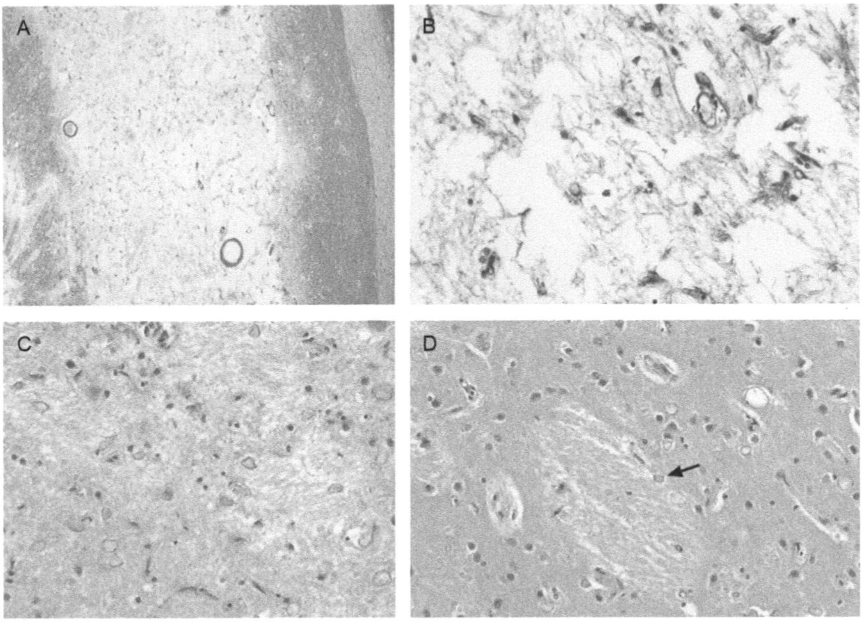

Figure 8.10 Histopathology of NFT. (A and B) Posterior putamen. Cavitary lesion with loss of almost all elements. (C and D) Anterior putamen. (C) Milder lesion with rarefaction and vacuolated nuclei. (D) Loss of glial nuclei from myelin-depleted intrinsic fiber bundle (arrow). Hematoxylin and eosin, original magnifications: (A) 10×; (B) 75×; (C) 75×; (D) 75×. *(Modified from Mancuso et al. (2005) with permission).* (For color version of this figure, the reader is referred to the online version of this chapter.)

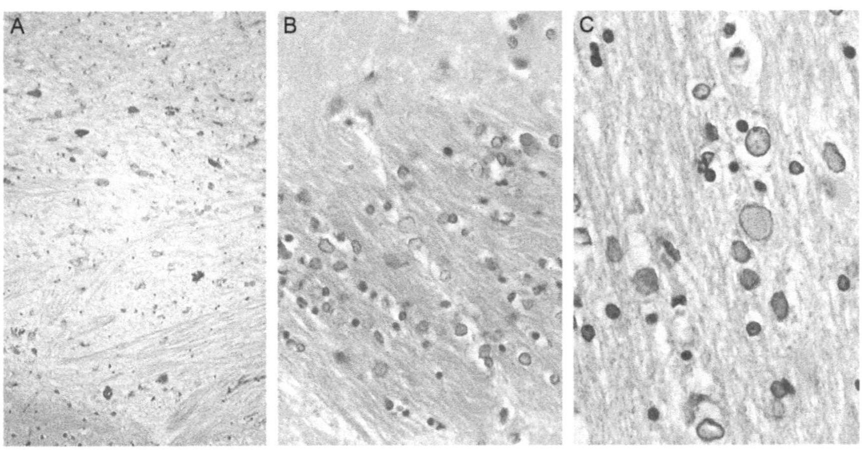

Figure 8.11 Additional features of NFT. (A) Pallidum. Marked loss of fibers and neurons in most lateral portion of GPe. (B) Caudate. Vacuolated nuclei within intrinsic fiber bundle. (C) External capsule. Variably sized vacuolated nuclei. Hematoxylin and eosin, original magnifications: (A) 75×; (B) 150×; (C) 220×. *(Modified from Mancuso et al. (2005) with permission).* (For color version of this figure, the reader is referred to the online version of this chapter.)

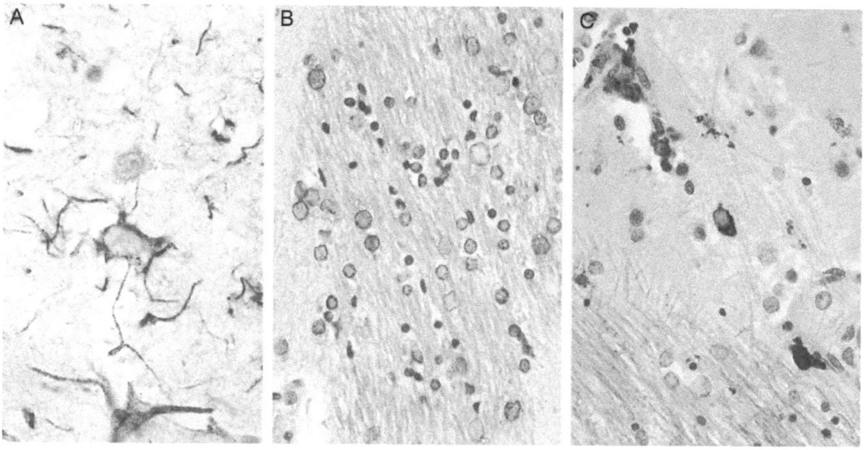

Figure 8.12 Vacuolated nuclei in NFT. (A) Putamen. GFAP-immunoreactive astrocytic cytoplasm around large vacuolated nucleus. (B) Caudate. Several CA II-immunoreactive oligodendroglial cells (nuclear and cytoplasmic) within intrinsic fibers. (C) Putamen. Mildly swollen nucleus within CD-68-immunoreactive microglial cell in center of field. Original magnifications: (A) 150×; (B) 150×; (C) 150×. *(Modified from Mancuso et al. (2005) with permission).* (For color version of this figure, the reader is referred to the online version of this chapter.)

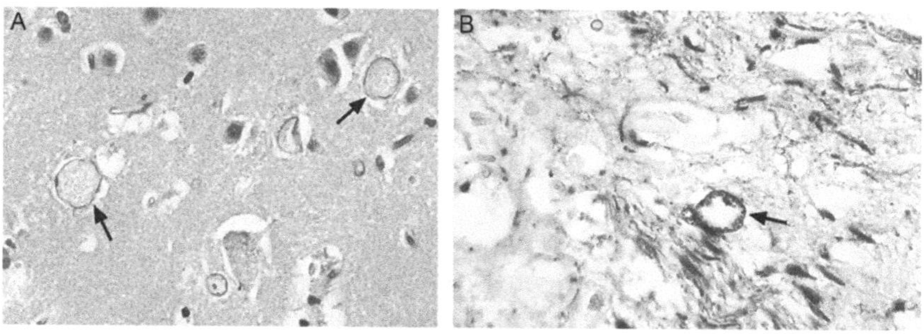

Figure 8.13 Additional vacuolated nuclei. Anterior putamen. (A) Markedly vacuolated nuclei (arrows). (B) Some partially immunoreactive for neurofilaments (arrow). Original magnifications: (A) Hematoxylin and eosin, 150×; (B) 150×. *(Modified from Mancuso et al. (2005) with permission).* (For color version of this figure, the reader is referred to the online version of this chapter.)

ACP pathology has been described in detail by Kaneko et al. (2002, 2012), Oide, Yoshida, Kaneko, Ohta, and Arima (2006), and Hattori et al. (2012).

8.1. Histologic findings

Pervasive pyramidal neuron dropout is seen in the caudate nucleus and putamen with accompanying widespread astrocytosis, along with enlarged and

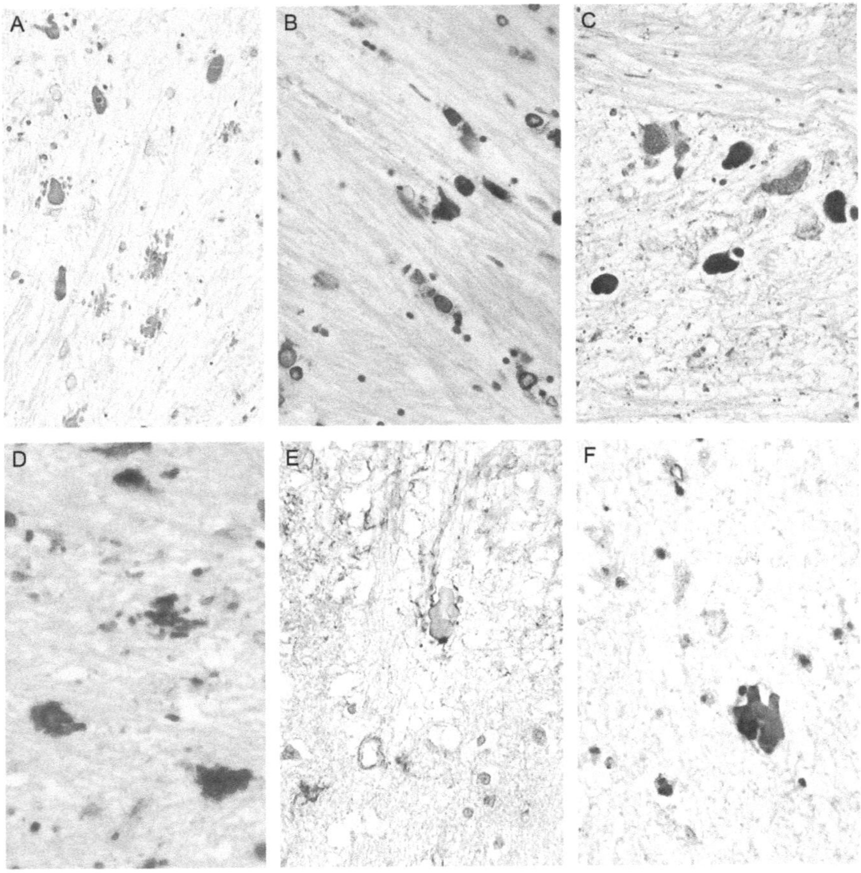

Figure 8.14 Hyaline deposits. Pallidum. (A) Pleomorphic hyaline deposits, some containing nuclei, within the same area of GPe as in Fig. 8.6A. (B) Prussian blue reactivity. (C) PTAH reactivity. (D) Polyclonal antiferritin immunoreactivity. (E) Rare association of hyaline deposits with GFAP-immunoreactive astrocyte. (F) HO-2-immunoreactivity. Original magnifications: (A) Hematoxylin and eosin, 150×; (B) Perls' stain, 220×; (C) 220×; (D) 250×; (E) 220×; (F) 250×. *(Modified from Mancuso et al. (2005) with permission).* (For color version of this figure, the reader is referred to the online version of this chapter.)

sometimes bizarrely distorted astrocytes. This deformation is consistent with the morphology of Alzheimer type 1 astrocytes.

8.2. Spheroids

GFAP-positive and synaptophysin- and neurofilament-negative eosinophilic globular structures described as "grumose or foamy spheroid bodies" appeared to represent ballooned, iron-overloaded astrocytes. These structures measured approximately 10–60 μm in diameter. The astrocytic origin of these structures distinguishes them from spheroids, which are derived from degenerating neurons. Globular bodies are positive for 4-HNE and ubiquitin. Ultrastructurally, globular structures have dense bodies and glial fibril-like elements by electron microscopy.

8.3. Iron deposition

Examination of several cases in different stages of disease has indicated that iron deposition begins in the caudate, putamen, thalamus, and dentate nucleus but subsequently spreads throughout the cortex with the frontal and temporal cortices being most profoundly affected. Iron deposition also occurs in the retina (Wolkow et al., 2011) and cerebellum (Oide et al., 2006). Iron deposition is most prominently detected with Perls' stain, although Turnbull staining is also positive to a lesser extent. Iron is seen prominently localized to astrocytic end-feet in association with brain capillaries.

No significant tau or synuclein pathology has been reported in patients with ACP.

Figures 8.15 and 8.16 demonstrate key pathologic findings in ACP.

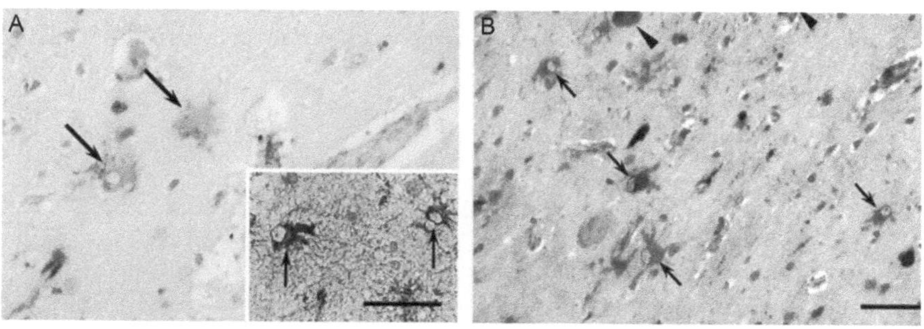

Figure 8.15 Findings in the cerebral cortex in aceruloplasminemia. Iron accumulates in the cytoplasm of astrocytes (A, arrows) (Perls' stain). Enlarged astrocytes (A, inset) and globular structures were clearly stained with anti-GFAP antibody. Anti-ferritin light-chain antibody strongly reacts to deformed astrocytes (F, arrows) and globular structures (F, arrowheads). Scale bar: 50 μm. *(Modified from Kaneko et al. (2012) with permission).* (For color version of this figure, the reader is referred to the online version of this chapter.)

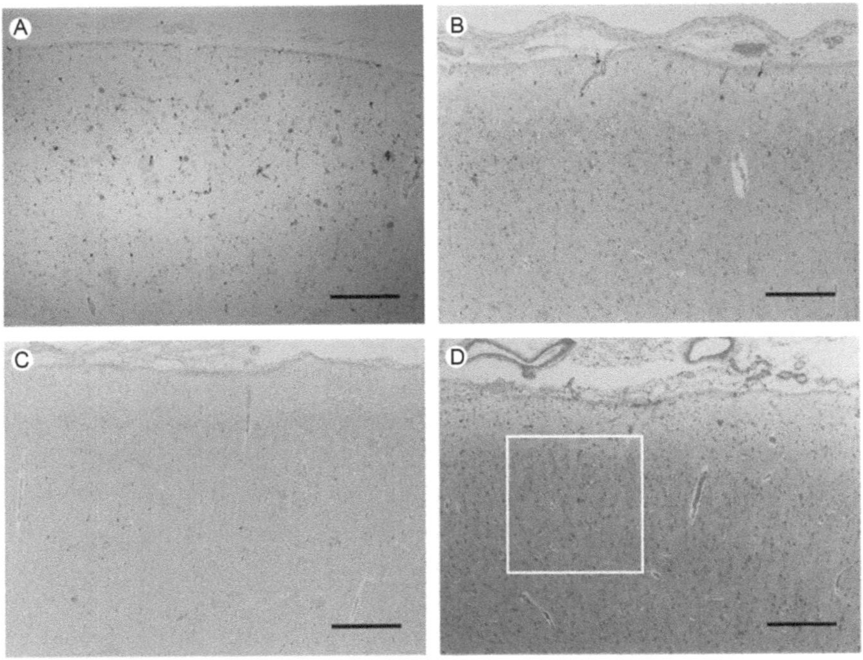

Figure 8.16 Low-magnification findings of frontal cortical sections. (A–C) Berlin blue-stained for iron detection. (D) Klüver–Barrera-stained specimen. (A and D) Present case; (B) elder sister of the present case (60-year-old woman); (C) control patient (50-year-old man). Remarkable iron deposition is seen in the present case (A). Weak (B) or little (C) iron deposition is observed in other cases. Klüver–Barrera-stained specimen for pyramidal cell count (D). Pyramidal cells in the cortical third layer inside the white square (1 mm^2) were counted. Ten serial areas were counted and averaged. Bar, 500 μm. *(Modified from Kaneko et al. (2012) with permission).* (For color version of this figure, the reader is referred to the online version of this chapter.)

9. CONCLUSIONS

Despite differences in clinical features, neuroimaging, and pathologic findings in subtypes of NBIA, common features argue for a common nosology among these monogenic neurodegenerative diseases. However, whether ACP and NFT should be considered in the same category as the other forms of NBIA remains controversial, as these two disorders appear to be more clearly related to a primary disturbance in iron trafficking. Additional insight into pathogenic mechanisms is needed before this question can be resolved.

Pathologically, patients with NBIA show a degeneration of both neurons and astrocytes with relative sparing of oligodendrocytes. There are limited activation of microglia and little evidence of an inflammatory infiltrate. Iron deposition occurs primarily in the globus pallidus in most forms of NBIA. Iron accumulation occurs to a lesser extent in the substantia nigra in several NBIA subtypes, except in BPAN where iron is most evident in the substantia nigra. Spheroid bodies are common to many forms of NBIA, including PKAN, PLAN, MPAN, and likely BPAN. They may represent degenerating neurons and/or the accumulation of protein and lipid storage material and damaged organelles (mitochondria and vesicles).

Several forms of NBIA show pathologic similarities to more common neurodegenerative diseases. In particular, PKAN, PLAN, MPAN, and BPAN all feature neurofibrillary tangles, although without amyloid plagues and all have been associated with dementia. α-Synuclein-containing Lewy bodies are not only prominent in MPAN but also widely seen in PLAN and BPAN, while the brains of PKAN patients lack Lewy body deposition. In addition, the inconsistent association of these neuropathologic features with disease severity suggests that these findings represent downstream or secondary effects.

Is the deposition of iron in crucial brain regions such as the globus pallidus then singly responsible for disease? Likely not. Although this hypothesis is attractive, with free-radical formation catalyzed by free iron via the Fenton reaction, disease onset can precede iron deposition in PKAN, while in PLAN, KRS or FAHN patients might not develop iron deposition at all. Iron overload may contribute to the burden of disease and may even represent a viable therapeutic target, but there are likely other mechanisms that play a more central role in the neurodegeneration that occurs. What then might lead several monogenic disorders to exhibit a similar phenotype and a common pattern of secondary iron deposition? Genes leading to NBIA have been implicated in mitochondrial function, lipid metabolism, and autophagy. Perhaps, an inappropriate activation of genes with iron-response elements may occur, but further studies will be needed to answer these and other questions surrounding these enigmatic disorders.

REFERENCES

Baraibar, M. A., Muhoberac, B. B., Garringer, H. J., Hurley, T. D., & Vidal, R. (2010). Unraveling of the E-helices and disruption of 4-fold pores are associated with iron mishandling in a mutant ferritin causing neurodegeneration. *Journal of Biological Chemistry, 285*(3), 1950–1956.

Bras, J., Verloes, A., Schneider, S. A., Mole, S. E., & Guerreiro, R. J. (2012). Mutation of the parkinsonism gene ATP13A2 causes neuronal ceroid-lipofuscinosis. *Human Molecular Genetics*, *21*(12), 2646–2650.

Brüggemann, N., Hagenah, J., Reetz, K., Schmidt, A., Kasten, M., Buchmann, I., et al. (2010). Recessively inherited parkinsonism: Effect of ATP13A2 mutations on the clinical and neuroimaging phenotype. *Archives of Neurology*, *67*(11), 1357–1363.

Crosiers, D., Ceulemans, B., Meeus, B., Nuytemans, K., Pals, P., Van Broeckhoven, C., et al. (2011). Juvenile dystonia-parkinsonism and dementia caused by a novel ATP13A2 frameshift mutation. *Parkinsonism & Related Disorders*, *17*(2), 135–138.

Eidelberg, D., Sotrel, A., Joachim, C., Selkoe, D., Forman, A., Pendlebury, W. W., et al. (1987). Adult onset Hallervorden-Spatz disease with neurofibrillary pathology. A discrete clinicopathological entity. *Brain*, *110*(Pt 4), 993–1013.

Gregory, A., Westaway, S. K., Holm, I. E., Kotzbauer, P. T., Hogarth, P., Sonek, S., et al. (2008). Neurodegeneration associated with genetic defects in phospholipase A(2). *Neurology*, *71*(18), 1402–1409.

Hartig, M. B., Iuso, A., Haack, T., Kmiec, T., Jurkiewicz, E., Heim, K., et al. (2011). Absence of an orphan mitochondrial protein, c19orf12, causes a distinct clinical subtype of neurodegeneration with brain iron accumulation. *American Journal of Human Genetics*, *89*(4), 543–550.

Hattori, A., Miyajima, H., Tomosugi, N., Tatsumi, Y., Hayashi, H., & Wakusawa, S. (2012). Clinicopathological study of Japanese patients with genetic iron overload syndromes. *Pathology International*, *62*(9), 612–618.

Hautot, D., Pankhurst, Q. A., Morris, C. M., Curtis, A., Burn, J., & Dobson, J. (2007). Preliminary observation of elevated levels of nanocrystalline iron oxide in the basal ganglia of neuroferritinopathy patients. *Biochimica et Biophysica Acta*, *1772*(1), 21–25.

Hayflick, S. J., Hartman, M., Coryell, J., Gitschier, J., & Rowley, H. (2006). Brain MRI in neurodegeneration with brain iron accumulation with and without PANK2 mutations. *American Journal of Neuroradiology*, *27*(6), 1230–1233.

Hogarth, P., Gregory, A., Kruer, M. C., Sanford, L., Wagoner, W., Natowicz, M. R., et al. (2013). New NBIA subtype: Genetic, clinical, pathologic, and radiographic features of MPAN. *Neurology*, *80*(3), 268–275.

Kaneko, K., Hineno, A., Yoshida, K., Ohara, S., Morita, H., & Ikeda, S. (2012). Extensive brain pathology in a patient with aceruloplasminemia with a prolonged duration of illness. *Human Pathology*, *43*(3), 451–456.

Kaneko, K., Yoshida, K., Arima, K., Ohara, S., Miyajima, H., Kato, T., et al. (2002). Astrocytic deformity and globular structures are characteristic of the brains of patients with aceruloplasminemia. *Journal of Neuropathology and Experimental Neurology*, *61*(12), 1069–1077.

Kruer, M. C., Hiken, M., Gregory, A., Malandrini, A., Clark, D., Hogarth, P., et al. (2011). Novel histopathologic findings in molecularly-confirmed pantothenate kinase-associated neurodegeneration. *Brain*, *134*(Pt 4), 947–958.

Kuo, Y. M., Duncan, J. L., Westaway, S. K., Yang, H., Nune, G., Xu, E. Y., et al. (2005). Deficiency of pantothenate kinase 2 (Pank2) in mice leads to retinal degeneration and azoospermia. *Human Molecular Genetics*, *14*(1), 49–57.

Kuo, Y. M., Hayflick, S. J., & Gitschier, J. (2007). Deprivation of pantothenic acid elicits a movement disorder and azoospermia in a mouse model of pantothenate kinase-associated neurodegeneration. *Journal of Inherited Metabolic Disease*, *30*(3), 310–317.

Li, A., Paudel, R., Johnson, R., Courtney, R., Lees, A. J., Holton, J. L., et al. (2012). Pantothenate kinase-associated neurodegeneration is not a synucleinopathy. *Neuropathology and Applied Neurobiology*, (Epub ahead of print).

Malandrini, A., Bonuccelli, U., Parrotta, E., Ceravolo, R., Berti, G., & Guazzi, G. C. (1995). Myopathic involvement in two cases of Hallervorden-Spatz disease. *Brain & Development, 17*(4), 286–290.

Malandrini, A., Cavallaro, T., Fabrizi, G. M., Berti, G., Salvestroni, R., Salvadori, C., et al. (1995). Ultrastructure and immunoreactivity of dystrophic axons indicate a different pathogenesis of Hallervorden-Spatz disease and infantile neuroaxonal dystrophy. *Virchows Archiv, 427*(4), 415–421.

Malandrini, A., Rubegni, A., Battisti, C., Berti, G., & Federico, A. (2013). Electron-dense lamellated inclusions in 2 siblings with Kufor-Rakeb syndrome. *Movement Disorders,* (Epub ahead of print).

Malik, I., Turk, J., Mancuso, D. J., Montier, L., Wohltmann, M., Wozniak, D. F., et al. (2008). Disrupted membrane homeostasis and accumulation of ubiquitinated proteins in a mouse model of infantile neuroaxonal dystrophy caused by PLA2G6 mutations. *American Journal of Pathology, 172*(2), 406–416.

Mancuso, M., Davidzon, G., Kurlan, R. M., Tawil, R., Bonilla, E., Di Mauro, S., et al. (2005). Hereditary ferritinopathy: A novel mutation, its cellular pathology, and pathogenetic insights. *Journal of Neuropathology and Experimental Neurology, 64*(4), 280–294.

Oide, T., Yoshida, K., Kaneko, K., Ohta, M., & Arima, K. (2006). Iron overload and antioxidative role of perivascular astrocytes in aceruloplasminemia. *Neuropathology and Applied Neurobiology, 32*(2), 170–176.

Paisan-Ruiz, C., Bhatia, K. P., Li, A., Hernandez, D., Davis, M., Wood, N. W., et al. (2009). Characterization of PLA2G6 as a locus for dystonia-parkinsonism. *Annals of Neurology, 65*(1), 19–23.

Paisán-Ruiz, C., Guevara, R., Federoff, M., Hanagasi, H., Sina, F., Elahi, E., et al. (2010). Early-onset L-dopa-responsive parkinsonism with pyramidal signs due to ATP13A2, PLA2G6, FBXO7 and spatacsin mutations. *Movement Disorders, 25*(12), 1791–1800.

Paisán-Ruiz, C., Li, A., Schneider, S. A., Holton, J. L., Johnson, R., Kidd, D., et al. (2012). Widespread Lewy body and tau accumulation in childhood and adult onset dystonia-parkinsonism cases with PLA2G6 mutations. *Neurobiology of Aging, 33*(4), 814–823.

Potter, K. A., Kern, M. J., Fullbright, G., Bielawski, J., Scherer, S. S., Yum, S. W., et al. (2011). Central nervous system dysfunction in a mouse model of FA2H deficiency. *Glia, 59*(7), 1009–1021.

Santoro, L., Breedveld, G. J., Manganelli, F., Iodice, R., Pisciotta, C., Nolano, M., et al. (2011). Novel ATP13A2 (PARK9) homozygous mutation in a family with marked phenotype variability. *Neurogenetics, 12*(1), 33–39.

Schneider, S. A., Paisan-Ruiz, C., Quinn, N. P., Lees, A. J., Houlden, H., Hardy, J., et al. (2010). ATP13A2 mutations (PARK9) cause neurodegeneration with brain iron accumulation. *Movement Disorders, 25*(8), 979–984.

Schneider, S. A., & Bhatia, K. P. (2013). Excess iron harms the brain: The syndromes of neurodegeneration with brain iron accumulation (NBIA). *Journal of Neural Transmission, 120*(4), 695–703.

Schultheis, P. J., Fleming, S. M., Clippinger, A. K., Lewis, J., Tsunemi, T., Giasson, B., et al. (2013). Atp13a2-deficient mice exhibit neuronal ceroid lipofuscinosis, limited α-synuclein accumulation and age-dependent sensorimotor deficits. *Human Molecular Genetics, 22*(10), 2067–2082.

Shevell, M. (1992). Racial hygiene, active euthanasia, and Julius Hallervorden. *Neurology, 42*(11), 2214–2219.

Wada, H., Yasuda, T., Miura, I., Watabe, K., Sawa, C., Kamijuku, H., et al. (2009). Establishment of an improved mouse model for infantile neuroaxonal dystrophy that shows early disease onset and bears a point mutation in Pla2g6. *American Journal of Pathology, 175*(6), 2257–2263.

Wolkow, N., Song, Y., Wu, T. D., Qian, J., Guerquin-Kern, J. L., & Dunaief, J. L. (2011). Aceruloplasminemia: Retinal histopathologic manifestations and iron-mediated melanosome degradation. *Archives of Ophthalmology*, *129*(11), 1466–1474.

Zhou, B., Westaway, S. K., Levinson, B., Johnson, M. A., Gitschier, J., & Hayflick, S. J. (2001). A novel pantothenate kinase gene (PANK2) is defective in Hallervorden-Spatz syndrome. *Nature Genetics*, *28*(4), 345–349.

Zöller, I., Meixner, M., Hartmann, D., Büssow, H., Meyer, R., Gieselmann, V., et al. (2008). Absence of 2-hydroxylated sphingolipids is compatible with normal neural development but causes late-onset axon and myelin sheath degeneration. *Journal of Neuroscience*, *28*(39), 9741–9754.

CHAPTER NINE

Imaging of Iron

Petr Dusek[*,†,1], **Monika Dezortova**[‡], **Jens Wuerfel**[†,§]

[*]Department of Neurology and Center of Clinical Neuroscience, Charles University in Prague, 1st Faculty of Medicine and General University Hospital, Prague, Czech Republic
[†]Institut für interventionelle und diagnostische Neuroradiologie, Universitätsmedizin Göttingen, Göttingen, Germany
[‡]MR Unit, Department of Diagnostic and Interventional Radiology, Institute for Clinical and Experimental Medicine, Prague, Czech Republic
[§]NeuroCure Clinical Research Center, Charité University Medicine and Max Delbrück Center for Molecular Medicine, Berlin, Germany
[1]Correspondence author: e-mail address: pdusek@gmail.com

Contents

International Review of Neurobiology, Volume 110
ISSN 0074-7742
http://dx.doi.org/10.1016/B978-0-12-410502-7.00010-7

Abstract

Magnetic resonance imaging (MRI) enables a noninvasive *in vivo* quantification of iron in various organs. Several techniques have been developed that detect signal alterations derived mainly from the magnetic properties of ferritin and hemosiderin, the major iron storage compounds. High magnetic susceptibility of ferritin shortens the transversal relaxation time of nearby water protons and thus induces a focal signal extinction of iron-rich areas in T2-weighted (T2w) MRI. T2w tissue contrast is additionally influenced by other factors such as water content, myelin density, and the presence of other metals. Therefore, more specific methods are needed with higher specificity to iron. These *in vivo* techniques can be divided into three groups: relaxometry, magnetic field correlation imaging and phase-based contrast covering susceptibility-weighted imaging, and quantitative susceptibility mapping. The differential diagnosis of various neurological disorders is aided by characteristic patterns of iron depositions. Reliable estimates of cerebral tissue iron concentration are equally important in studying physiological age-related as well as pathological conditions in neurodegenerative, neuroinflammatory, and vascular diseases. In the future, monitoring changes in iron storage and content may serve as sensitive biomarker for diagnosis as well as treatment monitoring.

1. INTRODUCTION

Several systemic and neurological disorders are associated with dysregulation of iron metabolism (Dusek, Jankovic, & Le, 2012; Fleming & Ponka, 2012; Schneider et al., 2013). The ability to noninvasively quantify the iron content in affected organs is important not only for diagnostic purposes but also in the assessment of disease progression and treatment effects. Furthermore, with the advent of new chelating agents, there is increasing need for methods able to reliably measure changes of iron content in tissues (Stankiewicz et al., 2007).

Computed tomography can detect iron deposits in areas with increased density (Boltshauser et al., 1987; Tennison, Bouldin, & Whaley, 1988; Van Kirk, Larsen, & O'Connor, 1986), while transcranial sonography reveals hyperechogenicity in the same locations (Berg, 2006; Kostic et al., 2012; Liman, Wellmer, Rostasy, Bahr, & Kermer, 2012; Zecca et al., 2005). Positron emission tomography with radioactively labeled iron (e.g., ^{52}Fe-citrate) may be used to determine pharmacokinetic properties of iron circulation and metabolism (Bruehlmeier et al., 2000). Magnetic resonance imaging (MRI) emerged as the most powerful and sensitive technique for detecting even small iron quantities, utilizing changes in relaxation times and thus the inherent image contrast, when paramagnetic ions are present.

The advent of high-field MRI and iron-sensitive imaging techniques such as T2*-weighted imaging (T2*w), susceptibility-weighted imaging (SWI), or field-dependent relaxivity increase (FDRI) made MRI an indispensable tool in clinical diagnostic routine as well as in research studies of disorders with iron dysregulation (Haacke et al., 2005). This chapter is focused on the detection and quantification of cerebral iron deposits with a special focus on neurodegeneration.

Recently, iron and metal ion metabolism and the development of specific, iron particle-containing contrast agents have gained increasing interest, for example, iron contrast labeling of cells enabled the *in vivo* observation of their transport and their precise localization in the body.

2. IRON METABOLISM AND ITS CELLULAR DISTRIBUTION

Brain and systemic iron metabolism are separated by the blood–brain barrier (BBB) and blood–cerebrospinal fluid barrier. The adult human brain contains approximately 60 mg of iron, which begins to be deposited in the brain parenchyma shortly after birth and is gradually accumulated mainly within the first three decades of life (Hallgren & Sourander, 1958). In the developing brain, iron colocalizes with the myelinogenic foci; the most prominent increase in cerebral iron uptake coincides with the peak of myelination (Todorich, Pasquini, Garcia, Paez, & Connor, 2009). Iron is not distributed uniformly throughout the brain, with differences in iron content among various cell types and brain regions. Thus, high iron concentrations are particularly present in oligodendrocytes, while neurons or astrocytes show a considerably lower iron concentration (Connor, Menzies, St Martin, & Mufson, 1990). Iron content is not equally distributed in oligodendrocytes throughout the brain, but seems to be particularly high intracortically in comparison to white matter myelin (Fukunaga et al., 2010). Here, it seems important to recognize that myelinated fibers are present also in deep gray matter nuclei (Pott et al., 2009). When developmental myelination is accomplished, there is no further iron accumulation in oligodendrocytes. In contrast, microglia and astrocytes continue to accumulate iron during adulthood and senescence (Connor et al., 1990). Microglia might serve as brain iron capacitor, bearing the ability to promptly accumulate and release iron. In the brain, microglia iron levels vary considerably (Connor & Menzies, 1990).

Systemic iron, while crossing the BBB, preferentially enters endothelial cells bound to transferrin (TF). The influx of iron into the central nervous system (CNS) is regulated through TF receptor expression on cerebral capillary endothelial cells in the brain parenchyma and the choroid plexus (Fishman, Rubin, Handrahan, Connor, & Fine, 1987; Taylor & Morgan, 1991). Endothelial cells forming the BBB express the iron storage protein ferritin, may accumulate substantial amounts of iron, and may thus serve as an iron buffer for the CNS (Connor et al., 2011).

After its release from endothelial cells, iron is oxidized into the ferric state. This process is facilitated by ceruloplasmin localized at astrocytic perivascular foot processes (Attieh, Mukhopadhyay, Seshadri, Tripoulas, & Fox, 1999). Various transporter proteins can thereafter bind and distribute ferric ions throughout the brain. Only a small amount of the absorbed iron stays nonbound, forming the labile iron pool. Intracellular storage and detoxification of free iron are mediated by the ferritin complex that is composed of two subunits, ferritin heavy polypeptide 1 and ferritin light polypeptide. In catecholaminergic neurons of the substantia nigra (SN) and the locus coeruleus, the major iron storage molecule is neuromelanin, an aggregate derived from oxidized excess catechols contained in organelles of lysosomal type (Tribl et al., 2006; Zecca et al., 2001). Most of cellular iron is not localized in the cytosol, but is dispatched to mitochondria and the endosomal/lysosomal compartment. Mitochondrial iron may be stored in mitochondrial ferritin expressed by metabolically highly active cells and protecting mitochondria from oxidative damage (Levi & Rovida, 2009; Li, Jankovic, & Le, 2011).

The principal mechanism of iron release from ferritin is lysosomal degradation. Lysosomes recycle also other iron-rich structures of cytochromes, hemoglobin, or mitochondria (Johansson et al., 2010). Consequently, lysosomes contain higher quantities of iron compared to other organelles. In postmitotic cells such as neurons, autophagy, that is, recycling of worn-out proteins and organelles, is important for the cellular homeostasis. Neuronal lysosomes accumulate high amounts of degraded iron by this mechanism (Terman & Kurz, 2013). Agglomerates of denatured and incompletely degraded ferritin molecules processed by lysosome form the brown insoluble compound called hemosiderin (Schenck & Zimmerman, 2004). Other autophagic intralysosomal pigments like lipofuscin and neuromelanin also progressively accumulate in nondividing cells. They are composed of indigestible oxidized proteins and lipid residues and also contain high amounts of iron. Lipofuscin accumulates predominantly in large neurons and in brain regions involved in motor function (Sulzer et al., 2008).

3. IRON EFFECT ON MR

MRI is very sensitive to the presence of iron in many biochemical forms. Thus, iron may be considered an intrinsic contrast agent. The presence of iron in the brain, liver, and other organs greatly influences the contrast of T2w MRI. Principally, magnetic crystals can cause changes in the local susceptibility and thus shorten T2 relaxation times of nearby water protons. Magnetic susceptibility is a physical quantity that describes the degree of magnetization in response to an applied magnetic field. Materials are divided into several categories according to their behavior in the magnetic field: diamagnetic, paramagnetic, and ferromagnetic. Ferromagnetic materials possess permanent magnetization independent of an external magnetic field, and their susceptibility is positive. Paramagnetic materials induce magnetization to an external magnetic field causing "positive" susceptibility. The opposite behavior can be observed with diamagnetic materials, that is, the applied external magnetic field is decreased with subsequent "negative susceptibility effect" and signal extinction (Duyn, 2013). The ability of an ion to influence a proton's relaxation rate depends on its concentration and relaxivity (Schenck, 2003).

Qualitative correlations between low MR signal intensity on spin echo T2w images and brain regions containing high iron concentrations have been described since the beginning of MR imaging (Doyle et al., 1982; Drayer et al., 1986). The signal extinction or "dark appearance" on T2w images caused by a shortening of the transversal relaxation times pertains especially to deep gray matter nuclei of the brain. The predominant chemical form of iron deposits in the brain is ferritin (Stark et al., 1983; Vymazal, Brooks, et al., 1996). Besides transversal relaxation, which strongly correlates with iron content in the human brain (Gelman, Ewing, Gorell, Spickler, & Solomon, 2001; Ogg & Steen, 1998; Vymazal et al., 1999), ferritin has also an effect on longitudinal relaxation times, however much weaker, and therefore produces only minor changes on T1w images (Gossuin, Burtea, et al., 2004; Vymazal et al., 1992, 1995; Vymazal, Zak, Bulte, Aisen, & Brooks, 1996). Due to the shortening of the T1 relaxation time, such changes are usually visualized as hyperintensities on T1w MRI. The effect of ferritin clustering on longitudinal relaxation is also much weaker compared to its effect on transversal relaxation (Gossuin, Gillis, Muller, & Hocq, 2007; Wood, Fassler, & Meade, 2004).

There are two major forms of iron in the human body, heme and nonheme iron. Heme iron forms a part of hemoglobin and is essential for its capacity to oxygen transportation. Nonheme iron is present in transporter molecules such as TF as well as storage molecules such as ferritin and hemosiderin. Apart from ferritin, which has been extensively studied, little is known about the magnetic properties and their contribution to MR contrast of other iron forms.

Further interest in iron metabolism and imaging of iron deposits has been raised by the development of superparamagnetic iron oxide (SPIO) contrast agents. SPIO particles were developed as MRI contrast agents. Commercially available dextran-coated SPIO were successfully used for cell labeling even in human applications (Saudek et al., 2010).

Strong susceptibility effects and "blooming" on T2w and T2*w caused by SPIO nanoparticles increase the MR contrast and thus enable the visualization of even single cell. Only at high intracellular concentrations, the cell metabolism as well as the immunologic properties might be altered. A number of studies investigated cell transplantations including not only stem cells but also tumor cells and immune cells and the visualization of their transport, spreading, and circulation in the body and long-term survival observation. Two commercially available SPIO contrast agents were clinically approved for human MRI applications: Ferucarbotran (Resovist®, Bayer Healthcare, Germany) with a particle size range of 45–60 nm and ferumoxides (Endorem®, Guerbet, France) in Europe, distributed as Feridex® (Berlex) in the United States, with a particle size of about 120–180 nm. Both contrast agents were successfully applied for cell labeling (Jendelova et al., 2003; Jirak et al., 2004). Unfortunately, the production of both contrast agents has been stopped for commercial reasons. Hence, today, no SPIO contrast agents are available clinical studies.

3.1. Heme iron

Approximately 70 % of total body iron is present as heme iron in hemoglobin (Schenck & Zimmerman, 2004). The heme group is also necessary for proper functioning of several enzymes, e.g. cytochrome-c oxidase or catalase. Oxygenated and deoxygenated hemoglobin differ in their magnetic properties: Oxyhemoglobin is diamagnetic and therefore does not significantly influence transversal tissue relaxivity. In contrast, deoxyhemoglobin carries strong paramagnetic properties (Brooks, Vymazal, Bulte, Baumgarner, & Tran, 1995). Consequently, under normoxic condition, the influence of deoxyhemoglobin on tissue T2* or phase contrast is weak (Marques, Maddage, Mlynarik, & Gruetter, 2009; Punwani et al., 1997).

3.2. Transport molecules: transferrin and the labile iron pool

In the human brain parenchyma, intracellular free iron water ions are virtually not present except very low concentrations in lysosomes. Cytosol harbors low concentration of various iron-containing enzymes and proteins as well as the chaperon-bound cytoplasmic iron, the so-called labile iron pool. All these compounds are considered too scarce for affecting the MR contrast (Schenck, 2003). TF-bound iron is only weakly magnetic, and the TF concentration is considered insufficient to cause alternations of the MR signal (Schweser, Deistung, Lehr, & Reichenbach, 2011). TF levels are typically 10–50 times lower compared to ferritin in the cortex and the basal ganglia. The difference is less pronounced in white matter since it contains 2.5–3.5 times more TF than gray matter (Connor, Snyder, Beard, Fine, & Mufson, 1992). TF is mostly associated with oligodendrocytes, although astrocytes may also contain considerable amounts in elderly patients. Although both of them are dominantly present in oligodendrocytes, the distribution of ferritin and TF in brain does not show a major overlap (Connor et al., 1990).

3.3. Storage molecules: ferritin and hemosiderin

Ferritin consists of a spherical protein shell surrounding an inner cavity, which is able to accumulate and store up to 4500 iron atoms as superparamagnetic crystalline ferric oxyhydroxide, known as ferrihydrite ($5Fe_2O_3{\cdot}9H_2O$). The number of iron atoms stored within the ferritin molecule is called the loading factor. Ferritin is the major iron storage molecule in glial cells, particularly in oligodendrocytes (Gerber & Connor, 1989). Ferritin levels in the basal ganglia and the cortex closely match with the distribution of iron. This is in agreement with ferritin being the major iron storage molecule. It is estimated that approximately 20–25% of body iron is stored by ferritin. Consequently, ferritin and hemosiderin are the most important sources of iron-related signal changes in cerebral MRI, since they contain 33–88 % of nonheme iron of the brain (Schenck, 2003).

The behavior of ferritin in the magnetic field has been extensively studied in solutions as well as in tissues (for review, see Gossuin, Muller, & Gillis, 2004). Early *in vitro* studies showed that aqueous solution of ferritin decreases the transverse relaxation time (T2) proportionally to the concentration of ferritin (Koenig, Brown, Gibson, Ward, & Peters, 1986). The reciprocal of the T2 decay time constant is called transverse relaxation rate R2 (1/T2) and is more often used to characterize this effect. Relaxivity is defined as relaxation rate enhancement caused by an increase of 1 mg(Fe)/g in the iron concentration, indicated in s^{-1}/(mg(Fe)/g). Relaxation rates increase linearly with the loading factor of ferritin, provided that the protein concentration remains constant. Thus, the contribution to relaxation rate per iron atom is constant, regardless of low or high ferritin loading, and R2 may be used as a measure of tissue iron concentration (Vymazal, Zak, et al., 1996). R2 of aqueous solution of ferritin is linearly dependent on the magnetic field strength (Vymazal, Brooks, et al., 1996), and iron-dependent contrast becomes much more prominent as the field strength increases. Therefore, in high-field scanners, iron-rich nuclei appear much darker compared to scanner with lower field that may lead an inexperienced rater to overestimate the iron content (Fig. 9.1). R2 is also influenced by the temperature of the ferritin solution. Low temperature leads to increased R2 (Gossuin, Roch, Muller, & Gillis, 2000), but this effect does not affect ferritin measurements *in vivo* since temperature of human body is constant. The nearly

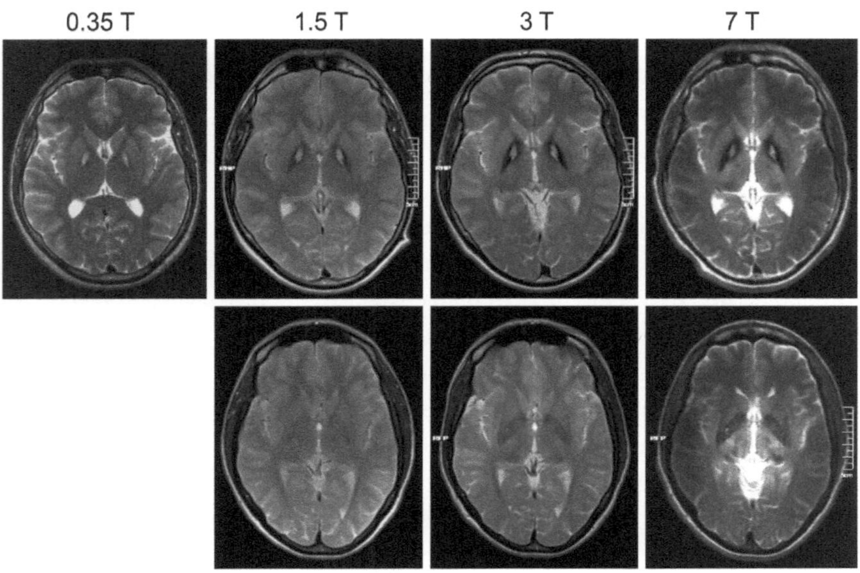

Figure 9.1 Effect of magnetic field strength on the transversal relaxation time in globus pallidus. Top row: T2-weighted images of a 32-year-old patient with pantothenate kinase-associated neurodegeneration (PKAN) examined at 0.35 T (TR/TE = 5040/99 ms), 1.5 T (TR/TE = 3000/75.9 ms), 3 T (TR/TE = 3000/75.9 ms), and 7 T (TR/TE = 6760/75.9 ms) field strengths. Bottom row: A 40-year-old volunteer examined by the same sequences and magnetic field strengths (except 0.35 T). Note the darkening of globus pallidus with the magnetic field strength increase.

linear relationship between field strength (B0) and R2 of ferritin is in contrast to the exponential behavior predicted by the presumption that the ferritin core is ferro- or ferrimagnetic. The unusual magnetic properties of ferritin may contribute to this effect, since its nanometric size of the grains gives ferritin superparamagnetic qualities, while the crystal structure of ferrihydrite is antiferromagnetic at lower temperatures (Brooks, Vymazal, Goldfarb, Bulte, & Aisen, 1998; Haacke et al., 2005; Vymazal, Zak, et al., 1996).

The tissue relaxivity of ferritin, that is, within the brain parenchyma, is significantly increased in comparison to aqueous solutions (Gossuin et al., 2004; Vymazal et al., 1995), at 1.5 T by factor three (Engelhardt et al., 1994). Furthermore, there are differences in transverse relaxation rate of ferritin within various body organs (Gossuin, Burtea, et al., 2004). The relaxation rate in liver and spleen increases linearly with the magnetic field strengths, whereas it shows an exponential behavior in cerebral gray matter (Hocq et al., 2009). This finding may be interpreted as a result of slower diffusion due to the increased compartmentalization and/or resulting from clustering of ferritin in tissues. It could be confirmed by *in vitro* study that increasing the size of individual ferritin clusters enhanced the transversal relaxivity. Moreover, with a critical cluster size, the linear dependence of R2 on the magnetic field is turned into an exponential dependence (Gossuin et al., 2007). Another study examined transversal relaxivity of liposomal fraction, that is, ferritin clusters coated by biological membranes, which closely simulates conditions in tissues. Liposomal ferritin in this study exhibited six times stronger transversal relaxivity than not-aggregated ferritin (Wood et al., 2004). Very tightly aggregated ferritin clusters, resembling hemosiderin granules, may possess stronger superparamagnetic properties due to the crystal structure and the spatial relationship, that is, distance between iron ions. The degree of ferritin clustering as well as the ferritin/hemosiderin ratio may be specific to the individual organ and thus explain differences in tissue relaxivity. In fact, the transverse relaxation rate is predominantly influenced by hemosiderin, but it is relatively insensitive to the dispersed not-aggregated ferritin (Haque et al., 2003). In iron-overloaded liver, biexponential T2 signal decay has been observed, suggestive of multiple sources causing the signal decay (Bulte, Miller, Vymazal, Brooks, & Frank, 1997). A method based on multiple spin echo (MSE) sequence was thus used to decompose the tissue signal decay into two major components: deriving from dispersed soluble fraction of ferritin iron and from aggregated insoluble hemosiderin iron. In a phantom model as well as in human liver, this method exhibited a better correlation with the tissue iron concentration compared to T2* relaxometry (Jensen et al., 2010). In conclusion, the linear dependence of ferritin R2 on the magnetic field strength seems to be caused by not-aggregated ferritin in solution. Human tissues contain highly aggregated ferritin and hemosiderin, which distort the linear dependence observed *in vitro*.

4. POTENTIAL CONFOUNDING FACTORS IN IRON IMAGING

Tissue relaxation is influenced by the concentration of iron as well as its chemical form. However, the relationship between tissue iron content and its transversal relaxation times is even more complex. Multiple sources and mechanisms besides iron concentration additionally contribute to the contrast in T2w (Adisetiyo et al., 2012; Duyn et al., 2007; Lee et al., 2012). It was estimated that R2 contrast is dominated by iron only in those tissues with iron concentration exceeding 5 mg Fe/100 g fresh weight (House et al., 2007). Several researchers reported only weak correlation between tissue R2 and iron concentration determined biochemically (Brooks, Luthert, Gadian, & Marsden, 1989; Chen et al., 1989, 1993).

4.1. Water and myelin content

The water proton content is a major contributor to T2w signal intensities (Kamman, Go, Brouwer, & Berendsen, 1988). The slight difference in water content in white matter (72% water) compared to gray matter (84% water) may rule their different T2w contrasts. Indeed, higher R2 value in white matter compared to gray matter has been observed even when iron concentrations did not differ (Haacke et al., 2005; Whittall et al., 1997). The effect of water on R2 is related to the tissue diffusion coefficient (Bulte, Vymazal, Brooks, Pierpaoli, & Frank, 1993). Thus, R2 is not reliable in quantifying iron of pathological processes with increased water content, such as gliosis, edema, or axonal and neuronal loss.

Densely myelinated fiber pathways with low water content, such as the anterior commissure, the internal capsule, or the fornix, present with reduced signal intensities on T2w despite low iron content (Curnes, Burger, Djang, & Boyko, 1988). Several studies documented the contribution of myelin to magnetic susceptibility contrast, which becomes particularly dominant at ultrahigh magnetic fields (Duyn, 2013). Even after complete iron extraction from the brain parenchyma, the contrast between white and gray matter remained, suggestive of diamagnetic properties caused by iron-free myelin (Fukunaga et al., 2010). This hypothesis was confirmed in demyelinating mouse models (Lee et al., 2012; Liu, Li, Johnson, & Wu, 2011) and human neonates (Zhong et al., 2011). Both studies showed markedly reduced white versus gray matter contrast in the absence of myelin. However, iron measurement was not performed in these studies. Thus, it cannot be ruled out that differences in iron quantities in mature white and gray brain matter significantly contribute to their T2w contrast. Furthermore, heterogeneity in transversal relaxation of different white matter tracts was not sufficiently explained by variations in iron concentration (Li et al., 2009). This was especially apparent in the optic radiation, a brain area that contains very low amounts of iron, but shows very high transversal relaxation rates. R2 heterogeneity correlated well with the myelin content and its microstructure. Thus, it was hypothesized that a higher fraction of bound water could increase R2 in strongly myelinated fiber tracts. Finally, both iron content and myelin density were shown to be independent predictors of transversal relaxation rate in frontal and occipital white and gray matter. R2 correlated with the iron content in gray matter but not in white matter, confirming that R2 is not a reliable parameter measuring iron in white matter (Langkammer, Krebs, et al., 2012).

Moreover, there is a susceptibility contrast derived from the white matter microstructure and its orientation relative to B0, since white matter fiber orientation relative to the main magnetic field also modulates transversal relaxivity. This may be explained by small magnetic field inhomogeneities created by diamagnetic anisotropic susceptibility sources along the fiber direction in the myelin sheath (Bender & Klose, 2010; Denk, Hernandez Torres, MacKay, & Rauscher, 2011; He & Yablonskiy, 2009; Lee et al., 2011; Li, Wu, Avram, & Liu, 2012; Wisnieff et al., 2013). These orientation-dependent effects may lead to an increase of up to 50% of the transversal relaxation rate at 7 T in those fibers oriented perpendicular to B0 in comparison to fibers running parallel to B0 (Sati et al., 2012). In conclusion, iron concentration is not the dominant source of susceptibility-weighted contrast in white matter.

4.2. Metal ions

Other ions with paramagnetic or diamagnetic properties, such as copper, chromium, and manganese, can potentially affect tissue signal and mimic the presence of iron. Accumulation of these elements has been documented histochemically within the basal ganglia (Casanova & Araque, 2003; Harder et al., 2008). However, in the case of copper, its relatively small magnetic moment renders T2 relaxation rate too small to significantly influence the tissue contrast at lower or conventional field strengths (Vymazal, Bulte, Frank, Di Chiro, & Brooks, 1993). Moreover, the tissue copper concentration is at least two to four times lower than that of iron (Litwin et al., 2013; Schenck, 1996). Manganese also seems not to affect significantly the MR signal under physiological conditions (Schenck, 2003).

Only under pathological conditions, seen, for example, in Wilson or other liver diseases, the concentrations of these metals may increase high enough to contribute to the MR signal. Accumulation of manganese as notable in the basal ganglia of patients suffering from chronic liver disease, long-term parenteral nutrition, or manganese intoxications causes much higher relative changes in T1w compared to T2w (Fig. 9.2). It usually becomes apparent as a strong T1w hyperintensity and simultaneous weaker T2 signal drop particularly in the globus pallidus (GP) (Vymazal, Babis, et al., 1996). In contrast to signal abnormalities caused by iron accumulation, changes in relaxation times based on manganese are reversible after liver transplantation without recurrence even up to 15 years (Herynek et al., 2001; Herynek, Wagnerova, Hejlova, Dezortova, & Hajek, 2012). The combined effect of manganese and iron on T1 and T2 relaxation times has been also studied and found to be nonlinear and complicated (Zhang et al., 2009).

Calcium is an example of a diamagnetic element influencing the tissue T2 relaxation. *In vitro* measurements in the deep gray matter nuclei found calcium concentrations in the range of iron concentrations (Riederer et al., 1989). Calcium is also subject to age-related changes and accumulations (Casanova & Araque, 2003). Thus, in tissues with high calcium content, such as the GP, its diamagnetic susceptibility effect can be potentially significant and confound iron measurements.

4.3. Chemical exchange

Chemical exchange of protons between macromolecules and free water may contribute to phase contrast between gray and white matter (Luo, He, d'Avignon, Ackerman, & Yablonskiy, 2010; Shmueli, Dodd, Li, & Duyn, 2011; Zhong, Leupold, von Elverfeldt, & Speck, 2008). Through this

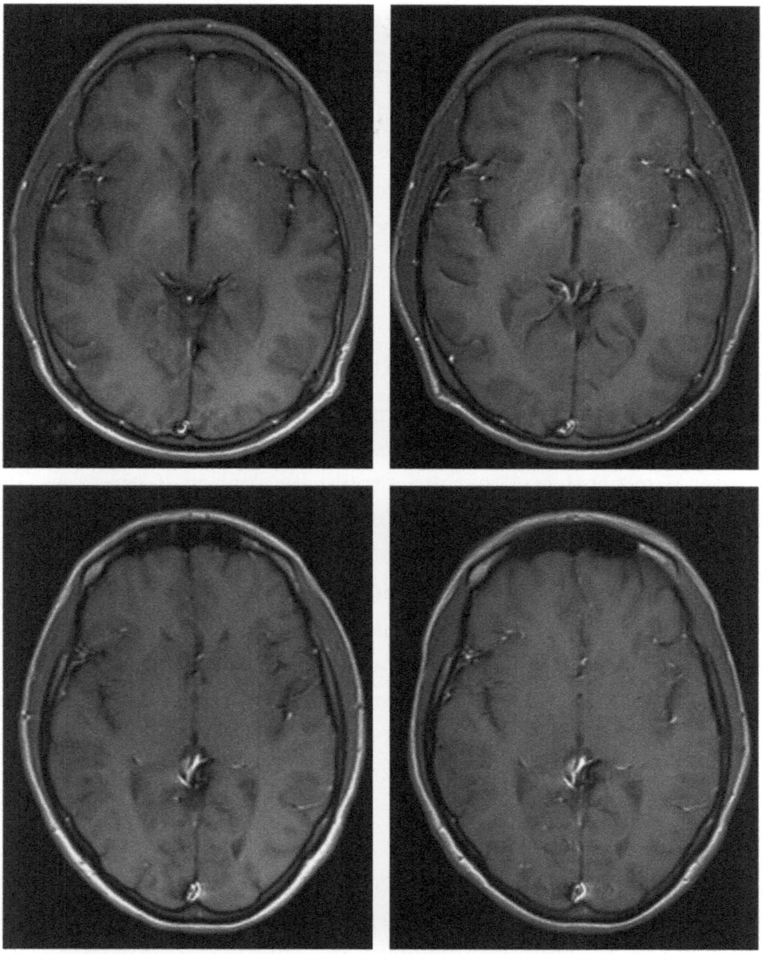

Figure 9.2 T1 hyperintense signal in globus pallidus reflecting manganese deposits in hepatic encephalopathy. Top row: T1-weighted images (TR/TE = 200/6.9 ms) of a 25-year-old patient with chronic hepatic encephalopathy at 1.5 T (left) and 3 T (right) field strength. Bottom row: A 40-year-old volunteer examined by the same sequence and magnetic field strengths, that is, 1.5 T (left) and 3 T (right).

mechanism, white matter has a positive phase shift relative to gray matter. The direction of the gray–white matter exchange-induced phase differences is reciprocal to the overall gray–white matter phase difference usually observed *in vivo*. Thus, chemical exchange cannot be the main source of gray–white matter phase contrast, but may contribute to it (Shmueli et al., 2011).

5. MRI TECHNIQUES SENSITIVE FOR IRON DETECTION

With the knowledge of the confounds mentioned earlier, there is a growing interest in developing and validating MR methods sensitive and specific for tissue iron content. In order to overcome impeding factors, several novel imaging techniques aimed at measuring tissue iron concentrations have been proposed.

5.1. T2, T2*, and T2' relaxometry

MR relaxometry, that is, the determination of relaxation times, is most frequently used for iron quantification. However, there are still discussions which T2 parameter and method should be applied. Generally, T2* is more sensitive to low iron content, but in case of high iron concentrations, T2 seems more accurate as it is less sensitive to field inhomogeneities.

T2 relaxometry is commonly based on a multi-spin echo sequence proposed by Meiboom and Gill (1958) already in 1958 as a modification to the Hahn spin echo sequence. The sequence is known as CPMG (Carr–Purcell–Meiboom–Gill) and employs multiple equidistant phase-shifted refocusing 180° pulses, each followed by an echo. It belongs to the standard sequences available on clinical MR systems. It enables the direct calculation of T2 or R2 (1/T2) maps with only minor influence of external magnetic field inhomogeneities.

On the other hand, T2* relaxation and thus its inverse R2* parameter refer to the decay of transverse magnetization caused by a combination of spin–spin relaxation and local magnetic field inhomogeneity. R2', defined as difference between R2* and R2, is directly related to the local field inhomogeneity induced by the presence of iron as well as other sources. Therefore, R2* and R2' parameters are theoretically more sensitive to the iron content compared to R2 (Gelman et al., 1999, 2001; Graham, Paley, Grunewald, Hoggard, & Griffiths, 2000). In reality, however, R2' has limited sensitivity due to its small magnitude. Compared to R2, R2*/R2' parameters are biased by other background sources of local magnetic field variations not related to the tissue iron content (Hikita et al., 2005; Reichenbach et al., 1997). Global magnetic field inhomogeneities exert less influence on R2* the smaller the voxel size is; therefore, reducing the slice thickness is desirable (Young, Cox, Bryant, & Bydder, 1988).

Another relaxometry method is based on the adiabatic T2rho relaxation time constant (Nestrasil et al., 2010; Wheaton et al., 2004). However, this technique is rarely used.

In comparison to T2 measurements, $T2^*$ relaxometry applying a multiple gradient echo sequence with different TEs enables the direct measurement of the signal intensity decay caused by dephasing of the transverse magnetization. It reflects the net loss of spin coherence irrespective to its origin. More efficient way of $R2^*$/R2' measurement is through acquiring multiple echoes during one acquisition. These multisection measurements during single acquisition can be performed by gradient echo sampling of free induction decay and echo (GESFIDE) (Ma & Wehrli, 1996) or a sequence called partially refocused interleaved multiple echo (PRIME) (Graham et al., 2000). Multiecho sequences are able to provide data that suffice to construct multisection maps of $R2^*$, R2', and R2 in one single acquisition (Gelman et al., 1999).

A postmortem validation study showed that $R2^*$ relaxation rate in gray matter linearly correlates with brain iron concentration determined chemically (Langkammer et al., 2010). In the SN of Parkinson's disease patients, R2' showed the best correlation with biochemically measured iron concentration in comparison to other methods (Ordidge, Gorell, Deniau, Knight, & Helpern, 1994). R2' also better reflects iron content of the frontal white matter, which is typically overestimated based on R2 values (Gelman et al., 1999). In direct comparison with phase values, $R2^*$ showed better correlation with iron concentration in the human brain *in vivo* (Yan, Sun, Yan, Wang, & Lou, 2012).

5.2. Field-dependent relaxometry

It is known that transversal relaxation times depend on the magnetic field strength. In case of iron, field dependence of the relaxivity of ferritin solutions shows a linear increase with the magnetic field strength with the slope depending on the ferritin loading factor (Dezortova et al., 2012; Gossuin et al., 2000; Vymazal, Zak, et al., 1996). Contrary, SPIO nanoparticles showed more or less constant relaxivity values in the fields higher than 0.5 T (Bulte et al., 1993; Rohrer, Bauer, Mintorovitch, Requardt, & Weinmann, 2005). The use of different magnetic fields is not preferentially focused on iron quantification but enables determination of the form in which the iron is deposited. The limitation of this method is a necessity of measuring R2 in MR systems with different field strengths, preventing its routine application in clinical practice.

The comparison of *in vivo* and *in vitro* experiments using magnetic fields with different strengths is challenging. There are significant differences between ferritin-induced relaxation in aqueous solutions and that in tissues. It can be explained by variable clustering of ferritin or presence of more iron forms with different magnetic properties (Gossuin et al., 2000). It was shown that ferritin clusters form quadratic dependence of R2 on the field strength. Very similar behavior was described in brain nuclei *in vitro*, whereas linear dependence on the magnetic field was seen for liver and spleen samples (Gossuin et al., 2007). The nonlinear properties in the shape of the field-dependent relaxation curve were also described *in vivo* in the brain nuclei and frontal white matter of healthy volunteers and pantothenate kinase-associated neurodegeneration patients where higher content of iron in GP was accompanied by different shape of field-dependent relaxation curve. Iron was quantified based on T2 relaxometry, and additionally concentrations of antiferromagnetic/paramagnetic (ferritin-like) and ferrimagnetic (hemosiderin-like) iron were estimated (Dezortova et al., 2012). The method of field-dependent relaxometry in two different magnetic fields has been elaborated by Bartzokis and coworkers (Bartzokis, Aravagiri, Oldendorf, Mintz, & Marder, 1993). Their method is known as FDRI, and it has been shown that FDRI results correlate well with postmortem brain iron concentrations (Bartzokis et al., 1993, Bartzokis, Sultzer, et al., 1994, Bartzokis et al., 1997, 1999; Bartzokis & Tishler, 2000; Pfefferbaum, Adalsteinsson, Rohlfing, & Sullivan, 2009).

5.3. Magnetic field correlation imaging

Magnetic field correlation (MFC) imaging can supposedly quantify microscopic magnetic field inhomogeneities with the size ranging from 1 to 100 μm independently of molecular relaxation mechanisms unrelated to iron. It has a more direct physical relationship to magnetic field inhomogeneities compared to R2', although it can be influenced by water diffusion irregularities. Voxel in the vicinity of large veins, air/tissue, or bone–tissue interfaces may be also affected by the contribution of macroscopic field gradients (Jensen et al., 2006, 2009). MFC uses asymmetric spin echo pulse sequence with multiple refocusing pulse time shifts. In this sequence, the time between the initial 90° excitation pulse and a 180° refocusing pulse is shifted from the half value of the time to echo (TE/2) by up to 16 ms.

In contrast to R2, iron concentration estimated by MFC correlated with postmortem measurements not only in deep gray matter nuclei but also in white matter and thalamus (Adisetiyo et al., 2012). No comparisons were made with $R2^*$ or R2' so far.

5.4. Susceptibility-weighted imaging

SWI, similarly to $T2^*$ imaging, relies on spin dephasing caused by local field inhomogeneities, but in addition to magnitude information uses also phase information. The local changes in the frequency of precessing protons, that is, phase of the $T2^*$w gradient echo signal, bring additional, independent information regarding tissue contrast. Typically, a fully flow-compensated, three-dimensional, fast low-angle shot gradient echo sequence is used in SWI (Haacke et al., 2007; Reichenbach, Venkatesan, Schillinger, Kido, & Haacke, 1997). Phase images give a mapping of the field changes independent of the magnitude response, and they are thus a robust means to measure local changes in susceptibility. In SWI, the phase image is used as a mask to enhance the contrast of the magnitude image (Ogg, Langston, Haacke, Steen, & Taylor, 1999; Reichenbach, Venkatesan, Schillinger, Kido, & Haacke, 1997). Paramagnetic substances, such as ferritin or hemosiderin, increase the magnetic field, and their presence leads to a positive phase shift relative to the surrounding parenchyma. Diamagnetic substances, such as calcium or myelin, cause a negative phase shift (Gupta et al., 2001; Yamada, Imakita, Sakuma, & Takamiya, 1996). A limitation of this method with regards to iron measurement is that phase values are influenced not only by the iron content but also by the difference in the magnetic susceptibility of the surrounding tissue. Background field low-spatial-frequency effects from air/tissue and tissue/bone interfaces lead to unwanted phase information and must be removed by a high-pass phase filter (Chen et al., 2010; Hammond et al., 2008). Symmetric central 64×64 filter was recommended as it removes low-spatial-frequency components, but it may cause also some loss of the phase contrast for the structures of interest (Wang et al., 2000).

Susceptibility can be quantified through its volume content, often indicated in parts per million (ppm) (Reichenbach, Venkatesan, Yablonskiy, et al., 1997). The concept of phase mapping and the normal range of phase shifts between different tissues are established (Haacke et al., 2007). Phase images appear on a scale from $-\pi$ to $+\pi$ with linear dependence on the concentration of paramagnetic substances (Wang et al., 2000). Alternatively, it is possible to correlate phase units with the ppm field deviations. It is expected that most phase-dependent tissue contrast in the basal ganglia derives from the ferritin concentration. However, the phase value of SWI in some brain regions does not correlate always well with the iron content (Pfefferbaum et al., 2009; Yan, Sun, Yan, et al., 2012). Especially in the GP, iron content estimated by the phase image was much lower compared to R2* relaxation or referential chemical measurement. Presumably, GP is rich not only in iron but also in calcium, which may shift the phase to the opposite direction, interfering with the iron quantification. Due to diamagnetic properties of myelin, which in analogy to calcium cancel out iron effect, phase images cannot be used for iron measurement in white matter unless the contribution of myelin is corrected for (Langkammer, Krebs, et al., 2012). Susceptibility from deoxygenated hemoglobin is another potential confounding factor in tissue iron quantification. Particularly in hypoxemic states, hemoglobin contributes to the contrast between gray and white matter in phase images because of differences in local cerebral blood volume and may thus obscure the estimates of nonheme iron concentration (Rauscher, Sedlacik, Barth, Mentzel, & Reichenbach, 2005).

5.5. Quantitative susceptibility mapping

Quantitative susceptibility mapping (QSM) is a technique that, similary to SWI, uses phase information of three-dimensional gradient echo sequence to produce maps of magnetic properties of tissues (de Rochefort, Brown, Prince, & Wang, 2008; Liu, Liu et al., 2011; Liu, Spincemaille, de Rochefort, Kressler, & Wang, 2009; Schweser et al., 2011; Shmueli et al., 2009; Wharton, Schafer, & Bowtell, 2010). Through advanced postprocessing, QSM solves the magnetic field to susceptibility source inverse problem and generates images where voxel intensity is linearly proportional to the underlying tissue apparent magnetic susceptibility. Although dual-echo sequences are prefered, susceptibility maps can be retrospectively calculated from the single-echo GRE measurements routinely acquired during SWI. In comparison to SWI, QSM provides a quantitative and spatially specific image contrast, which is differentially sensitive to myelin and iron content (Reichenbach, 2012). QSM provides maps of relative susceptibility differences rather than absolute values, and all points in the map must be related to a reference region, typically cerebrospinal fluid or occipital white matter (Cheng, Neelavalli, & Haacke, 2009).

One of the QSM computation methods uses a multiangle susceptibility approach in order to overcome anisotropic magnetic susceptibility by averaging the phase values acquired in three different head positions with respect to B0. This method is called calculation of susceptibility through multiple orientation sampling; it requires performing three scans and subsequent coregistration (Liu et al., 2009). Therefore, this method is time-consuming and thus not practical for clinical routine applications.

Morphology enabled dipole inversion method only requires a single-angle acquisition, so it is a more practical solution. This approach compares the structural consistency between the susceptibility map and the magnitude image. Specifically, voxels with known zero susceptibility and voxels that are part of edges in the susceptibility map but not in the edges of the magnitude image are used as spatial priors (de Rochefort et al., 2010; Liu et al., 2012; Liu, Liu, et al., 2011; Schweser et al., 2011; Schweser, Deistung, Sommer, & Reichenbach, 2013). The method of QSM computation using threshold-based k-space division also requires only a single-angle acquisition (Wharton & Bowtell, 2010; Wharton et al., 2010).

In a postmortem study, a very close correlation between iron concentration determined by QSM and chemical analysis was found in gray matter; however, poorer correlation was found in white matter (Langkammer, Schweser, et al., 2012). In another QSM study, the phase contrast correlated well with $R2^*$ and with the iron content in deep gray matter nuclei, while poor correlations were found in white matter (Deistung et al., 2013). Performing QSM and $R2^*$ images at the same time can theoretically disentangle the contributions of iron and myelin through their differential effects on these contrasts. An increasing concentration of both iron and myelin causes an additive increase of $R2^*$, while the paramagnetic property of the former increases and the diamagnetic property of the latter decreases magnetic susceptibility (Deistung et al., 2013). Similarly, the combination of $T2^*$w imaging and QSM may overcome confounds caused by the diamagnetic effects of calcium (Schweser, Deistung, Lehr, & Reichenbach, 2010).

6. QUANTITATIVE IRON MEASUREMENT USING MRI

Since the iron content is directly related to the pathophysiology of certain diseases, the ability to quantitatively assess regional brain iron is important for monitoring disease progression and treatment efficacy. Several postmortem studies examined and quantified iron in the brain parenchyma, using biochemical (atomic absorption spectrophotometry, inductively coupled plasma spectroscopy, instrumental neutron activation, colorimetry, and X-ray fluorescence) and histochemical (Prussian blue and Perls stain) assays (see Haacke et al., 2005 for review). In general, basal ganglia contain more iron than cortical gray matter. Highest concentrations of iron were found in GP (21 μg/100 g fresh weight), exceeding iron concentration of the liver (Hallgren & Sourander, 1958). Other iron-rich brain structures are SN, red nucleus, and putamen. All methods described earlier are in good agreement regarding the iron concentration in the basal ganglia. On the other hand, differing results were obtained regarding the white matter iron concentrations.

A majority of *in vivo* quantitative MRI studies related the estimates of iron content to previously published referential biochemical values (Adisetiyo et al., 2012; Bartzokis, Sultzer, et al., 1994, Bartzokis et al.,1997; Bilgic, Pfefferbaum, Rohlfing, Sullivan, & Adalsteinsson, 2012; Gelman et al., 1999, 2001; Graham et al., 2000; Hasan, Walimuni, Kramer, & Narayana, 2012; Hikita et al., 2005; Pfefferbaum et al., 2009; Schweser et al., 2011; Shmueli et al., 2009; Vymazal et al., 1999; Wharton & Bowtell, 2010; Xu, Wang, & Zhang, 2008; Yan, Sun, Yan, et al., 2012; Yao et al., 2009), mostly taken from the study by Hallgren and Sourander (1958), while *in vitro* studies directly compared MRI and biochemical measurements in the same brains (Hopp et al., 2010; Langkammer et al., 2010; Langkammer, Schweser, et al., 2012; Yao et al., 2009; Zheng, Nichol, Liu, Cheng, & Haacke, 2013) (Table 9.1). For quantitative iron measurements, mean signal intensities across all pixels in defined regions of interest (ROIs) are usually calculated. Most studies use iron estimates in several representative ROIs for correlation: GP, putamen, caudate, SN, red nucleus, thalamus, cortical gray, and white matter. Precise delineation of ROIs and the exclusion of large vessels and possible partial volume areas from the measurement are necessary prerequisites for reliable determinations of iron levels (Fig. 9.3). Applying this method, it is possible to examine two aspects of iron accumulation: the extent of hypointensities reflecting the position of iron deposits and the magnitude of hypointense changes reflecting the iron concentration.

The theoretical paramagnetic contribution of iron stored in ferritin/hemosiderin to tissue susceptibility χ_{Fe} is approximately $1.27 \times 10^{-6}\ \rho . c_{Fe}$, where ρ is tissue density in g/cm^3 and c_{Fe} is concentration of iron in fresh tissue in mg/g (Schenck, 1992). Susceptibility of iron is often indicated in ppm, that is, $\chi_{Fe} = 1.27$ ppm susceptibility per mg iron/cm^3 of fresh tissue.

Table 9.1 Comparison of MRI studies that performed correlation analysis between iron measurement in various brain regions using MRI and reference values determined chemically

Authors	Method	Field strength	Subject nr.	Structures	Correlation coefficient	Reference
In vivo studies						
Gelman et al., 2001	R1	3 T	12	CA, GP, PUT, TH, SN, FWM, PFC	$r=0.98$ (excluding FWM, TH)	HS
Vymazal et al., 1999	R1/R2	1.5 T	18	CA, GP, PUT, TH, SN, RN, PFC	R1 $r=0.88$ R2 $r=0.83$	HS
Hasan et al., 2012	R2	3 T	86	CA, GP, PUT, TH, PFC, FC, PC, TC, OC, CRB, FWM	$r=0.91$	HS
Hikita et al., 2005	R2/R2' (MSE/GESFIDE)	1.5 T	13	CA, GP, PUT, FWM, PFC	R2 (MSE) $r=0.92$ R2 (GESFIDE) $r=0.87$ R2' (GESFIDE) $r=0.33$	HS
Gelman et al., 1999	R2/R2'	3 T	6	CA, GP, PUT, RN, SN, FWM, PFC	R2 $r=0.92$ R2' $r=0.90$	HS
Graham et al., 2000	R2'	1.5 T	14	CA, GP, PUT, SN, FWM	$r=0.99$	HS

Continued

Table 9.1 Comparison of MRI studies that performed correlation analysis between iron measurement in various brain regions using MRI and reference values determined chemically—cont'd

Authors	Method	Field strength	Subject nr.	Structures	Correlation coefficient	Reference
Adisetiyo et al., 2012	MFC/R2	3 T	26	CA, GP, PUT, TH, FWM	MFC $r=0.75$ R2 $r=0.92$	HS
Bartzokis, Sultzer, et al., 1994	FDRI	0.5/1.5 T	76	CA, GP, PUT, FWM	$r=0.99$	HS
Bartzokis et al., 1997	FDRI	0.5/1.5 T	13	CA, GP, PUT, FWM	$r=0.98$	HS
Xu et al., 2008	SWI phase	1.5 T	78	CA, GP, PUT, RN, SN, TH, FWM	$r=0.8$	HS
Pfefferbaum et al., 2009; Bilgic et al., 2012	FDRI/SWI phase/QSM	1.5/3 T	23	CA, GP, PUT, RN, SN, TH, DN, FWM	FDRI rho = 0.95 Phase rho = 0.82 QSM $r=0.88$	HS
Yao et al., 2009	$R2^*$/SWI phase	1.5/3/7 T	9	CA, GP, PUT, TH	$R2^*$ 1.5 T $r=0.79$ $R2^*$3 T $r=0.87$ $R2^*$7 T $r=0.92$ SWI N.S.	HS
Yan, Sun, Yan, et al., 2012	SWI phase/SWI magnitude/$R2^*$	3 T	24	CA, GP, PUT, RN, SN, TH, FWM	$R2^*$ $r=0.98$ Phase $r=0.69$ Magnitude $r=0.98$	HS
Shmueli et al., 2009	SWI phase/$R2^*$/QSM	7 T	1	PUT, RN, SN	Phase $r=0.46$ $R2^*$ $r=0.83$ QSM $r=0.92$	HS

Schweser et al., 2011	QSM (myelin corrected)	3 T	5	CA, GP, PUT, RN, SN, TH, DN, MO, FWM, PFC, MC	$r=0.86$ $r=0.94$ (excluding RN)	HS
Wharton & Bowtell, 2010	QSM	7 T	5	CA, GP, PUT, RN, SN, TH, IC PFC, PC	$r=0.96$	HS
In vitro studies						
Yao et al., 2009	$R2^*$	1.5/3/ 7 T	2	CA, GP, PUT, TH	$r=0.93$	ICPMS
Langkammer et al., 2010	$R2/R2^*$	3 T	7	CA, GP, PUT, TH, CC, FWM, TWM, OWM	R2 $r=0.82$ $R2^*$ $r=0.95$	ICPMS
Hopp et al., 2010	SWI phase	1.5 T	5	n.a.	$r=0.9$	XRF
Langkammer, Schweser, et al., 2012	QSM	3 T	13	CA, GP, PUT, TH, CC, FWM, TWM, OWM	$r=0.87$	ICPMS
Zheng et al., 2013	QSM	3 T	1	CN, GP, PUT	Left hemisphere $r=0.87$ Right hemisphere $r=0.74$	XRF

In vivo studies used previously published referential values, mostly from the article by Hallgren and Sourander. Correlation coefficients are not directly comparable since mentioned studies used different field strengths, employed various numbers of subjects, and measured iron content in different ROIs.

HS, Hallgren and Sourander; XRF, X-ray fluorescence; ICPMS, inductively coupled plasma mass spectrometry; CA, head of the caudate; GP, globus pallidus, PUT, putamen; TH, thalamus; SN, substantia nigra; RN, red nucleus; FWM, frontal white matter; TWM, temporal white matter; OWM, occipital white matter; IC, internal capsule; PFC, prefrontal cortex; FC, frontal cortex; PC, parietal cortex; TC, temporal cortex; OC, occipital cortex; MO, medulla oblongata; CRB, cerebellum; DN, dentate nucleus.

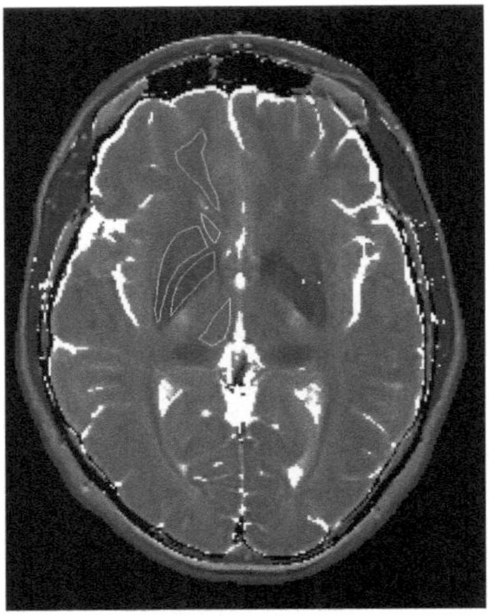

Figure 9.3 Example of MR image with delineated ROIs in basal ganglia on a T2-weighted image. T2 map of a 40-year-old volunteer acquired by the CPMG sequence with 32 echoes, spacing 13.2 ms, and TR=3000 ms. ROIs are marked in caudate head, putamen, globus pallidus, thalamus, and frontal white matter region.

However, the iron contribution to the tissue susceptibility determined from postmortem brains and *in vivo* human studies is significantly lower than the predicted value (Zheng et al., 2013), except one study where susceptibility was corrected for myelin content (Schweser et al., 2011). Besides the myelin contribution, another possible explanation for the lower than expected iron contribution to the tissue susceptibility is the presence of forms of iron that are nonparamagnetic and thus MR-invisible (Zheng et al., 2013).

Supposing that cerebral iron is present as ferritin iron only, its concentration can also be calculated from the age dependence of T2 relaxation times (or R2) and iron content (Hallgren & Sourander, 1958; Schenker, Meier, Wichmann, Boesiger, & Valavanis, 1993). For the GP area measured at 1.5 T MR system, the iron concentration in μg/ml corresponds to the equation 37/T2 (in seconds) −358. Other authors published similar equations relating $R2^*$ parameter to iron content in basal ganglia (Deistung et al., 2013; Peran et al., 2009). Nevertheless, multifield examinations indicate that besides iron in the form of ferritin, there can be iron deposits in other forms with different magnetic properties present. It results in unreliable total iron content estimates due to strong susceptibility effects of these forms of iron (Dezortova et al., 2012).

A good correlation between MRI estimates and iron concentration was confirmed both *in vivo* and in postmortem studies. A good correlation was also observed between different MR methods in estimating brain iron content (Bilgic et al., 2012; Daugherty & Raz, 2013). Results of relaxometry in different fields and MFC significantly correlated also with atlas-based R2 brain maps of healthy subjects (Ge et al., 2007; Hasan et al., 2012). On the other hand, weak correlations were observed between SWI and multifield methods (Pfefferbaum et al., 2009) and between SWI and iron content determined chemically *in vitro* (Yao et al., 2009). Insufficient correlation is caused especially by large differences between phase signal in GP and expected values based on iron concentration measured *in vitro* (Yan, Sun, Yan, et al., 2012). In a postmortem validation study, R2* maps were superior to R2 maps and phase imaging with regard to iron content determined histochemically (Walsh et al., 2013a). When relationship between R2* and iron content was examined under different field strengths, best linear correlation was achieved using 7 T (Yao et al., 2009) confirming the advantage of ultrahigh field MRI in measuring iron concentrations (see Table 9.1).

7. IMAGING OF IRON IN PHYSIOLOGICAL AND PATHOLOGICAL CONDITIONS

MRI has brought tremendous progress to our understanding of brain iron metabolism during aging as well as in various pathological conditions. Detecting changes of the iron content in subcortical brain structures contributes to the differential diagnoses of several genetic and neurodegenerative diseases. MRI has also proven capable of monitoring the treatment effects of iron chelators (Boddaert et al., 2007; Zorzi et al., 2011).

7.1. Age-related iron increase

Increasing degree of T2w hypointensity that is seen in several brain regions in MR images with increasing age (Schenker et al., 1993) parallels postmortem neuropathologic findings of age-related iron accumulation. Increased iron content in basal ganglia was associated with worse performance in motor and cognitive tasks in the elderly population (Pujol et al., 1992; Schroder, Figueiredo, & de Lima, 2013). There are disputes about the rate of iron accumulation in individual deep gray matter nuclei in relation to age. In general, brain iron accumulation follows a fast pace up until third to fourth decade, whereas in late adulthood and old age, slow, if any, iron accumulation is usually observed. Life span differences in iron content are regionally specific. In SN, red nucleus, and GP, there is relatively rapid increase of iron content during early adulthood, whereas in the putamen and head of the caudate, slow iron accumulation is seen in the elderly patients (Milton, Atlas, Lexa, Mozley, & Gur, 1991; Wang, Li, Wei, Li, & Dai, 2012). Thus, putamen yields greatest differences in iron concentration between adult and elderly subjects (Bilgic et al., 2012), and recent meta-analysis confirmed robust age-related changes in putamen (Daugherty & Raz, 2013). Iron accumulation in putamen is not homogenous with its posterior part having much higher iron concentration. It has been suggested to divide putamen into four subregions for the purpose of iron content measurements. Based on iron accumulation in these subregions, four stages of deposition patterns have been postulated: stage 1 (no significant iron accumulation), stage 2 (iron deposits in the posterior outer subregion), stage 3 (iron deposits in both posterior subregions), and stage 4 (iron present all over putamen) (Haacke et al., 2007; Harder et al., 2008). Contrary to putamen, iron accumulation in GP follows an exponential curve with a relatively fast pace in early adulthood and very slow accumulation with nearly reaching a plateau thereafter (Hallgren & Sourander, 1958; Pujol et al., 1992). Iron first accumulates in the most medial part of GP internus and later extends laterally, eventually reaching GP externus. At the same time, iron accumulation in GP follows the posteroanterior direction (Aquino et al., 2009).

The fast iron accumulation in childhood has been associated with the production of energy and constituents for the extensive myelination process. The cause of increasing iron content in the basal ganglia after the fourth decade is far less understood. It is unclear whether this slow accumulation is a physiological finding in elderly or an early biomarker of neurodegenerative, vascular, or demyelinating diseases since older population is known to be heterogeneous and may have unnoticed preclinical Alzheimer or small vessel disease. Increased iron content in basal ganglia has been connected with white matter disruption in multiple sclerosis (MS) (Bakshi, Dmochowski, Shaikh, & Jacobs, 2001; Burgetova, Seidl, Krasensky, Horakova, & Vaneckova, 2010; Ge et al., 2007; Khalil et al., 2009, 2011; Langkammer et al., 2013), age-related white matter hyperintensities (Gebril, Simpson, Kirby, Brayne, & Ince, 2011; Yan, Sun, Chen, et al., 2013), cerebral autosomal dominant arteriopathy with subcortical infarcts and leukoencephalopathy (CADASIL) (Liem et al., 2012), and rare genetic syndromes from the neurodegeneration with brain iron accumulation (NBIA) group (Dusek & Schneider, 2012).

Importantly, iron deposits in the GP were proportional to the severity of white matter hyperintensities, and both were also associated with increasing age (Yan, Sun, Chen, et al., 2012). There are several possible explanations for these findings. Firstly, this association may reflect toxic effect of accumulated iron on oligodendrocytes in the white matter. From this perspective, it remains difficult to explain why iron accumulation occurs in demyelinating disorders primarily affecting oligodendrocytes. Secondly, iron deposits may be a marker of increased myelin turnover since myelination is dependent on iron supply. Lastly, iron accumulation and white matter disruption may have a common cause. Iron accumulation in basal ganglia has been found also after severe ischemic–anoxic insults (Dietrich & Bradley, 1988). Thus, increased iron content and white matter disease in the normally aging population might be a consequence of chronic brain ischemia due to small vessel disease. In order to disentangle the relation between age-related iron and white matter changes, longitudinal studies with repeated measurements of iron deposits, white matter disease, and cerebral perfusion are needed.

7.2. Gender and hemispheric differences

Aging is the main risk factor for most neurodegenerative disorders (Rhodes & Ritz, 2008), and it was suggested that age-related iron accumulation is one of the factors connecting aging and neurodegeneration (Bartzokis et al., 2007). Along this line, it was suggested that the lower incidence of neurodegenerative disorders in females might be explained by higher lifelong body iron losses through menstruation and subsequent lower brain iron content that was confirmed by MRI (Bartzokis et al., 1997, 2007). However, lower brain iron levels in females were not confirmed by other studies (Burgetova et al., 2010; Xu et al., 2008). Interestingly, higher iron content determined chemically was documented in the left hemisphere suggesting relationship with motor lateralization (Haacke et al., 2010; Langkammer et al., 2010; Supprian, Hofmann, Warmuth-Metz, Franzek, & Becker, 1997; Xu et al., 2008). Other studies, however, did not find hemispheric differences (Bilgic et al., 2012; Vymazal et al., 1999). Altogether, these findings indicate that variations in iron content related to hemispheric dominance and gender are less robust than age-related differences. Also, their association with increased risk of neurodegenerative changes remains speculative. Nevertheless, researchers performing imaging studies should take these potential effects into consideration.

7.3. Imaging of iron in neurological disorders

MRI has become an invaluable tool in differential diagnosis of disorders from the NBIA group (for review, see Kruer et al., 2012; McNeill et al., 2008; Schipper, 2012). MRI is very sensitive in detecting even microscopic hemosiderin deposits resulting from cerebral microbleeds resulting from amyloid angiopathy or arterial hypertension. Excellent correlation between microbleeds and small hypointense lesions in T2, T2*, or SWI contrasts was documented by postmortem MRI and histochemistry studies (De Reuck et al., 2011; Fazekas et al., 1999; Schrag et al., 2010; Tanaka, Ueno, Nakayama, Takano, & Takebayashi, 1999; Tatsumi, Shinohara, & Yamamoto, 2008) (Table 9.2). Another postmortem MR and histochemistry studies documented good correlation between hypointense regions in T2, T2*, and SWI contrasts and iron staining in white matter lesions and basal ganglia in MS patients (Bagnato et al., 2011; Yao et al., 2012). Yet another postmortem studies showed that T2* hypointense signal in deep layers of motor cortex of patients with amyotrophic lateral sclerosis (ALS) (Kwan et al., 2012) and in striatum of patients with CADASIL (Liem et al., 2012) reflects iron deposits in perivascular space and in glia (Table 9.2).

Signs of iron accumulation in neuroimaging *in vivo* have been reported in almost every neurodegenerative disorder including PD and other parkinsonian syndromes (Gorell et al., 1995; Gupta, Saini, Kesavadas, Sarma, & Kishore, 2010; Wallis et al., 2008), Huntington disease (Bartzokis, Cummings, Perlman, Hance, & Mintz, 1999; Dumas et al., 2012; Vymazal et al., 2007), Alzheimer disease (Bartzokis, Sultzer, et al., 1994; Bartzokis & Tishler, 2000; Wang, Zhu, Wei, Li, & Li, 2013), ALS (Kwan et al., 2012), and even in essential tremor (Novellino et al., 2013), suggesting that iron may be involved in some essential process involved in neurodegeneration such as apoptosis (Sian-Hulsmann, Mandel, Youdim, & Riederer, 2011). On the other hand, it is not possible to exclude that iron increase is just a consequence of cellular loss (Sastry & Arendash, 1995).

Detection of increased iron content in these disorders has mostly scientific significance. In parkinsonian disorders, however, there is also an emerging clinical application of iron-sensitive MRI sequences since it can help in the diagnosis of PD and differential diagnosis of parkinsonian syndromes. The combination of several MRI parameters in SN can discriminate between PD patients and controls with an excellent accuracy. SN of PD patients exhibits decreased fractional anisotropy reflecting microstructural

Table 9.2 List of postmortem studies correlating MRI changes with histochemical examination for iron detection

Study	MRI method	Field strength	Histochemistry method	Subjects nr.	Diagnosis	Major findings
Walsh et al., 2013a	R2, R2*, SWI	4.7 T	Perls stain	4	MS	R2* has strongest (>R2, >SWI) correlation with iron staining in BG MS lesion appearance was variable in regard to iron content
Bagnato et al., 2011/Yao et al., 2012	R2*, SWI	7 T/ 11.7 T	Perls stain Turnbull blue stain Immunostaining for ferritin, L-ferritin, oligodendrocyte, microglia, macrophage, astrocyte, axon	2	MS	Iron invariably resulted in R2* hyperintensity but not vice versa Sources of iron included oligodendrocytes in NAWM, microbleeds in the center, and microglia at the rims of WM lesions R2* reduction corresponded to loss of both iron and myelin Negative phase shift in phase images corresponded to focal iron deposits
Fazekas et al., 1999	T2*	1.5 T	H–E, Masson trichrome, Kluver–Barrera, Congo red	11	IH	62% of MR hypointensities: focal accumulation of hemosiderin containing macrophages adjacent to small blood vessels in BG 38% of MR hypointensities: without specific pathological correlate Several small microbleeds were MR negative

Continued

Table 9.2 List of postmortem studies correlating MRI changes with histochemical examination for iron detection—cont'd

Study	MRI method	Field strength	Histochemistry method	Subjects nr.	Diagnosis	Major findings
Tanaka et al., 1999	T2	1 T	H–E, Masson trichrome, Kluver–Barrera	3	IH	Majority of T2 hypointense lesions reflect hemosiderin pigments within the perivascular space One T2 hypointense lesion reflects organized miliary pseudoaneurysm
Tatsumi et al., 2008	T2, T2*	1.5 T	H–E, Berlin blue	1	IH	8/9 T2* hypointense lesions were identified as perivascular hemosiderin-laden macrophages 1/9 was identified as vascular pseudocalcification
Schrag et al., 2010	T2*, SWI	3 T	H–E, Prussian blue Immunostaining for HO-1, CD68, complement C6, Abeta1-42, CD3, CD 20, MAP-2	8	AD	38 hypointense lesions identified in SWI images: 16 represent old hematomas with perivascular hemosiderin deposits, 10 represent acute hematoma, 7 represent small cavities ringed by hemosiderin deposits, 3 represent small hemosiderin/hematoidin granules, 2 vessel wall abnormalities Acute hematomas were noted to propagate along perivascular spaces activating microglia and lymphocytes and inducing apoptosis

De Reuck et al., 2011	T2, T2*	7 T	H–E, Perls stain, luxol fast blue	20	AD	96% microbleeds were detected as hypointense lesions in T2* 3 cortical microbleeds were missed 85% small minibleeds (diameter 200–500 μm) were detected as hypointense lesions in T2* 25 minibleeds were missed 41 hyposignals in T2* in striatum and DWM were not related to minibleeds In basal ganglia, many T2* hyposignals correlated to iron/calcium deposits in walls of deep perforating vessels
Liem et al., 2012	T2, T2*	7 T	H–E, Perls stain, luxol fast blue, von Kossa	3	CADASIL	T2* hypointensities in putamen, caudate, and GP correspond to iron deposited perivascularly and in glial cells throughout the parenchyma T2* hypointensities did not match the calcium deposits
Kwan et al., 2012	R2*	7 T	H–E, Perls stain, Bielschowsky silver stain, luxol fast blue, immunostain for GFAP, TDP-43, FUS, ubiquitin	2	ALS	R2* hyperintense rim in deep layers of motor cortex correlated with iron deposits in microglia

H–E, hematoxylin–eosin; GFAP, glial fibrillary acidic protein; MAP-2, microtubule-associated protein 2; HO-1, heme oxygenase-1; TDP43, TAR DNA-binding protein 43 kDa; FUS, fused in sarcoma gene; WM, white matter; NAWM, normal-appearing white matter; BG, basal ganglia; MS, multiple sclerosis; IH, intracranial hemorrhage; AD, Alzheimer disease; CADASIL, cerebral autosomal dominant arteriopathy with subcortical infarcts and leukoencephalopathy; ALS, amyotrophic lateral sclerosis.

disintegration and increased R2* reflecting enhanced tissue iron stores (Du et al., 2011; Graham et al., 2000; Michaeli et al., 2007; Peran et al., 2010). QSM imaging of SN also confirmed increased iron content in PD (Lotfipour et al., 2012). Moreover, according to several MRI studies, iron deposition in the SN correlates with the disease severity as well as with the motor score in the contralateral body side (Gorell et al., 1995; Jin et al., 2011) but not with the disease duration (Wallis et al., 2008). Correlation between disease severity and SN iron content suggests direct relationship between neurodegeneration and iron accumulation and may be used for monitoring of PD progression. T2*w images obtained by a 7 T MRI scanner offer an excellent contrast of the SN pars compacta and may enable radiological diagnosis of PD by the "naked eye" (Cho et al., 2011). The pattern of deposits in iron-sensitive MRI is also very helpful in differential diagnosis of parkinsonian syndromes. T2 hypointensities in putamen have low sensitivity but high specificity for atypical parkinsonism (Arabia et al., 2010). Increased putaminal and GP iron is the most consistent finding in MRI studies, and its visualization may thus be helpful in the differential diagnosis between PD and multiple system atrophy (MSA) or progressive supranuclear palsy (PSP) (Boelmans et al., 2012; Gupta et al., 2010; Vymazal et al., 1999; Wang et al., 2012). More specifically, patients with MSA have signs of significant increases of iron deposits in the posterolateral putamen and adjacent lateral aspect of the GP, whereas PSP patients in the anterior and medial aspects of the GP and thalamus (Han et al., 2013). We suggest that iron-sensitive MRI sequences should be included in the routine differential diagnosis scanning protocol of parkinsonian patients.

8. CONCLUSIONS

Several MRI techniques have been developed and proposed for iron detection. T2 and T2* relaxometry, use of different magnetic fields, MFC imaging, and QSM showed good correlation with *in vitro* iron measurement in the gray matter. All of these methods have inherent advantages and disadvantages for different research purposes, and none of them is preferred over the others. Nevertheless, limited understanding of iron substances' behavior in magnetic field, insufficient knowledge about exact proportions of different iron forms in tissues, and sensitivity to other confounding factors still prevent these techniques from usage in routine clinical practice (Brittenham, Badman, & National Institute of Diabetes and Digestive and

Kidney Diseases (NIDDK) Workshop, 2003). Availability of ultrahigh field scanners that possess high accuracy for iron detection will boost the utility of MRI not only in research but also in clinical practice. There is a need for further postmortem MRI histopathology correlation studies in order to validate the use of MRI in iron measurement in pathological states with potential confounding factors such as myelin disruption or accumulation of other metals. Research should also focus on studying behavior of nonferritin iron substances, such as lipofuscin or neuromelanin, in the magnetic field and their contribution to MRI contrast. Lastly, the utility of MRI in assessment of disorders with disturbed iron metabolism would be tremendously increased, if we understood better the role of iron accumulation in their pathophysiology.

ACKNOWLEDGMENTS

This work was supported by the Czech Ministry of Education, research project PRVOUK-P26/LF1/4, and by the Ministry of Health, Czech Republic: 00023001 IKEM Institution support.

REFERENCES

Adisetiyo, V., Jensen, J. H., Ramani, A., Tabesh, A., Di Martino, A., Fieremans, E., et al. (2012). In vivo assessment of age-related brain iron differences by magnetic field correlation imaging. *Journal of Magnetic Resonance Imaging, 36*(2), 322–331.

Aquino, D., Bizzi, A., Grisoli, M., Garavaglia, B., Bruzzone, M. G., Nardocci, N., et al. (2009). Age-related iron deposition in the basal ganglia: Quantitative analysis in healthy subjects. *Radiology, 252*(1), 165–172.

Arabia, G., Morelli, M., Paglionico, S., Novellino, F., Salsone, M., Giofre, L., et al. (2010). An magnetic resonance imaging T2*-weighted sequence at short echo time to detect putaminal hypointensity in Parkinsonisms. *Movement Disorders, 25*(16), 2728–2734.

Attieh, Z. K., Mukhopadhyay, C. K., Seshadri, V., Tripoulas, N. A., & Fox, P. L. (1999). Ceruloplasmin ferroxidase activity stimulates cellular iron uptake by a trivalent cation-specific transport mechanism. *Journal of Biological Chemistry, 274*(2), 1116–1123.

Bagnato, F., Hametner, S., Yao, B., van Gelderen, P., Merkle, H., Cantor, F. K., et al. (2011). Tracking iron in multiple sclerosis: A combined imaging and histopathological study at 7 Tesla. *Brain, 134*(Pt 12), 3599–3612.

Bakshi, R., Dmochowski, J., Shaikh, Z. A., & Jacobs, L. (2001). Gray matter T2 hypointensity is related to plaques and atrophy in the brains of multiple sclerosis patients. *Journal of Neurological Sciences, 185*(1), 19–26.

Bartzokis, G., Aravagiri, M., Oldendorf, W. H., Mintz, J., & Marder, S. R. (1993). Field dependent transverse relaxation rate increase may be a specific measure of tissue iron stores. *Magnetic Resonance in Medicine, 29*(4), 459–464.

Bartzokis, G., Beckson, M., Hance, D. B., Marx, P., Foster, J. A., & Marder, S. R. (1997). MR evaluation of age-related increase of brain iron in young adult and older normal males. *Magnetic Resonance Imaging, 15*(1), 29–35.

Bartzokis, G., Cummings, J. L., Markham, C. H., Marmarelis, P. Z., Treciokas, L. J., Tishler, T. A., et al. (1999). MRI evaluation of brain iron in earlier- and later-onset Parkinson's disease and normal subjects. *Magnetic Resonance Imaging, 17*(2), 213–222.

Bartzokis, G., Cummings, J., Perlman, S., Hance, D. B., & Mintz, J. (1999). Increased basal ganglia iron levels in Huntington disease. *Archives of Neurology, 56*(5), 569–574.

Bartzokis, G., Sultzer, D., Mintz, J., Holt, L. E., Marx, P., Phelan, C. K., et al. (1994). In vivo evaluation of brain iron in Alzheimer's disease and normal subjects using MRI. *Biological Psychiatry, 35*(7), 480–487.

Bartzokis, G., & Tishler, T. A. (2000). MRI evaluation of basal ganglia ferritin iron and neurotoxicity in Alzheimer's and Huntingon's disease. *Cell and Molecular Biology (Noisy-le-grand), 46*(4), 821–833.

Bartzokis, G., Tishler, T. A., Lu, P. H., Villablanca, P., Altshuler, L. L., Carter, M., et al. (2007). Brain ferritin iron may influence age- and gender-related risks of neurodegeneration. *Neurobiology of Aging, 28*(3), 414–423.

Bender, B., & Klose, U. (2010). The in vivo influence of white matter fiber orientation towards B(0) on T2* in the human brain. *NMR in Biomedicine, 23*(9), 1071–1076.

Berg, D. (2006). In vivo detection of iron and neuromelanin by transcranial sonography—A new approach for early detection of substantia nigra damage. *Journal of Neural Transmission, 113*(6), 775–780.

Bilgic, B., Pfefferbaum, A., Rohlfing, T., Sullivan, E. V., & Adalsteinsson, E. (2012). MRI estimates of brain iron concentration in normal aging using quantitative susceptibility mapping. *NeuroImage, 59*(3), 2625–2635.

Boddaert, N., Le Quan Sang, K. H., Rotig, A., Leroy-Willig, A., Gallet, S., Brunelle, F., et al. (2007). Selective iron chelation in Friedreich ataxia: Biologic and clinical implications. *Blood, 110*(1), 401–408.

Boelmans, K., Holst, B., Hackius, M., Finsterbusch, J., Gerloff, C., Fiehler, J., et al. (2012). Brain iron deposition fingerprints in Parkinson's disease and progressive supranuclear palsy. *Movement Disorders, 27*(3), 421–427.

Boltshauser, E., Lang, W., Janzer, R., Briner, J., Spiess, H., Kleihues, P., et al. (1987). Computed tomography in Hallervorden-Spatz disease. *Neuropediatrics, 18*(2), 81–83.

Brittenham, G. M., Badman, D. G., & National Institute of Diabetes and Digestive and Kidney Diseases (NIDDK) Workshop (2003). Noninvasive measurement of iron: Report of an NIDDK workshop. *Blood, 101*(1), 15–19.

Brooks, D. J., Luthert, P., Gadian, D., & Marsden, C. D. (1989). Does signal-attenuation on high-field T2-weighted MRI of the brain reflect regional cerebral iron deposition? Observations on the relationship between regional cerebral water proton T2 values and iron levels. *Journal of Neurology, Neurosurgery, and Psychiatry, 52*(1), 108–111.

Brooks, R. A., Vymazal, J., Bulte, J. W., Baumgarner, C. D., & Tran, V. (1995). Comparison of T2 relaxation in blood, brain, and ferritin. *Journal of Magnetic Resonance Imaging, 5*(4), 446–450.

Brooks, R. A., Vymazal, J., Goldfarb, R. B., Bulte, J. W., & Aisen, P. (1998). Relaxometry and magnetometry of ferritin. *Magnetic Resonance in Medicine, 40*(2), 227–235.

Bruehlmeier, M., Leenders, K. L., Vontobel, P., Calonder, C., Antonini, A., & Weindl, A. (2000). Increased cerebral iron uptake in Wilson's disease: A 52Fe-citrate PET study. *Journal of Nuclear Medicine, 41*(5), 781–787.

Bulte, J. W., Miller, G. F., Vymazal, J., Brooks, R. A., & Frank, J. A. (1997). Hepatic hemosiderosis in non-human primates: Quantification of liver iron using different field strengths. *Magnetic Resonance in Medicine, 37*(4), 530–536.

Bulte, J. W., Vymazal, J., Brooks, R. A., Pierpaoli, C., & Frank, J. A. (1993). Frequency dependence of MR relaxation times II. Iron oxides. *Journal of Magnetic Resonance Imaging, 3*(4), 641–648.

Burgetova, A., Seidl, Z., Krasensky, J., Horakova, D., & Vaneckova, M. (2010). Multiple sclerosis and the accumulation of iron in the Basal Ganglia: Quantitative assessment of brain iron using MRI t(2) relaxometry. *European Neurology, 63*(3), 136–143.

Casanova, M. F., & Araque, J. M. (2003). Mineralization of the basal ganglia: Implications for neuropsychiatry, pathology and neuroimaging. *Psychiatry Research*, *121*(1), 59–87.

Chen, J. C., Hardy, P. A., Clauberg, M., Joshi, J. G., Parravano, J., Deck, J. H., et al. (1989). T2 values in the human brain: Comparison with quantitative assays of iron and ferritin. *Radiology*, *173*(2), 521–526.

Chen, J. C., Hardy, P. A., Kucharczyk, W., Clauberg, M., Joshi, J. G., Vourlas, A., et al. (1993). MR of human postmortem brain tissue: Correlative study between T2 and assays of iron and ferritin in Parkinson and Huntington disease. *American Journal of Neuroradiology*, *14*(2), 275–281.

Chen, Z., Johnston, L. A., Kwon, D. H., Oh, S. H., Cho, Z. H., & Egan, G. F. (2010). An optimised framework for reconstructing and processing MR phase images. *NeuroImage*, *49*(2), 1289–1300.

Cheng, Y. C., Neelavalli, J., & Haacke, E. M. (2009). Limitations of calculating field distributions and magnetic susceptibilities in MRI using a Fourier based method. *Physics in Medicine and Biology*, *54*(5), 1169–1189.

Cho, Z. H., Oh, S. H., Kim, J. M., Park, S. Y., Kwon, D. H., Jeong, H. J., et al. (2011). Direct visualization of Parkinson's disease by in vivo human brain imaging using 7.0 T magnetic resonance imaging. *Movement Disorders*, *26*(4), 713–718.

Connor, J. R., & Menzies, S. L. (1990). Altered cellular distribution of iron in the central nervous system of myelin deficient rats. *Neuroscience*, *34*(1), 265–271.

Connor, J. R., Menzies, S. L., St Martin, S. M., & Mufson, E. J. (1990). Cellular distribution of transferrin, ferritin, and iron in normal and aged human brains. *Journal of Neuroscience Research*, *27*(4), 595–611.

Connor, J. R., Ponnuru, P., Wang, X. S., Patton, S. M., Allen, R. P., & Earley, C. J. (2011). Profile of altered brain iron acquisition in restless legs syndrome. *Brain*, *134*(Pt 4), 959–968.

Connor, J. R., Snyder, B. S., Beard, J. L., Fine, R. E., & Mufson, E. J. (1992). Regional distribution of iron and iron-regulatory proteins in the brain in aging and Alzheimer's disease. *Journal of Neuroscience Research*, *31*(2), 327–335.

Curnes, J. T., Burger, P. C., Djang, W. T., & Boyko, O. B. (1988). MR imaging of compact white matter pathways. *American Journal of Neuroradiology*, *9*(6), 1061–1068.

Daugherty, A., & Raz, N. (2013). Age-related differences in iron content of subcortical nuclei observed in vivo: A meta-analysis. *NeuroImage*, *70*, 113–121.

De Reuck, J., Auger, F., Cordonnier, C., Deramecourt, V., Durieux, N., Pasquier, F., et al. (2011). Comparison of 7.0-T $T(2)^*$-magnetic resonance imaging of cerebral bleeds in post-mortem brain sections of Alzheimer patients with their neuropathological correlates. *Cerebrovascular Diseases*, *31*(5), 511–517.

de Rochefort, L., Brown, R., Prince, M. R., & Wang, Y. (2008). Quantitative MR susceptibility mapping using piece-wise constant regularized inversion of the magnetic field. *Magnetic Resonance in Medicine*, *60*(4), 1003–1009.

de Rochefort, L., Liu, T., Kressler, B., Liu, J., Spincemaille, P., Lebon, V., et al. (2010). Quantitative susceptibility map reconstruction from MR phase data using bayesian regularization: Validation and application to brain imaging. *Magnetic Resonance in Medicine*, *63*(1), 194–206.

Deistung, A., Schafer, A., Schweser, F., Biedermann, U., Turner, R., & Reichenbach, J. R. (2013). Toward in vivo histology: A comparison of quantitative susceptibility mapping (QSM) with magnitude-, phase-, and $R2^*$-imaging at ultra-high magnetic field strength. *NeuroImage*, *65*, 299–314.

Denk, C., Hernandez Torres, E., MacKay, A., & Rauscher, A. (2011). The influence of white matter fibre orientation on MR signal phase and decay. *NMR in Biomedicine*, *24*(3), 246–252.

Dezortova, M., Herynek, V., Krssak, M., Kronerwetter, C., Trattnig, S., & Hajek, M. (2012). Two forms of iron as an intrinsic contrast agent in the basal ganglia of PKAN patients. *Contrast Media & Molecular Imaging*, 7(6), 509–515.

Dietrich, R. B., & Bradley, W. G., Jr. (1988). Iron accumulation in the basal ganglia following severe ischemic-anoxic insults in children. *Radiology*, *168*(1), 203–206.

Doyle, F. H., Pennock, J. M., Banks, L. M., McDonnell, M. J., Bydder, G. M., Steiner, R. E., et al. (1982). Nuclear magnetic resonance imaging of the liver: Initial experience. *American Journal of Roentgenology*, *138*(2), 193–200.

Drayer, B., Burger, P., Darwin, R., Riederer, S., Herfkens, R., & Johnson, G. A. (1986). MRI of brain iron. *American Journal of Roentgenology*, *147*(1), 103–110.

Du, G., Lewis, M. M., Styner, M., Shaffer, M. L., Sen, S., Yang, Q. X., et al. (2011). Combined $R2^*$ and diffusion tensor imaging changes in the substantia nigra in Parkinson's disease. *Movement Disorders*, *26*(9), 1627–1632.

Dumas, E. M., Versluis, M. J., van den Bogaard, S. J., van Osch, M. J., Hart, E. P., van Roon-Mom, W. M., et al. (2012). Elevated brain iron is independent from atrophy in Huntington's Disease. *NeuroImage*, *61*(3), 558–564.

Dusek, P., Jankovic, J., & Le, W. (2012). Iron dysregulation in movement disorders. *Neurobiology of Disease*, *46*(1), 1–18.

Dusek, P., & Schneider, S. A. (2012). Neurodegeneration with brain iron accumulation. *Current Opinion in Neurology*, *25*(4), 499–506.

Duyn, J. (2013). MR susceptibility imaging. *Journal of Magnetic Resonance*, *229*, 198–207.

Duyn, J. H., van Gelderen, P., Li, T. Q., de Zwart, J. A., Koretsky, A. P., & Fukunaga, M. (2007). High-field MRI of brain cortical substructure based on signal phase. *Proceedings of the National Academy of Sciences of the United States of America*, *104*(28), 11796–11801.

Engelhardt, R., Langkowski, J. H., Fischer, R., Nielsen, P., Kooijman, H., Heinrich, H. C., et al. (1994). Liver iron quantification: Studies in aqueous iron solutions, iron overloaded rats, and patients with hereditary hemochromatosis. *Magnetic Resonance Imaging*, *12*(7), 999–1007.

Fazekas, F., Kleinert, R., Roob, G., Kleinert, G., Kapeller, P., Schmidt, R., et al. (1999). Histopathologic analysis of foci of signal loss on gradient-echo $T2^*$-weighted MR images in patients with spontaneous intracerebral hemorrhage: Evidence of microangiopathy-related microbleeds. *American Journal of Neuroradiology*, *20*(4), 637–642.

Fishman, J. B., Rubin, J. B., Handrahan, J. V., Connor, J. R., & Fine, R. E. (1987). Receptor-mediated transcytosis of transferrin across the blood-brain barrier. *Journal of Neuroscience Research*, *18*(2), 299–304.

Fleming, R. E., & Ponka, P. (2012). Iron overload in human disease. *New England Journal of Medicine*, *366*(4), 348–359.

Fukunaga, M., Li, T. Q., van Gelderen, P., de Zwart, J. A., Shmueli, K., Yao, B., et al. (2010). Layer-specific variation of iron content in cerebral cortex as a source of MRI contrast. *Proceedings of the National Academy of Sciences of the United States of America*, *107*(8), 3834–3839.

Ge, Y., Jensen, J. H., Lu, H., Helpern, J. A., Miles, L., Inglese, M., et al. (2007). Quantitative assessment of iron accumulation in the deep gray matter of multiple sclerosis by magnetic field correlation imaging. *American Journal of Neuroradiology*, *28*(9), 1639–1644.

Gebril, O. H., Simpson, J. E., Kirby, J., Brayne, C., & Ince, P. G. (2011). Brain iron dysregulation and the risk of ageing white matter lesions. *Neuromolecular Medicine*, *13*(4), 289–299.

Gelman, N., Ewing, J. R., Gorell, J. M., Spickler, E. M., & Solomon, E. G. (2001). Interregional variation of longitudinal relaxation rates in human brain at 3.0 T: Relation to estimated iron and water contents. *Magnetic Resonance in Medicine*, *45*(1), 71–79.

Gelman, N., Gorell, J. M., Barker, P. B., Savage, R. M., Spickler, E. M., Windham, J. P., et al. (1999). MR imaging of human brain at 3.0 T: Preliminary report on transverse relaxation rates and relation to estimated iron content. *Radiology*, *210*(3), 759–767.

Gerber, M. R., & Connor, J. R. (1989). Do oligodendrocytes mediate iron regulation in the human brain? *Annals of Neurology, 26*(1), 95–98.

Gorell, J. M., Ordidge, R. J., Brown, G. G., Deniau, J. C., Buderer, N. M., & Helpern, J. A. (1995). Increased iron-related MRI contrast in the substantia nigra in Parkinson's disease. *Neurology, 45*(6), 1138–1143.

Gossuin, Y., Burtea, C., Monseux, A., Toubeau, G., Roch, A., Muller, R. N., et al. (2004). Ferritin-induced relaxation in tissues: An in vitro study. *Journal of Magnetic Resonance Imaging, 20*(4), 690–696.

Gossuin, Y., Gillis, P., Muller, R. N., & Hocq, A. (2007). Relaxation by clustered ferritin: A model for ferritin-induced relaxation in vivo. *NMR in Biomedicine, 20*(8), 749–756.

Gossuin, Y., Muller, R. N., & Gillis, P. (2004). Relaxation induced by ferritin: A better understanding for an improved MRI iron quantification. *NMR in Biomedicine, 17*(7), 427–432.

Gossuin, Y., Roch, A., Muller, R. N., & Gillis, P. (2000). Relaxation induced by ferritin and ferritin-like magnetic particles: The role of proton exchange. *Magnetic Resonance in Medicine, 43*(2), 237–243.

Graham, J. M., Paley, M. N., Grunewald, R. A., Hoggard, N., & Griffiths, P. D. (2000). Brain iron deposition in Parkinson's disease imaged using the PRIME magnetic resonance sequence. *Brain, 123*(Pt 12), 2423–2431.

Gupta, R. K., Rao, S. B., Jain, R., Pal, L., Kumar, R., Venkatesh, S. K., et al. (2001). Differentiation of calcification from chronic hemorrhage with corrected gradient echo phase imaging. *Journal of Computer Assisted Tomography, 25*(5), 698–704.

Gupta, D., Saini, J., Kesavadas, C., Sarma, P. S., & Kishore, A. (2010). Utility of susceptibility-weighted MRI in differentiating Parkinson's disease and atypical parkinsonism. *Neuroradiology, 52*(12), 1087–1094.

Haacke, E. M., Ayaz, M., Khan, A., Manova, E. S., Krishnamurthy, B., Gollapalli, L., et al. (2007). Establishing a baseline phase behavior in magnetic resonance imaging to determine normal vs. abnormal iron content in the brain. *Journal of Magnetic Resonance Imaging, 26*(2), 256–264.

Haacke, E. M., Cheng, N. Y., House, M. J., Liu, Q., Neelavalli, J., Ogg, R. J., et al. (2005). Imaging iron stores in the brain using magnetic resonance imaging. *Magnetic Resonance Imaging, 23*(1), 1–25.

Haacke, E. M., Miao, Y., Liu, M., Habib, C. A., Katkuri, Y., Liu, T., et al. (2010). Correlation of putative iron content as represented by changes in $R2^*$ and phase with age in deep gray matter of healthy adults. *Journal of Magnetic Resonance Imaging, 32*(3), 561–576.

Hallgren, B., & Sourander, P. (1958). The effect of age on the non-haemin iron in the human brain. *Journal of Neurochemistry, 3*(1), 41–51.

Hammond, K. E., Lupo, J. M., Xu, D., Metcalf, M., Kelley, D. A., Pelletier, D., et al. (2008). Development of a robust method for generating 7.0 T multichannel phase images of the brain with application to normal volunteers and patients with neurological diseases. *NeuroImage, 39*(4), 1682–1692.

Han, Y. H., Lee, J. H., Kang, B. M., Mun, C. W., Baik, S. K., Shin, Y. I., et al. (2013). Topographical differences of brain iron deposition between progressive supranuclear palsy and parkinsonian variant multiple system atrophy. *Journal of Neurological Sciences, 325*(1–2), 29–35.

Haque, T. L., Miki, Y., Kanagaki, M., Takahashi, T., Yamamoto, A., Konishi, J., et al. (2003). MR contrast of ferritin and hemosiderin in the brain: Comparison among gradient-echo, conventional spin-echo and fast spin-echo sequences. *European Journal of Radiology, 48*(3), 230–236.

Harder, S. L., Hopp, K. M., Ward, H., Neglio, H., Gitlin, J., & Kido, D. (2008). Mineralization of the deep gray matter with age: A retrospective review with susceptibility-weighted MR imaging. *American Journal of Neuroradiology, 29*(1), 176–183.

Hasan, K. M., Walimuni, I. S., Kramer, L. A., & Narayana, P. A. (2012). Human brain iron mapping using atlas-based T2 relaxometry. *Magnetic Resonance in Medicine*, *67*(3), 731–739.

He, X., & Yablonskiy, D. A. (2009). Biophysical mechanisms of phase contrast in gradient echo MRI. *Proceedings of the National Academy of Sciences of the United States of America*, *106*(32), 13558–13563.

Herynek, V., Babis, M., Trunecka, P., Filip, K., Vymazal, J., Dezortova, M., et al. (2001). Chronic liver disease: Relaxometry in the brain after liver transplantation. *Magnetic Resonance Materials in Physics Biology and Medicine*, *12*(1), 10–15.

Herynek, V., Wagnerova, D., Hejlova, I., Dezortova, M., & Hajek, M. (2012). Changes in the brain during long-term follow-up after liver transplantation. *Journal of Magnetic Resonance Imaging*, *35*(6), 1332–1337.

Hikita, T., Abe, K., Sakoda, S., Tanaka, H., Murase, K., & Fujita, N. (2005). Determination of transverse relaxation rate for estimating iron deposits in central nervous system. *Neuroscience Research*, *51*(1), 67–71.

Hocq, A., Brouette, N., Saussez, S., Luhmer, M., Gillis, P., & Gossuin, Y. (2009). Variable-field relaxometry of iron-containing human tissues: A preliminary study. *Contrast Media & Molecular Imaging*, *4*(4), 157–164.

Hopp, K., Popescu, B. F., McCrea, R. P., Harder, S. L., Robinson, C. A., Haacke, M. E., et al. (2010). Brain iron detected by SWI high pass filtered phase calibrated with synchrotron X-ray fluorescence. *Journal of Magnetic Resonance Imaging*, *31*(6), 1346–1354.

House, M. J., St Pierre, T. G., Kowdley, K. V., Montine, T., Connor, J., Beard, J., et al. (2007). Correlation of proton transverse relaxation rates (R2) with iron concentrations in postmortem brain tissue from Alzheimer's disease patients. *Magnetic Resonance in Medicine*, *57*(1), 172–180.

Jendelova, P., Herynek, V., DeCroos, J., Glogarova, K., Andersson, B., Hajek, M., et al. (2003). Imaging the fate of implanted bone marrow stromal cells labeled with superparamagnetic nanoparticles. *Magnetic Resonance in Medicine*, *50*(4), 767–776.

Jensen, J. H., Chandra, R., Ramani, A., Lu, H., Johnson, G., Lee, S. P., et al. (2006). Magnetic field correlation imaging. *Magnetic Resonance in Medicine*, *55*(6), 1350–1361.

Jensen, J. H., Szulc, K., Hu, C., Ramani, A., Lu, H., Xuan, L., et al. (2009). Magnetic field correlation as a measure of iron-generated magnetic field inhomogeneities in the brain. *Magnetic Resonance in Medicine*, *61*(2), 481–485.

Jensen, J. H., Tang, H., Tosti, C. L., Swaminathan, S. V., Nunez, A., Hultman, K., et al. (2010). Separate MRI quantification of dispersed (ferritin-like) and aggregated (hemosiderin-like) storage iron. *Magnetic Resonance in Medicine*, *63*(5), 1201–1209.

Jin, L., Wang, J., Zhao, L., Jin, H., Fei, G., Zhang, Y., et al. (2011). Decreased serum ceruloplasmin levels characteristically aggravate nigral iron deposition in Parkinson's disease. *Brain*, *134*(Pt 1), 50–58.

Jirak, D., Kriz, J., Herynek, V., Andersson, B., Girman, P., Burian, M., et al. (2004). MRI of transplanted pancreatic islets. *Magnetic Resonance in Medicine*, *52*(6), 1228–1233.

Johansson, A. C., Appelqvist, H., Nilsson, C., Kagedal, K., Roberg, K., & Ollinger, K. (2010). Regulation of apoptosis-associated lysosomal membrane permeabilization. *Apoptosis: An International Journal on Programmed Cell Death*, *15*(5), 527–540.

Kamman, R. L., Go, K. G., Brouwer, W., & Berendsen, H. J. (1988). Nuclear magnetic resonance relaxation in experimental brain edema: Effects of water concentration, protein concentration, and temperature. *Magnetic Resonance in Medicine*, *6*(3), 265–274.

Khalil, M., Enzinger, C., Langkammer, C., Tscherner, M., Wallner-Blazek, M., Jehna, M., et al. (2009). Quantitative assessment of brain iron by $R(2)^*$ relaxometry in patients with clinically isolated syndrome and relapsing-remitting multiple sclerosis. *Multiple Sclerosis*, *15*(9), 1048–1054.

Khalil, M., Langkammer, C., Ropele, S., Petrovic, K., Wallner-Blazek, M., Loitfelder, M., et al. (2011). Determinants of brain iron in multiple sclerosis: A quantitative 3 T MRI study. *Neurology*, 77(18), 1691–1697.

Koenig, S. H., Brown, R. D., III., Gibson, J. F., Ward, R. J., & Peters, T. J. (1986). Relaxometry of ferritin solutions and the influence of the Fe3+ core ions. *Magnetic Resonance in Medicine*, *3*(5), 755–767.

Kostic, V. S., Svetel, M., Mijajlovic, M., Pavlovic, A., Jecmenica-Lukic, M., & Kozic, D. (2012). Transcranial sonography in pantothenate kinase-associated neurodegeneration. *Journal of Neurology*, *259*(5), 959–964.

Kruer, M. C., Boddaert, N., Schneider, S. A., Houlden, H., Bhatia, K. P., Gregory, A., et al. (2012). Neuroimaging features of neurodegeneration with brain iron accumulation. *American Journal of Neuroradiology*, *33*(3), 407–414.

Kwan, J. Y., Jeong, S. Y., Van Gelderen, P., Deng, H. X., Quezado, M. M., Danielian, L. E., et al. (2012). Iron accumulation in deep cortical layers accounts for MRI signal abnormalities in ALS: Correlating 7 tesla MRI and pathology. *PLoS One*, 7(4), e35241.

Langkammer, C., Krebs, N., Goessler, W., Scheurer, E., Ebner, F., Yen, K., et al. (2010). Quantitative MR imaging of brain iron: A postmortem validation study. *Radiology*, *257*(2), 455–462.

Langkammer, C., Krebs, N., Goessler, W., Scheurer, E., Yen, K., Fazekas, F., et al. (2012). Susceptibility induced gray-white matter MRI contrast in the human brain. *NeuroImage*, *59*(2), 1413–1419.

Langkammer, C., Liu, T., Khalil, M., Enzinger, C., Jehna, M., Fuchs, S., et al. (2013). Quantitative susceptibility mapping in multiple sclerosis. *Radiology*, *267*(2), 551–559.

Langkammer, C., Schweser, F., Krebs, N., Deistung, A., Goessler, W., Scheurer, E., et al. (2012). Quantitative susceptibility mapping (QSM) as a means to measure brain iron? A post mortem validation study. *NeuroImage*, *62*(3), 1593–1599.

Lee, J., Shmueli, K., Kang, B. T., Yao, B., Fukunaga, M., van Gelderen, P., et al. (2012). The contribution of myelin to magnetic susceptibility-weighted contrasts in high-field MRI of the brain. *NeuroImage*, *59*(4), 3967–3975.

Lee, J., van Gelderen, P., Kuo, L. W., Merkle, H., Silva, A. C., & Duyn, J. H. (2011). T2*-based fiber orientation mapping. *NeuroImage*, *57*(1), 225–234.

Levi, S., & Rovida, E. (2009). The role of iron in mitochondrial function. *Biochimica et Biophysica Acta*, *1790*(7), 629–636.

Li, X., Jankovic, J., & Le, W. (2011). Iron chelation and neuroprotection in neurodegenerative diseases. *Journal of Neural Transmission*, *118*(3), 473–477.

Li, W., Wu, B., Avram, A. V., & Liu, C. (2012). Magnetic susceptibility anisotropy of human brain in vivo and its molecular underpinnings. *NeuroImage*, *59*(3), 2088–2097.

Li, T. Q., Yao, B., van Gelderen, P., Merkle, H., Dodd, S., Talagala, L., et al. (2009). Characterization of $T(2)^*$ heterogeneity in human brain white matter. *Magnetic Resonance in Medicine*, *62*(6), 1652–1657.

Liem, M. K., Lesnik Oberstein, S. A., Versluis, M. J., Maat-Schieman, M. L., Haan, J., Webb, A. G., et al. (2012). 7 T MRI reveals diffuse iron deposition in putamen and caudate nucleus in CADASIL. *Journal of Neurology, Neurosurgery, and Psychiatry*, *83*(12), 1180–1185.

Liman, J., Wellmer, A., Rostasy, K., Bahr, M., & Kermer, P. (2012). Transcranial ultrasound in neurodegeneration with brain iron accumulation (NBIA). *European Journal of Paediatric Neurology*, *16*(2), 175–178.

Litwin, T., Gromadzka, G., Szpak, G. M., Jablonka-Salach, K., Bulska, E., & Czlonkowska, A. (2013). Brain metal accumulation in Wilson's disease. *Journal of Neurological Sciences*, *329*(1–2), 55–58.

Liu, C., Li, W., Johnson, G. A., & Wu, B. (2011). High-field (9.4 T) MRI of brain dysmyelination by quantitative mapping of magnetic susceptibility. *NeuroImage*, *56*(3), 930–938.

Liu, J., Liu, T., de Rochefort, L., Ledoux, J., Khalidov, I., Chen, W., et al. (2012). Morphology enabled dipole inversion for quantitative susceptibility mapping using structural consistency between the magnitude image and the susceptibility map. *NeuroImage*, *59*(3), 2560–2568.

Liu, T., Liu, J., de Rochefort, L., Spincemaille, P., Khalidov, I., Ledoux, J. R., et al. (2011). Morphology enabled dipole inversion (MEDI) from a single-angle acquisition: Comparison with COSMOS in human brain imaging. *Magnetic Resonance in Medicine*, *66*(3), 777–783.

Liu, T., Spincemaille, P., de Rochefort, L., Kressler, B., & Wang, Y. (2009). Calculation of susceptibility through multiple orientation sampling (COSMOS): A method for conditioning the inverse problem from measured magnetic field map to susceptibility source image in MRI. *Magnetic Resonance in Medicine*, *61*(1), 196–204.

Lotfipour, A. K., Wharton, S., Schwarz, S. T., Gontu, V., Schafer, A., Peters, A. M., et al. (2012). High resolution magnetic susceptibility mapping of the substantia nigra in Parkinson's disease. *Journal of Magnetic Resonance Imaging*, *35*(1), 48–55.

Luo, J., He, X., d'Avignon, D. A., Ackerman, J. J., & Yablonskiy, D. A. (2010). Protein-induced water 1H MR frequency shifts: Contributions from magnetic susceptibility and exchange effects. *Journal of Magnetic Resonance*, *202*(1), 102–108.

Ma, J., & Wehrli, F. W. (1996). Method for image-based measurement of the reversible and irreversible contribution to the transverse-relaxation rate. *Journal of Magnetic Resonance Series B*, *111*(1), 61–69.

Marques, J. P., Maddage, R., Mlynarik, V., & Gruetter, R. (2009). On the origin of the MR image phase contrast: An in vivo MR microscopy study of the rat brain at 14.1 T. *NeuroImage*, *46*(2), 345–352.

McNeill, A., Birchall, D., Hayflick, S. J., Gregory, A., Schenk, J. F., Zimmerman, E. A., et al. (2008). T2* and FSE MRI distinguishes four subtypes of neurodegeneration with brain iron accumulation. *Neurology*, *70*(18), 1614–1619.

Meiboom, S., & Gill, D. (1958). Modified spin-echo method for measuring nuclear relaxation times. *Review of Scientific Instruments*, *29*(8), 688–691.

Michaeli, S., Oz, G., Sorce, D. J., Garwood, M., Ugurbil, K., Majestic, S., et al. (2007). Assessment of brain iron and neuronal integrity in patients with Parkinson's disease using novel MRI contrasts. *Movement Disorders*, *22*(3), 334–340.

Milton, W. J., Atlas, S. W., Lexa, F. J., Mozley, P. D., & Gur, R. E. (1991). Deep gray matter hypointensity patterns with aging in healthy adults: MR imaging at 1.5 T. *Radiology*, *181*(3), 715–719.

Nestrasil, I., Michaeli, S., Liimatainen, T., Rydeen, C. E., Kotz, C. M., Nixon, J. P., et al. (2010). T1rho and T2rho MRI in the evaluation of Parkinson's disease. *Journal of Neurology*, *257*(6), 964–968.

Novellino, F., Cherubini, A., Chiriaco, C., Morelli, M., Salsone, M., Arabia, G., et al. (2013). Brain iron deposition in essential tremor: A quantitative 3-Tesla magnetic resonance imaging study. *Movement Disorders*, *28*(2), 196–200.

Ogg, R. J., Langston, J. W., Haacke, E. M., Steen, R. G., & Taylor, J. S. (1999). The correlation between phase shifts in gradient-echo MR images and regional brain iron concentration. *Magnetic Resonance Imaging*, *17*(8), 1141–1148.

Ogg, R. J., & Steen, R. G. (1998). Age-related changes in brain T1 are correlated with iron concentration. *Magnetic Resonance in Medicine*, *40*(5), 749–753.

Ordidge, R. J., Gorell, J. M., Deniau, J. C., Knight, R. A., & Helpern, J. A. (1994). Assessment of relative brain iron concentrations using T2-weighted and T2*-weighted MRI at 3 Tesla. *Magnetic Resonance in Medicine*, *32*(3), 335–341.

Peran, P., Cherubini, A., Assogna, F., Piras, F., Quattrocchi, C., Peppe, A., et al. (2010). Magnetic resonance imaging markers of Parkinson's disease nigrostriatal signature. *Brain*, *133*(11), 3423–3433.

Peran, P., Cherubini, A., Luccichenti, G., Hagberg, G., Demonet, J. F., Rascol, O., et al. (2009). Volume and iron content in basal ganglia and thalamus. *Human Brain Mapping, 30*(8), 2667–2675.

Pfefferbaum, A., Adalsteinsson, E., Rohlfing, T., & Sullivan, E. V. (2009). MRI estimates of brain iron concentration in normal aging: Comparison of field-dependent (FDRI) and phase (SWI) methods. *NeuroImage, 47*(2), 493–500.

Pott, F., Gingele, S., Clarner, T., Dang, J., Baumgartner, W., Beyer, C., et al. (2009). Cuprizone effect on myelination, astrogliosis and microglia attraction in the mouse basal ganglia. *Brain Research, 1305*, 137–149.

Pujol, J., Junque, C., Vendrell, P., Grau, J. M., Marti-Vilalta, J. L., Olive, C., et al. (1992). Biological significance of iron-related magnetic resonance imaging changes in the brain. *Archives of Neurology, 49*(7), 711–717.

Punwani, S., Cooper, C. E., Clemence, M., Penrice, J., Amess, P., Thornton, J., et al. (1997). Correlation between absolute deoxyhaemoglobin [dHb] measured by near infrared spectroscopy (NIRS) and absolute R2' as determined by magnetic resonance imaging (MRI). *Advances in Experimental Medicine and Biology, 413*, 129–137.

Rauscher, A., Sedlacik, J., Barth, M., Mentzel, H. J., & Reichenbach, J. R. (2005). Magnetic susceptibility-weighted MR phase imaging of the human brain. *American Journal of Neuroradiology, 26*(4), 736–742.

Reichenbach, J. R. (2012). The future of susceptibility contrast for assessment of anatomy and function. *NeuroImage, 62*(2), 1311–1315.

Reichenbach, J. R., Venkatesan, R., Schillinger, D. J., Kido, D. K., & Haacke, E. M. (1997). Small vessels in the human brain: MR venography with deoxyhemoglobin as an intrinsic contrast agent. *Radiology, 204*(1), 272–277.

Reichenbach, J. R., Venkatesan, R., Yablonskiy, D. A., Thompson, M. R., Lai, S., & Haacke, E. M. (1997). Theory and application of static field inhomogeneity effects in gradient-echo imaging. *Journal of Magnetic Resonance Imaging*, 7(2), 266–279.

Rhodes, S. L., & Ritz, B. (2008). Genetics of iron regulation and the possible role of iron in Parkinson's disease. *Neurobiology of Disease, 32*(2), 183–195.

Riederer, P., Sofic, E., Rausch, W. D., Schmidt, B., Reynolds, G. P., Jellinger, K., et al. (1989). Transition metals, ferritin, glutathione, and ascorbic acid in parkinsonian brains. *Journal of Neurochemistry, 52*(2), 515–520.

Rohrer, M., Bauer, H., Mintorovitch, J., Requardt, M., & Weinmann, H. J. (2005). Comparison of magnetic properties of MRI contrast media solutions at different magnetic field strengths. *Investigative Radiology, 40*(11), 715–724.

Sastry, S., & Arendash, G. W. (1995). Time-dependent changes in iron levels and associated neuronal loss within the substantia nigra following lesions within the neostriatum/globus pallidus complex. *Neuroscience, 67*(3), 649–666.

Sati, P., Silva, A. C., van Gelderen, P., Gaitan, M. I., Wohler, J. E., Jacobson, S., et al. (2012). In vivo quantification of T(2) anisotropy in white matter fibers in marmoset monkeys. *NeuroImage, 59*(2), 979–985.

Saudek, F., Jirak, D., Girman, P., Herynek, V., Dezortova, M., Kriz, J., et al. (2010). Magnetic resonance imaging of pancreatic islets transplanted into the liver in humans. *Transplantation, 90*(12), 1602–1606.

Schenck, J. F. (1992). Health and physiological effects of human exposure to whole-body four-tesla magnetic fields during MRI. *Annals of the New York Academy of Sciences, 649*, 285–301.

Schenck, J. F. (1996). The role of magnetic susceptibility in magnetic resonance imaging: MRI magnetic compatibility of the first and second kinds. *Medical Physics, 23*(6), 815–850.

Schenck, J. F. (2003). Magnetic resonance imaging of brain iron. *Journal of the Neurological Sciences, 207*(1–2), 99–102.

Schenck, J. F., & Zimmerman, E. A. (2004). High-field magnetic resonance imaging of brain iron: Birth of a biomarker? *NMR in Biomedicine, 17*(7), 433–445.

Schenker, C., Meier, D., Wichmann, W., Boesiger, P., & Valavanis, A. (1993). Age distribution and iron dependency of the T2 relaxation time in the globus pallidus and putamen. *Neuroradiology*, *35*(2), 119–124.

Schipper, H. M. (2012). Neurodegeneration with brain iron accumulation—Clinical syndromes and neuroimaging. *Biochimica et Biophysica Acta*, *1822*(3), 350–360.

Schneider, S. A., Dusek, P., Hardy, J., Westenberger, A., Jankovic, J., & Bhatia, K. P. (2013). Genetics and Pathophysiology of Neurodegeneration with Brain Iron Accumulation (NBIA). *Current Neuropharmacology*, *11*(1), 59–79.

Schrag, M., McAuley, G., Pomakian, J., Jiffry, A., Tung, S., Mueller, C., et al. (2010). Correlation of hypointensities in susceptibility-weighted images to tissue histology in dementia patients with cerebral amyloid angiopathy: A postmortem MRI study. *Acta Neuropathologica*, *119*(3), 291–302.

Schroder, N., Figueiredo, L. S., & de Lima, M. N. (2013). Role of brain iron accumulation in cognitive dysfunction: Evidence from animal models and human studies. *Journal of Alzheimer's Disease*, *34*(4), 797–812.

Schweser, F., Deistung, A., Lehr, B. W., & Reichenbach, J. R. (2010). Differentiation between diamagnetic and paramagnetic cerebral lesions based on magnetic susceptibility mapping. *Medical Physics*, *37*(10), 5165–5178.

Schweser, F., Deistung, A., Lehr, B. W., & Reichenbach, J. R. (2011). Quantitative imaging of intrinsic magnetic tissue properties using MRI signal phase: An approach to in vivo brain iron metabolism? *NeuroImage*, *54*(4), 2789–2807.

Schweser, F., Deistung, A., Sommer, K., & Reichenbach, J. R. (2013). Toward online reconstruction of quantitative susceptibility maps: Superfast dipole inversion. *Magnetic Resonance in Medicine*, *69*(6), 1581–1593.

Shmueli, K., de Zwart, J. A., van Gelderen, P., Li, T. Q., Dodd, S. J., & Duyn, J. H. (2009). Magnetic susceptibility mapping of brain tissue in vivo using MRI phase data. *Magnetic Resonance in Medicine*, *62*(6), 1510–1522.

Shmueli, K., Dodd, S. J., Li, T. Q., & Duyn, J. H. (2011). The contribution of chemical exchange to MRI frequency shifts in brain tissue. *Magnetic Resonance in Medicine*, *65*(1), 35–43.

Sian-Hulsmann, J., Mandel, S., Youdim, M. B., & Riederer, P. (2011). The relevance of iron in the pathogenesis of Parkinson's disease. *Journal of Neurochemistry*, *118*(6), 939–957.

Stankiewicz, J., Panter, S. S., Neema, M., Arora, A., Batt, C. E., & Bakshi, R. (2007). Iron in chronic brain disorders: Imaging and neurotherapeutic implications. *Neurotherapeutics*, *4*(3), 371–386.

Stark, D. D., Bass, N. M., Moss, A. A., Bacon, B. R., McKerrow, J. H., Cann, C. E., et al. (1983). Nuclear magnetic resonance imaging of experimentally induced liver disease. *Radiology*, *148*(3), 743–751.

Sulzer, D., Mosharov, E., Talloczy, Z., Zucca, F. A., Simon, J. D., & Zecca, L. (2008). Neuronal pigmented autophagic vacuoles: Lipofuscin, neuromelanin, and ceroid as macroautophagic responses during aging and disease. *Journal of Neurochemistry*, *106*(1), 24–36.

Supprian, T., Hofmann, E., Warmuth-Metz, M., Franzek, E., & Becker, T. (1997). MRI T2 relaxation times of brain regions in schizophrenic patients and control subjects. *Psychiatry Research*, *75*(3), 173–182.

Tanaka, A., Ueno, Y., Nakayama, Y., Takano, K., & Takebayashi, S. (1999). Small chronic hemorrhages and ischemic lesions in association with spontaneous intracerebral hematomas. *Stroke*, *30*(8), 1637–1642.

Tatsumi, S., Shinohara, M., & Yamamoto, T. (2008). Direct comparison of histology of microbleeds with postmortem MR images: A case report. *Cerebrovascular Diseases*, *26*(2), 142–146.

Taylor, E. M., & Morgan, E. H. (1991). Role of transferrin in iron uptake by the brain: A comparative study. *Journal of Comparative Physiology B, 161*(5), 521–524.

Tennison, M. B., Bouldin, T. W., & Whaley, R. A. (1988). Mineralization of the basal ganglia detected by CT in Hallervorden-Spatz syndrome. *Neurology, 38*(1), 154–155.

Terman, A., & Kurz, T. (2013). Lysosomal iron, iron chelation, and cell death. *Antioxidants & Redox Signaling, 18*(8), 888–898.

Todorich, B., Pasquini, J. M., Garcia, C. I., Paez, P. M., & Connor, J. R. (2009). Oligodendrocytes and myelination: The role of iron. *Glia, 57*(5), 467–478.

Tribl, F., Marcus, K., Meyer, H. E., Bringmann, G., Gerlach, M., & Riederer, P. (2006). Subcellular proteomics reveals neuromelanin granules to be a lysosome-related organelle. *Journal of Neural Transmission, 113*(6), 741–749.

Van Kirk, M. P., Larsen, P. D., & O'Connor, P. S. (1986). New computed tomography scan finding in Hallervorden-Spatz syndrome. *Journal of Clinical Neuro-Ophthalmology, 6*(2), 86–90.

Vymazal, J., Babis, M., Brooks, R. A., Filip, K., Dezortova, M., Hrncarkova, H., et al. (1996). T1 and T2 alterations in the brains of patients with hepatic cirrhosis. *American Journal of Neuroradiology, 17*(2), 333–336.

Vymazal, J., Brooks, R. A., Baumgarner, C., Tran, V., Katz, D., Bulte, J. W., et al. (1996). The relation between brain iron and NMR relaxation times: An in vitro study. *Magnetic Resonance in Medicine, 35*(1), 56–61.

Vymazal, J., Brooks, R. A., Zak, O., McRill, C., Shen, C., & Di Chiro, G. (1992). T1 and T2 of ferritin at different field strengths: Effect on MRI. *Magnetic Resonance in Medicine, 27*(2), 368–374.

Vymazal, J., Bulte, J. W., Frank, J. A., Di Chiro, G., & Brooks, R. A. (1993). Frequency dependence of MR relaxation times I. Paramagnetic ions. *Journal of Magnetic Resonance Imaging, 3*(4), 637–640.

Vymazal, J., Hajek, M., Patronas, N., Giedd, J. N., Bulte, J. W., Baumgarner, C., et al. (1995). The quantitative relation between T1-weighted and T2-weighted MRI of normal gray matter and iron concentration. *Journal of Magnetic Resonance Imaging, 5*(5), 554–560.

Vymazal, J., Klempir, J., Jech, R., Zidovska, J., Syka, M., Ruzicka, E., et al. (2007). MR relaxometry in Huntington's disease: Correlation between imaging, genetic and clinical parameters. *Journal of Neurological Sciences, 263*(1–2), 20–25.

Vymazal, J., Righini, A., Brooks, R. A., Canesi, M., Mariani, C., Leonardi, M., et al. (1999). T1 and T2 in the brain of healthy subjects, patients with Parkinson disease, and patients with multiple system atrophy: Relation to iron content. *Radiology, 211*(2), 489–495.

Vymazal, J., Zak, O., Bulte, J. W., Aisen, P., & Brooks, R. A. (1996). T1 and T2 of ferritin solutions: Effect of loading factor. *Magnetic Resonance in Medicine, 36*(1), 61–65.

Wallis, L. I., Paley, M. N., Graham, J. M., Grunewald, R. A., Wignall, E. L., Joy, H. M., et al. (2008). MRI assessment of basal ganglia iron deposition in Parkinson's disease. *Journal of Magnetic Resonance Imaging, 28*(5), 1061–1067.

Walsh, A. J., Lebel, R. M., Eissa, A., Blevins, G., Catz, I., Lu, J. Q., et al. (2013a). Multiple sclerosis: Validation of MR imaging for quantification and detection of iron. *Radiology, 267*(2), 531–542.

Wang, Y., Butros, S. R., Shuai, X., Dai, Y., Chen, C., Liu, M., et al. (2012). Different iron-deposition patterns of multiple system atrophy with predominant parkinsonism and idiopathetic Parkinson diseases demonstrated by phase-corrected susceptibility-weighted imaging. *American Journal of Neuroradiology, 33*(2), 266–273.

Wang, D., Li, W. B., Wei, X. E., Li, Y. H., & Dai, Y. M. (2012). An investigation of age-related iron deposition using susceptibility weighted imaging. *PLoS One*, 7(11), e50706.

Wang, Y., Yu, Y., Li, D., Bae, K. T., Brown, J. J., Lin, W., et al. (2000). Artery and vein separation using susceptibility-dependent phase in contrast-enhanced MRA. *Journal of Magnetic Resonance Imaging*, *12*(5), 661–670.

Wang, D., Zhu, D., Wei, X. E., Li, Y. H., & Li, W. B. (2013). Using susceptibility-weighted images to quantify iron deposition differences in amnestic mild cognitive impairment and Alzheimer's disease. *Neurology India*, *61*(1), 26–34.

Wharton, S., & Bowtell, R. (2010). Whole-brain susceptibility mapping at high field: A comparison of multiple- and single-orientation methods. *NeuroImage*, *53*(2), 515–525.

Wharton, S., Schafer, A., & Bowtell, R. (2010). Susceptibility mapping in the human brain using threshold-based k-space division. *Magnetic Resonance in Medicine*, *63*(5), 1292–1304.

Wheaton, A. J., Borthakur, A., Corbo, M. T., Moonis, G., Melhem, E., & Reddy, R. (2004). T2rho-weighted contrast in MR images of the human brain. *Magnetic Resonance in Medicine*, *52*(6), 1223–1227.

Whittall, K. P., MacKay, A. L., Graeb, D. A., Nugent, R. A., Li, D. K., & Paty, D. W. (1997). In vivo measurement of T2 distributions and water contents in normal human brain. *Magnetic Resonance in Medicine*, *37*(1), 34–43.

Wisnieff, C., Liu, T., Spincemaille, P., Wang, S., Zhou, D., & Wang, Y. (2013). Magnetic susceptibility anisotropy: Cylindrical symmetry from macroscopically ordered anisotropic molecules and accuracy of MRI measurements using few orientations. *NeuroImage*, *70*, 363–376.

Wood, J. C., Fassler, J. D., & Meade, T. (2004). Mimicking liver iron overload using liposomal ferritin preparations. *Magnetic Resonance in Medicine*, *51*(3), 607–611.

Xu, X., Wang, Q., & Zhang, M. (2008). Age, gender, and hemispheric differences in iron deposition in the human brain: An in vivo MRI study. *NeuroImage*, *40*(1), 35–42.

Yamada, N., Imakita, S., Sakuma, T., & Takamiya, M. (1996). Intracranial calcification on gradient-echo phase image: Depiction of diamagnetic susceptibility. *Radiology*, *198*(1), 171–178.

Yan, S., Sun, J., Chen, Y., Selim, M., & Lou, M. (2013). Brain iron deposition in white matter hyperintensities: A 3-T MRI study. *Age (Dordr)*, *35*(5), 1927–1936.

Yan, S. Q., Sun, J. Z., Yan, Y. Q., Wang, H., & Lou, M. (2012). Evaluation of brain iron content based on magnetic resonance imaging (MRI): Comparison among phase value, $R2^*$ and magnitude signal intensity. *PLoS One*, 7(2), e31748.

Yao, B., Bagnato, F., Matsuura, E., Merkle, H., van Gelderen, P., Cantor, F. K., et al. (2012). Chronic multiple sclerosis lesions: Characterization with high-field-strength MR imaging. *Radiology*, *262*(1), 206–215.

Yao, B., Li, T. Q., Gelderen, P., Shmueli, K., de Zwart, J. A., & Duyn, J. H. (2009). Susceptibility contrast in high field MRI of human brain as a function of tissue iron content. *NeuroImage*, *44*(4), 1259–1266.

Young, I. R., Cox, I. J., Bryant, D. J., & Bydder, G. M. (1988). The benefits of increasing spatial resolution as a means of reducing artifacts due to field inhomogeneities. *Magnetic Resonance Imaging*, *6*(5), 585–590.

Zecca, L., Berg, D., Arzberger, T., Ruprecht, P., Rausch, W. D., Musicco, M., et al. (2005). In vivo detection of iron and neuromelanin by transcranial sonography: A new approach for early detection of substantia nigra damage. *Movement Disorders*, *20*(10), 1278–1285.

Zecca, L., Gallorini, M., Schunemann, V., Trautwein, A. X., Gerlach, M., Riederer, P., et al. (2001). Iron, neuromelanin and ferritin content in the substantia nigra of normal subjects at different ages: Consequences for iron storage and neurodegenerative processes. *Journal of Neurochemistry*, *76*(6), 1766–1773.

Zhang, N., Fitsanakis, V. A., Erikson, K. M., Aschner, M., Avison, M. J., & Gore, J. C. (2009). A model for the analysis of competitive relaxation effects of manganese and iron in vivo. *NMR in Biomedicine*, *22*(4), 391–404.

Zheng, W., Nichol, H., Liu, S., Cheng, Y. C., & Haacke, E. M. (2013). Measuring iron in the brain using quantitative susceptibility mapping and X-ray fluorescence imaging. *NeuroImage*, *78*, 68–74.

Zhong, K., Ernst, T., Buchthal, S., Speck, O., Anderson, L., & Chang, L. (2011). Phase contrast imaging in neonates. *NeuroImage*, *55*(3), 1068–1072.

Zhong, K., Leupold, J., von Elverfeldt, D., & Speck, O. (2008). The molecular basis for gray and white matter contrast in phase imaging. *NeuroImage*, *40*(4), 1561–1566.

Zorzi, G., Zibordi, F., Chiapparini, L., Bertini, E., Russo, L., Piga, A., et al. (2011). Iron-related MRI images in patients with pantothenate kinase-associated neurodegeneration (PKAN) treated with deferiprone: Results of a phase II pilot trial. *Movement Disorders*, *26*(9), 1756–1759.

CHAPTER TEN

The Role of Iron Imaging in Huntington's Disease

Simon J.A. van den Bogaard[1], Eve M. Dumas, Raymund A.C. Roos
Department of Neurology, Leiden University Medical Centre, Leiden, The Netherlands
[1]Corresponding author: e-mail address: s.j.a.van_den_bogaard@lumc.nl

Contents

Abstract

Huntington's disease (HD) is a devastating neurological disorder that affects the brain. The cause of HD is an expanded CAG trinucleotide repeat in the *Htt* gene. The *Htt* gene is responsible for the protein huntingtin, the exact functions of which have yet to be elucidated. The role of iron in the pathological cascade is usually mentioned in the context of an inability to regulate iron homeostasis, which generates a surplus of reactive iron and free radical toxicity, resulting in increased oxidative stress. In this review, we discuss the role of iron within the existing hypotheses of HD pathology, *ex vivo* findings in support of increased iron in HD, and finally *in vivo* MRI findings in manifest and premanifest HD. Both *in vivo* and *ex vivo* findings support the notion that excess iron is present in the brain of HD patients. There does not seem to be much evidence that iron accumulation is the initiator of the pathological cascade as little or no evidence can be found for very early increased iron in the HD brain, when neuronal cell loss is already extensive.

1. HUNTINGTON'S DISEASE

Huntington's disease (HD) is a devastating neurological disorder that affects the brain. The classical symptom triad consists of a movement disorder, cognitive decline, and psychiatric disturbances. The movement disorder is

International Review of Neurobiology, Volume 110
ISSN 0074-7742
http://dx.doi.org/10.1016/B978-0-12-410502-7.00011-9

characterized by chorea, dystonia, rigidity, hypokinesia, problems of balance, and oculomotor disturbances. Characteristics of cognitive decline are executive dysfunction, psychomotor slowing, and, in later stages, memory loss and global cognitive deterioration. Psychiatric symptoms include depression, irritability, apathy, aggression, and a high rate of suicide. In addition to these classical symptoms, weight loss, autonomic dysfunction, and sleep rhythm disturbance are also present (Aziz, Anguelova, Marinus, van Dijk, & Roos, 2010; Aziz et al., 2009; Novak & Tabrizi, 2010). Disease onset is usually in midlife, around 40 years of age, although it can range widely from age 2 to 80 years. The definition of disease onset is based on clinical measures, mostly motor assessments, and the genetic confirmation of an expanded CAG repeat. HD results in death within 15–20 years of onset (Novak & Tabrizi, 2010).

The cause of HD is an expanded Cytosine-Adenine-Guanine (CAG) trinucleotide repeat in the *Htt* gene on the short arm of chromosome 4 (The Huntington's Disease Collaborative Research Group, 1993). The *Htt* gene is responsible for the protein huntingtin, the exact functions of which have yet to be elucidated. The repeat size accounts for approximately 60–70% of the variability in disease onset. The length of the CAG expansion is inversely correlated with disease onset and rate of progression. CAG-repeat lengths of less than 28 are considered to be normal; CAG repeats of 28–35 are not related to disease expression, but are, however, considered to be unstable and to have the potential to expand further in offspring. If the number of repeats falls within the range 36–39, there is reduced penetrance, in some cases resulting in a milder disease course. Repeats of 40 and above are fully penetrant. When the age at onset is under 20 years, the so-called juvenile HD or Westphal variant, the number of CAG repeats exceeds 60 (Novak & Tabrizi, 2010; Sturrock & Leavitt, 2010). The result of the expansion is evident in neuronal dysfunction and neuronal loss in the brain.

Postmortem evidence indicates the pathological hallmark of the disease, namely, atrophy of the striatum and the external segment of the pallidum. As the disease progresses, the white matter, thalamus, cerebral cortex, and other subcortical gray matter structures also exhibit marked atrophy, resulting in whole brain volume loss. In particular, the medium spiny GABAergic projection neurons seem to be vulnerable to neurodegeneration in HD (Vonsattel, Keller, & Cortes Ramirez, 2011). *In vivo* magnetic resonance imaging (MRI) studies have revealed that the atrophy of the striatum and white matter occurs more than a decade prior to disease onset, the so-called premanifest or prodromal stage of the disease (Paulsen et al., 2010; Tabrizi et al., 2009).

2. THE PATHOLOGICAL CASCADE IN HD

There are many hypotheses that attempt to describe the pathological cascade in HD. First of all, direct toxic effects of the mutant huntingtin protein encoded by the *Htt* gene (mhtt). mhtt aggregates cannot be cleared by the ubiquitin–proteasome system. Second, transcription dysregulation of several genes, including those involved in neurotransmitter receptors, neuron structure, and stress responses, is considered. Third, mitochondrial dysfunction and energy metabolism pathways are described as a possible contributor within the pathological cascade. Fourth, excitotoxicity is described as neuronal overstimulation by glutamate or an increased sensitivity to glutamate. Finally, changes in axonal transport, intraneuronal vesicular transport, and synaptic dysfunction are thought to be involved (Estrada Sanchez, Mejia-Toiber, & Massieu, 2008; Ross & Tabrizi, 2011; Roze, Saudou, & Caboche, 2008).

It is, however, evident that many of these hypotheses are closely interlinked as the following examples clearly demonstrate dysregulation of the NMDA receptor by transcriptional dysregulation could lead to increased sensitivity of this particular receptor type, consistent with the excitotoxicity theory. The possible direct interaction of mhtt and mitochondrial membranes could lead to disrupted calcium homeostasis upon exposure to metabolic stress, leaving neurons more susceptible to damage, thus resulting in excitotoxicity. Furthermore, the altered trafficking of mitochondria potentially leads to impaired energy metabolism. Repressed transcription of transcriptional coactivators that regulate biogenesis of mitochondria and oxidative phosphorylation is also described (Roze et al., 2008; Sturrock & Leavitt, 2010).

To date, the exact sequence of events that lead to neuronal death is not fully understood. The role of iron is usually mentioned in the context of an inability to regulate iron homeostasis, which generates a surplus of reactive iron and free-radical toxicity, resulting in increased oxidative stress (Bartzokis et al., 2007; Zecca, Youdim, Riederer, Connor, & Crichton, 2004). In HD, increased levels of several markers of oxidative damage, such as DNA strand breaks, lipofuscin, and other markers of oxidative stress, have been reported (Browne & Beal, 2006). Oxidative damage does affect multiple cell systems; for instance, bioenergetic dysfunction appears to be linked to oxidative damage and increased free-radical generation. When the level of free radicals exceeds the capability of free-radical scavenging, this leads to a negative cycle of damage and increased sensitivity to damage, for example, excitotoxicity of neurons (Browne & Beal, 2006).

As we do not fully understand the order of events, Bartzokis et al. (2007) described a plausible relation between iron levels and the damage seen in the brain (Bartzokis et al., 2007). Iron levels are especially high in oligodendrocytes, which are important for myelination and remyelination. Thus when myelin defects are present, due to any of the previously stated pathophysiological processes, higher levels of oligodendrocytes and hence iron are present. Higher levels of oligodendrocytes in striatal regions in HD were already apparent from postmortem studies (Vonsattel et al., 2011). The raised level of iron is then implicated in the increased free-radical theory, the formation of radicals promoting further damage to neurons (Bartzokis et al., 2007).

3. *EX VIVO* FINDINGS OF IRON IN HD

Support for the implication of iron specifically in HD can be derived from a study of organelles. When assessing embryonic stem cells, several organelles were found to be huntingtin dependent. The specific role of one of these was proper regulation of iron homeostasis (Hilditch-Maguire et al., 2000). A study using zebra fish found that knockdown of *Htt* resulted in hypochromic blood due to a decrease in hemoglobin production and also an increase in transferrin receptor 1 transcript, suggestive of cellular iron starvation (Lumsden, Henshall, Dayan, Lardelli, & Richards, 2007). Making use of HD cell models, it was demonstrated that oxidative events are centered around the huntingtin inclusion bodies and that they are also CAG-repeat size dependent (Firdaus et al., 2006). The well-known animal model, R2/6 HD mice, has also been used to investigate iron accumulation in HD. R2/6 mice had increased staining for ferritin, an iron storage protein. This was specific for the striatum and was found to start 2–4 weeks postnatally. Other brain regions, such as the cortex and hippocampus, displayed increased ferritin from the age of 5–7 weeks. The ferritin was found primarily in the microglia. This corresponds with the presence of ferritin levels in human HD brain at various stages according to the Vonsattel grading system. One of the main conclusions of this chapter is that iron storage abnormalities occur early enough to be an important contributor to HD progression (Simmons et al., 2007). A study of the iron content, total iron-binding capacity, and ferritin in peripheral blood samples showed a decreased level of ferritin. The pathological implications of this finding are unknown (Morrison & Nevin, 1994).

Dexter et al. (1991) examined postmortem brain tissues in a variety of neurodegenerative disorders, including HD, and reported a 56% increase in iron in the caudate nucleus and 44% in the putamen. The latter was statistically not significant, probably due to the small sample size of 10 HD subjects (Dexter et al., 1991). A postmortem study by Chen et al. (1993) investigated iron and ferritin in pallidum and putamen in Parkinson's disease and HD. They described a threefold iron increase and a sixfold ferritin increase in HD, but could not find a relation to MRI T2 shortening in these areas (Chen et al., 1993). Later studies using newer MRI techniques have, however, found significant differences in iron content in HD compared to controls (see later section).

4. *IN VIVO* ASSESSMENT OF IRON ACCUMULATION IN HD USING MRI

MRI examination of iron can be performed using many different techniques including T2 relaxation, T2*, field-dependent relaxation rate increase (FDRI), and asymmetric spin echo.

Bartzokis et al. (1999) examined iron levels in 11 manifest HD participants and compared them to those in 27 age-matched controls. The technique of FDRI was applied, which makes use of two measurements of relaxation parameters at different field strengths. They demonstrated higher levels of iron in the putamen, caudate nucleus, and the pallidum. This was not a generalized phenomenon as there was no increase in iron in the white matter. The observed increase was present early in the disease course. An extension of this study by Bartzokis et al. (2007) showed that changes did occur not only in the basal ganglia but also in other brain areas. Frontal white matter iron levels were decreased, as were those in the genu of the corpus callosum. Iron levels were unchanged in the splenium of the corpus callosum, hippocampus, and thalamus. They concluded that myelin breakdown and changes in ferritin iron distribution occur in early manifest HD.

Vymazal et al. (2007), using T2 relaxometry as the outcome parameter, reported T2 shortening in the pallidum of HD patients compared to controls (Vymazal et al., 2007). The regions of interest were similar to those of Bartzokis et al. (1999). In this larger population of 34 HD participants, they did not find any abnormal iron levels in the caudate nucleus or putamen. A correlation was found between CAG-repeat length and iron in the left pallidum. Furthermore, there was an inverse correlation between CAG-repeat length and iron level in the left caudate. At first glance, the latter correlation was surprising as an increase in T2 and higher CAG repeats would be the opposite of what one might expect. The authors explain this by commenting on the form of iron in the different structures. It may be that the ferritin-bound iron is higher in the pallidum, which is mostly measured by T2 relaxometry. The level of low molecular iron might be higher in the caudate, not registered by the technique used in this study. The latter form is thought to be more "toxic" to the brain as it enables a greater degree of oxidative stress.

Premanifest gene carriers were examined by Jurgens et al. (2010), again using T2 relaxometry. They also demonstrated a higher amount of iron in the pallidum, suggesting that this iron accumulation starts very early, even before disease onset. Again, no findings in the other basal ganglia structures were found. The CAG-repeat length correlation was also positive in this study. Their population consisted of 17 premanifest gene carriers defined by the absence of "definite" motor signs, which corresponds to diagnostic confidence level 4 in the Unified Huntington's disease Rating Scale (UHDRS) motor examination. The mean UHDRS motor score was 3.5 ranging up to 10 points on this scale. This inclusion criterion was more liberal than that used nowadays in large-scale observational studies such as TRACK-HD, which has a cutoff point around 5 (Tabrizi et al., 2009). This could also explain the discrepancy with the results of the TRACK-HD group (Dumas et al., 2012), who did not find any significant differences in iron content in premanifest gene carriers compared to controls (Dumas et al., 2012). Dumas et al. (2012) did find increased iron levels in the caudate and putamen in manifest HD, but not in the pallidum (Fig. 10.1). These results are in contrast to those of Vymazal et al. (2007) and Jurgens et al. (2010), but in line with those presented by Bartzokis et al. (1999, 2007). The technique used by Dumas et al. (2012) is asymmetric spin echo, resulting in magnetic field correlation maps. It is hypothesized that this technique is more sensitive to iron accumulation and less influenced by other factors such as water content (Dumas et al., 2012). Furthermore, Dumas et al. (2012) reported that the iron content seems to be independent of atrophy of the striatal structures.

Whether or not iron is found to be elevated in the separate basal ganglia structures would seem to be at least partly dependent on the technique used. There does, however, seem to be accordance between all studies that iron is in fact elevated in basal ganglia structures. Possibly, the form in which iron is present in the different structures is indeed responsible for the differences between the four previously mentioned studies, as suggested by Vymazal

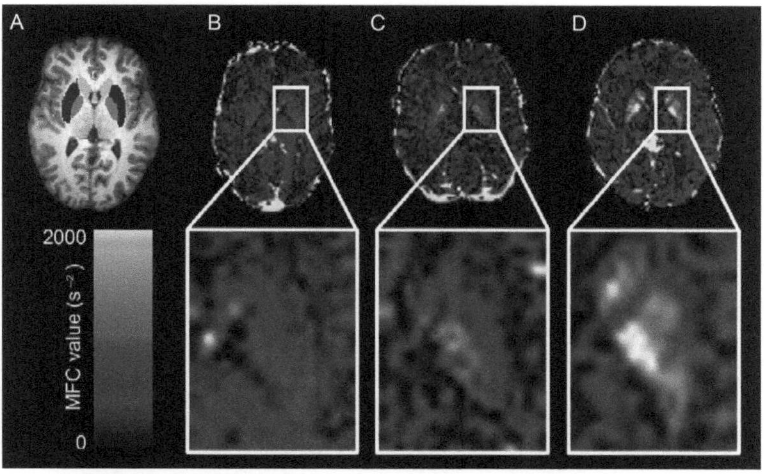

Figure 10.1 (A) Magnitude image showing segmentation of caudate nucleus (yellow), putamen (brown), and globus pallidus (orange). The corresponding MFC maps for a healthy control (B), premanifest gene carrier (C), and early Huntington's disease patient (D). High MFC values are found in the subcortical gray matter structures, known to correspond with high iron concentrations. Highest values are found in the early HD patient. The high MFC values found laterally near the tissue–skull interface are caused by macroscopic magnetic field inhomogeneities from the skull–tissue interface that could not be corrected properly. The distance to the areas of interest is so large that no interference with measurements in the deep gray-matter structures is expected. *Adapted from Dumas et al. (2012).* (For interpretation of the references to color in this figure legend, the reader is referred to the online version of this chapter.)

et al. (2007). Whether one considers iron to be elevated in premanifest HD depends on the interpretation of the correct cutoff point for clinical diagnosis. We would be inclined to state that iron is not significantly elevated in the premanifest stage, but is elevated in the early manifest stage or possibly during the transition from premanifest to manifest stage of the disease. Table 10.1 lists all the available *in vivo* MRI iron imaging studies.

5. CONCLUSION

Both *in vivo* and *ex vivo* findings support the notion that excess iron is present in the brain of HD patients. The pathological cascade of events in Huntington's disease is complex and not fully understood. Iron does, however, seem to have a place among the many pathways involved. Increased free-radical toxicity and impaired oxidative stress responses are likely to have

Table 10.1 Summary of the MRI studies available on Huntington's disease and iron quantification

	Disease stage examined	Sample size	MRI technique	Field strength	Findings
Bartzokis, Cummings, Perlman, et al. (1999); Bartzokis, Cummings, Markham, et al. (1999)	Manifest	11 HD 27 controls	FDRI	1.5 and 0.5 T	Elevated iron in caudate, putamen, and pallidum
Vymazal et al. (2007)	Manifest	34 HD 34 controls	T2 Relaxometry	1.5 T	T2 shortening in pallidum Correlation between CAG and pallidum T2 Inverse correlation between CAG and caudate T2
Bartzokis et al. (2007)	Manifest	11 HD 27 controls	FDRI	1.5 and 0.5 T	Increased ferritin iron levels caudate, putamen, pallidum Decreased ferritin in frontal white matter and genu corpus callosum Unchanged in hippocampus, thalamus, and splenium corpus callosum
Jurgens et al. (2010)	Premanifest	17 PMGC 15 controls	T2 hypointensity quantitation	3 T	Increased hypointensities in pallidum reflecting increased iron Correlation between more hypointensities and higher UHDRS motor score, lower SDMT score, higher CAG, greater probability of onset within 5 years
Dumas, Versluis, et al. (2012); Dumas, Jankovic, et al. (2012)	Premanifest and manifest	22 PMGC 27 HD 25 controls	Asymmetric spin echo MFC	3 T	Increased MFC in caudate and putamen in manifest HD, no findings in PMGC Atrophy is independent from iron levels

HD, Huntington's disease; FDRI, field-dependent relaxation rate increase; T, tesla, MFC, magnetic field correlation; PGMC, premanifest gene carrier; UHDRS, unified Huntington's disease rating scale; CAG, CAG repeat length.

an additional or even accelerating role within the pathological cascade. There does not seem to be much evidence that iron accumulation is the initiator of the cascade as little or no evidence can be found for very early increased iron in the HD brain, when neuronal cell loss is already extensive.

6. FUTURE PERSPECTIVES

It is debatable whether iron is a possible therapeutic target in HD. If iron is indeed involved in pathways in the pathological cascade, it might be worthwhile trying to reduce the amount and consequently the degree of oxidative stress, particularly when this process of neuronal damage becomes a vicious circle: the more damage, the more repair attempts by oligodendrocytes resulting in more oxidative stress, leading to more neuronal damage. At this time, however, this last assumption is purely speculative. Trying to stop a process that might only be a downstream process or a phenomenon secondary to the primary pathological events, would not be a particularly viable strategy. Reducing the amount of mhtt is considered to be a more upstream process. As suggested by Ross and Tabrizi (2011), however, we should keep an open mind about all possible therapeutic interventions (Ross & Tabrizi, 2011). Their very informative review of potential therapeutic strategies refers to the use of kynurenine 3-monooxygenase (KMO) inhibitors. KMO is a key microglial enzyme implicated in reactive oxygen species and the excitotoxicity of neurons. Mouse models show promising results with prolonged survival. Prolonged survival in this devastating disease that starts in midlife would constitute a considerable step forward.

REFERENCES

A novel gene containing a trinucleotide repeat that is expanded and unstable on Huntington's disease chromosomes (1993). The Huntington's Disease Collaborative Research Group. Cell, 72(6), 971–983.

Aziz, N. A., Anguelova, G. V., Marinus, J., van Dijk, J. G., & Roos, R. A. (2010). Autonomic symptoms in patients and pre-manifest mutation carriers of Huntington's disease. *European Journal of Neurology*, *17*(8), 1068–1074.

Aziz, N. A., Pijl, H., Frolich, M., van der Graaf, A. W., Roelfsema, F., & Roos, R. A. (2009). Increased hypothalamic-pituitary-adrenal axis activity in Huntington's disease. *Journal of Clinical Endocrinology and Metabolism*, *94*, 1223–1228.

Bartzokis, G., Cummings, J., Perlman, S., Hance, D. B., & Mintz, J. (1999). Increased basal ganglia iron levels in Huntington disease. *Archives of Neurology*, *56*(5), 569–574.

Bartzokis, G., Lu, P. H., Tishler, T. A., Fong, S. M., Oluwadara, B., Finn, J. P., et al. (2007). Myelin breakdown and iron changes in Huntington's disease: Pathogenesis and treatment implications. *Neurochemical Research*, *32*, 1655–1664.

Browne, S. E., & Beal, M. F. (2006). Oxidative damage in Huntington's disease pathogenesis. *Antioxidants & Redox Signaling*, *8*, 2061–2073.

Chen, J. C., Hardy, P. A., Kucharczyk, W., Clauberg, M., Joshi, J. G., Vourlas, A., et al. (1993). MR of human postmortem brain tissue: correlative study between T2 and assays of iron and ferritin in Parkinson and Huntington disease. *AJNR. American Journal of Neuroradiology, 14*, 275–281.

Dexter, D. T., Carayon, A., Javoy-Agid, F., Agid, Y., Wells, F. R., Daniel, S. E., et al. (1991). Alterations in the levels of iron, ferritin and other trace metals in Parkinson's disease and other neurodegenerative diseases affecting the basal ganglia. *Brain, 114*(Pt 4), 1953–1975.

Dumas, E. M., Versluis, M. J., van den Bogaard, S. J., van Osch, M. J., Hart, E. P., van Roon-Mom, W. M., et al. (2012). Elevated brain iron is independent from atrophy in Huntington's disease. *NeuroImage, 61*, 558–564.

Estrada Sanchez, A. M., Mejia-Toiber, J., & Massieu, L. (2008). Excitotoxic neuronal death and the pathogenesis of Huntington's disease. *Archives of Medical Research, 39*, 265–276.

Firdaus, W. J., Wyttenbach, A., Giuliano, P., Kretz-Remy, C., Currie, R. W., & Arrigo, A. P. (2006). Huntingtin inclusion bodies are iron-dependent centers of oxidative events. *FEBS Journal, 273*, 5428–5441.

Hilditch-Maguire, P., Trettel, F., Passani, L. A., Auerbach, A., Persichetti, F., & MacDonald, M. E. (2000). Huntingtin: An iron-regulated protein essential for normal nuclear and perinuclear organelles. *Human Molecular Genetics, 9*, 2789–2797.

Jurgens, C. K., Jasinschi, R., Ekin, A., Witjes-Ane, M. N., Middelkoop, H., van der Grond, J., et al. (2010). MRI T2 hypointensities in basal ganglia of premanifest Huntington's disease. *PLoS Currents, 2*.

Lumsden, A. L., Henshall, T. L., Dayan, S., Lardelli, M. T., & Richards, R. I. (2007). Huntingtin-deficient zebrafish exhibit defects in iron utilization and development. *Human Molecular Genetics, 16*(16), 1905–1920.

Morrison, P. J., & Nevin, N. C. (1994). Serum iron, total iron binding capacity and ferritin in early Huntington disease patients. *Irish Journal of Medical Science, 163*, 236–237.

Novak, M. J., & Tabrizi, S. J. (2010). Huntington's disease. *BMJ, 340*, c3109.

Paulsen, J. S., Nopoulos, P. C., Aylward, E., Ross, C. A., Johnson, H., Magnotta, V. A., et al. (2010). Striatal and white matter predictors of estimated diagnosis for Huntington disease. *Brain Research Bulletin, 82*, 201–207.

Ross, C. A., & Tabrizi, S. J. (2011). Huntington's disease: From molecular pathogenesis to clinical treatment. *Lancet Neurology, 10*, 83–98.

Roze, E., Saudou, F., & Caboche, J. (2008). Pathophysiology of Huntington's disease: From huntingtin functions to potential treatments. *Current Opinion in Neurology, 21*, 497–503.

Simmons, D. A., Casale, M., Alcon, B., Pham, N., Narayan, N., & Lynch, G. (2007). Ferritin accumulation in dystrophic microglia is an early event in the development of Huntington's disease. *Glia, 55*, 1074–1084.

Sturrock, A., & Leavitt, B. R. (2010). The clinical and genetic features of Huntington disease. *Journal of Geriatric Psychiatry and Neurology, 23*, 243–259.

Tabrizi, S. J., Langbehn, D. R., Leavitt, B. R., Roos, R. A., Durr, A., Craufurd, D., et al. (2009). Biological and clinical manifestations of Huntington's disease in the longitudinal TRACK-HD study: Cross-sectional analysis of baseline data. *Lancet Neurology, 8*, 791–801.

Vonsattel, J. P., Keller, C., & Cortes Ramirez, E. P. (2011). Huntington's disease—Neuropathology. *Handbook of Clinical Neurology, 100*, 83–100.

Vymazal, J., Klempir, J., Jech, R., Zidovska, J., Syka, M., Ruzicka, E., et al. (2007). MR relaxometry in Huntington's disease: Correlation between imaging, genetic and clinical parameters. *Journal of Neurological Sciences, 263*, 20–25.

Zecca, L., Youdim, M. B., Riederer, P., Connor, J. R., & Crichton, R. R. (2004). Iron, brain ageing and neurodegenerative disorders. *Nature Review Neuroscience, 5*, 863–873.

CHAPTER ELEVEN

Lysosomal Storage Disorders and Iron

Jose Miguel Bras[1]
Department of Molecular Neuroscience, Institute of Neurology, University College of London, London, United Kingdom
[1]Corresponding author: e-mail address: j.bras@ucl.ac.uk

Contents

Abstract

Lysosomal storage disorders are a group of about 50 rare metabolic diseases that result from defects in lysosomal function. The majority is recessively inherited and caused by mutations in genes encoding lysosomal proteins as the basis for its pathobiology. The lysosome plays a pivotal role in a cell's ability to recycle and degrade unwanted material. One of its functions relates to regulating iron levels throughout the body. Iron is a double-edged sword: It is absolutely required for an organism's survival, but high levels of iron quickly lead to cell death. In addition, recent results have put the lysosome on the map of pathways leading to common neurodegenerative diseases, such as Parkinson's disease and Alzheimer's disease. It is plausible that the mechanisms through which the lysosome acts in these diseases also involve iron and this would have significant implications in our understanding of the molecular etiology of these disorders.

International Review of Neurobiology, Volume 110
ISSN 0074-7742
http://dx.doi.org/10.1016/B978-0-12-410502-7.00012-0

1. OVERVIEW OF THE LYSOSOMAL COMPARTMENT AND ITS MAIN FUNCTIONS

Lysosomes are ubiquitous organelles that are present in animal cells and contain acid hydrolases. They constitute the primary degradative compartments of the cell.

Soon after their identification, in the 1950s by Christian de Duve and his coworkers (for a review, see de Duve, 2005), it became evident that the lysosome plays a pivotal role in heterophagy and autophagy. At the time, lysosomes were viewed as simple storage places for waste products, such as the age-pigment lipofuscin, and considered inactive. Later, the assumption of inactivity proved to be wrong, and lysosomes were recognized as integral parts of the lysosomal compartment and known to receive newly produced lysosomal enzymes by fusion with other lysosomes (Brunk & Ericsson, 1972). Lysosomes were also shown to have proton-pumping vacuolar ATPases, which are responsible for maintaining an acidic pH of 4.6–5 within the compartment, required for its adequate activity (Mellman, Fuchs, & Helenius, 1986).

Following the identification of endosomes (Helenius, Mellman, Wall, & Hubbard, 1983), it became clear that many endocytosed macromolecules, such as low-density lipoprotein, are delivered to lysosomes after their sequential passage through early and late endosomes (Mellman, 1996; Mukherjee, Ghosh, & Maxfield, 1997). The exact mechanism through which endosomes transfer their cargo to lysosomes was the subject of much controversy, with several theories being proposed. Electron microscopy and, more recently, time-lapse confocal microscopy experiments have shown that the fusion of endosomes with lysosomes creates hybrid organelles in which the bulk of the endocytosed cargo is degraded (Bright, Gratian, & Luzio, 2005). Subsequent reformation of lysosomes from these hybrid organelles is dependent on reuse or retrieval of membrane proteins by vesicular traffic (Bright et al., 2005; Pryor, Mullock, Bright, Gray, & Luzio, 2000).

Lysosomal hydrolases are synthesized in the endoplasmic reticulum, tagged with mannose-6-phosphate (M6P) in the cis-Golgi area, and packaged into transport vesicles (sometimes named primary lysosomes although they have a neutral pH) in the *trans*-Golgi network with the help of M6P receptors. They are then transferred to late endosomes, which arise from early endosomes containing endocytosed material. At this point, the lysosomal hydrolases are freed from M6P receptors and activated, while the M6P receptors are recirculated to the Golgi. The late endosomes then mature to lysosomes, which are characterized by their lack of M6P receptors, the presence of acid hydrolases, having an acidic pH, and containing endocytosed material to be degraded.

Some of these endocytosed ligands dissociate from their receptors in the acidic lumen of the early endosome, after which they are delivered to the lysosome for degradation, allowing for the recycling of the empty receptors to the cell surface. An example of such a ligand is the low-density lipoprotein. Others, as the epidermal growth factor, remain bound to their receptors (which are ubiquitylated (Haglund et al., 2003; Huang, Kirkpatrick, Jiang, Gygi, & Sorkin, 2006)) and, following endocytosis, are internalized from the surface of the endosome into luminal vesicles. These are then delivered to the lysosome upon fusion with the late endosomes (Luzio, Pryor, & Bright, 2007).

Phagocytosis is an essential process by which specialized cells engulf invading pathogens, apoptotic cells, and other foreign particles that are $>0.5\ \mu m$ in diameter. The engulfment of the target molecule forms a sealed intracellular component—the phagosome—which matures by multiple transient interactions with endosomal compartments, including lysosomes, to form a hybrid-like organelle termed the phagolysosome (Desjardins, 1995), whose primary function is to degrade the phagocytosed particle.

Lysosomes can also fuse with autophagosomes, in a process known as autophagy, to form autophagolysosomes where a variety of organelles and proteins are degraded into simple components, which in turn are reutilized by the anabolic machinery of the cell following transport to the cytosol (Klionsky, 2007; Terman, Gustafsson, & Brunk, 2007). This mechanism is important in many physiological processes (Levine & Klionsky, 2004), including response to starvation, cell growth, and innate immunity; one example is the removal of pathogenic bacteria when they have been released after endocytosis by nonphagocytic cells (Nakagawa et al., 2004).

Recently, it has also been shown that, in many cells, conventional lysosomes can also fuse with the plasma membrane in response to an increase in the concentration of cytosolic Ca^{2+} that triggers lysosome exocytosis (Rodriguez, Webster, Ortego, & Andrews, 1997). This lysosome exocytosis provides the extra membrane needed for plasma membrane wound repair (McNeil & Kirchhausen, 2005) and allows the formation of a parasitophorous vacuole, for example, when the trypanosomatid protozoan Leishmania infects cells (Wilson et al., 2008).

There is extensive evidence that lysosomes can fuse with late endosomes, the plasma membrane, phagosomes, and autophagosomes. Although much of these mechanistic events are now partially understood, several details are still elusive.

A better understanding of lysosome fusion would have considerable implications as the identification of better ways of upregulating autophagic pathways and the formation of autophagolysosomes would be greatly beneficial for novel approaches to the treatment of diseases where aggregate-prone cytosolic proteins cause a range of proteinopathies, such as Huntington's disease or Parkinson's disease.

2. IRON HOMEOSTASIS

2.1. Regulation of cellular iron uptake

Iron plays a pivotal role in mammalian metabolism, primarily because of the ease with which it can gain and lose electrons. This property makes iron a useful component of hemoglobin, myoglobin, cytochromes, and various nonheme enzymes.

However, though iron is essential for all eukaryotes, it can also be toxic, as the same property that permits iron to gain and lose electrons can also result in the donation of electrons to oxygen, causing the generation of superoxide anions and the hydroxyl radicals. These oxygen metabolites react readily with biological molecules, including proteins, lipids, and DNA. It is, thus, an absolute requirement of all mammalian species to tightly regulate the concentration of iron in biological fluids (De Domenico, McVey Ward, & Kaplan, 2008).

Since there is no known regulated form of iron excretion, the control of systemic iron levels occurs through the regulation of iron acquisition and storage. Loss of iron in humans occurs through exfoliation of enterocytes and skin cells and through menstruation and childbirth; however, these processes are independent of the iron levels of the body. Its balance is maintained by controlling iron acquisition, which may be increased in response to need.

Dietary iron absorption is a highly regulated process, which involves transport across the apical membrane of absorptive cells in the gut into the cells and across the basolateral membrane into blood plasma. In the plasma, iron is bound to the glycoprotein transferrin (Tf). Precursor erythrocytes can then bind iron-laden Tf, incorporating the iron into heme, which forms the largest pool of iron in humans (Munoz, Villar, &

Garcia-Erce, 2009; Sharp & Srai, 2007). Erythrocytes have a very well-defined life span, and when they are aged, they undergo phagocytosis by macrophages and are degraded within their lysosomal compartment (Dunn, Suryo Rahmanto, & Richardson, 2007). At this point, iron can be either stored or released back to the plasma, depending on the iron requirements of the body. If released, iron binds to Tf, an abundant plasma glycoprotein able to bind two atoms of iron with very high affinity (Aisen, Leibman, & Zweier, 1978). Therefore, little or no free, and potentially toxic, iron remains in circulation (Anderson & Vulpe, 2009). The complex transferrin–iron binds to the plasma membrane transferrin receptor 1 (TfR1) and is endocytosed (Richardson & Ponka, 1997). Here, iron is released and transported into the cytoplasmic pool of labile iron by divalent metal transporter 1, after which it is either incorporated and stored in ferritin or delivered to the mitochondria or elsewhere for synthesis of iron–sulfur complexes, heme, and iron-containing enzymes. Storage of excess iron in ferritin is essential to prevent iron-mediated oxidative processes, making ferritin a very important antioxidant (Arosio, Ingrassia, & Cavadini, 2009; Balla et al., 1992, 1993).

Both intestinal iron absorption and macrophage iron recycling are highly regulated, and disruption of either one leads to deficiencies in iron homeostasis and, potentially, to clinical phenotypes. This regulation occurs at both the cellular and systemic levels by transcriptional and posttranscriptional mechanisms. Ferritin and TfR1, which plays a vital role in iron metabolism (Levy, Jin, Fujiwara, Kuo, & Andrews, 1999), can be transcriptionally regulated by cytokines or differentiation factors. The mechanism of posttranscriptional control of ferritin and TfR1 involves iron-regulatory proteins (IRPs). When iron is limiting, IRPs bind iron-regulatory elements (IREs), which are stem-loop structures that are found in noncoding regions of messenger RNAs and encode proteins involved in iron metabolism. The binding of IRPs to IREs in the 5′-untranslated regions of mRNA prevents translation, whereas binding in the 3′-untranslated regions stabilizes the mRNA and prevents degradation (Galy, Ferring, Benesova, Benes, & Hentze, 2004; Rouault, 2006).

2.2. The role of lysosomes in intracellular iron metabolism

As mentioned previously, the lysosomal compartment is crucial for cell maintenance and has a variety of important functions, including participation in endocytic uptake of materials from the outside and autophagic degradation of damaged mitochondria and other organelles and numerous, mostly long-lived, proteins. Many of these proteins are ferruginous compounds, meaning that lysosomes of all cells contain low-mass redox-active iron, which explains the known lysosomal vulnerability to oxidative stress (Kurz, Terman, Gustafsson, & Brunk, 2008). Additionally, increased amounts of iron can accumulate in some lysosomes due to accumulation of iron-rich nondegradable materials, such as lipofuscin and hemosiderin (Jolly, Douglas, Davey, & Roiri, 1995).

Consequently, the sensitivity of individual lysosomes to oxidative stress may vary substantially. Nonetheless, lysosomes are very sensitive to oxidative stress and their membranes are easily peroxidized and permeabilized by the radicals that are formed secondary to Fenton-type reactions taking place in the lysosomes. The rupture of lysosomes with relocation of the lytic enzymes results in apoptosis or necrosis depending on the magnitude of this relocation (Brunk, Neuzil, & Eaton, 2001). Consequently, keeping the concentration of redox-active iron in lysosomes as low as possible is important for the survival of cells exposed to oxidative stress. The rapid transport of low-mass iron from the lysosomes to the cytosol is thus important, as well as methods of temporarily binding iron in a non-redox-active form. As mentioned before, ferritin is the main iron-storage protein, and by binding iron in a non-redox-active form, it is also an important antioxidant. When stored iron is needed for synthesis of iron-containing structures, it has to be released from ferritin. Research has shown that the major mechanism for this iron release involves lysosomal digestion (Yu, Persson, Eaton, & Brunk, 2003; Zhang et al., 2010).

Iron is an essential transition metal utilized in an extensive range of electron-transport mechanisms. It is now clear that the lysosome plays a pivotal role in the iron cellular homeostasis and that a disruption of this balance, be it by deregulation of iron levels or impaired lysosomal function, has pathogenic consequences.

3. LYSOSOMAL STORAGE DISORDERS

Current estimates place the number of soluble lysosomal hydrolases at greater than 60 and at least seven integral lysosomal membrane proteins (Eskelinen, Tanaka, & Saftig, 2003; Journet, Chapel, Kieffer, Roux, & Garin, 2002). In principle, mutations in the genes encoding any of these proteins could cause a lysosomal storage disorder (LSD), as long as the function of the protein is lost or altered. Over 40 LSDs that involve soluble hydrolases are known and a number of diseases have been identified that also involve the integral membrane proteins. In fact, it is plausible that mutations in any of the genes encoding these proteins cause a different type of LSD; however, given the essential nature of these proteins, some may not be compatible with life and, consequently, will be difficult to identify. A hypothetical example would be an individual having a homozygous truncating mutation in the gene encoding the ATPase responsible for maintaining the lysosome's pH. This would most likely lead to the inability of the lysosome to degrade any target and very rapidly lead to cell death, presumably at very early stages of development.

Historically, LSDs have been classified according to the accumulated substrate(s), primarily due to the fact that this accumulating compound was characterized years before the enzymatic defect was ever known. An example of this is the three types of Niemann–Pick disease, which were initially characterized as sphingomyelin storage disorders. It is now known that only Niemann–Pick A and B are caused by defective sphingomyelinase activity. Niemann–Pick C (NPC) is caused by the defective activity of a putative cholesterol transporter, the NPC1 protein, or by defects in the soluble lysosomal cholesterol-binding NPC2 protein. Perhaps a more useful classification of LSDs is one based on the characterization of the defective enzyme or protein.

The age of clinical onset and spectrum of clinical phenotype exhibited among different LSDs vary, depending on the level of protein function affected by specific mutations, the cell types where storage occurs, and the biochemistry of the stored material. Apart from lysosomal diseases involving substrate storage in the bone and cartilage (e.g., the mucopolysaccharidoses), most individuals born with these conditions appear normal at birth. The typical clinical presentation of an LSD is a neurodegenerative disease of infancy/early childhood onset (Wraith, 2002), but variants with a presentation in adulthood also occur (Nixon, Yang, & Lee, 2008; Shapiro, Logigian, Kolodny, & Pastores, 2008; Spada et al., 2006).

The majority of LSDs present with pathology of the central nervous system, and in some cases, neurodegeneration can occur in multiple brain regions (e.g., the thalamus, cortex, hippocampus, and cerebellum). A notable example of a nonneuronopathic form of LSD is type 1 Gaucher disease (beta-glucocerebrosidase deficiency leading to the accumulation of glucosylceramide), which is a relatively common form of LSD and particularly prevalent within the Ashkenazi Jewish population (Beutler, 2006; Germain, 2004). The major cell type affected by the excess glucosylceramide storage in this disease is the macrophage, whose dysfunction affects the production and turnover of cells belonging to the hematopoietic system. When the macrophage becomes unable to clear glucosylceramide, it accumulates in fibrils giving rise to typical structures visible on light microscopy and commonly known as Gaucher cells. Gaucher cells infiltrate into various organs and affect the immune system, bone strength, and spleen and liver functions (Beutler, 1988). Neurological symptoms can include seizures, dementia, and brainstem dysfunction. Among the peripheral symptoms are enlargement of the spleen and liver (hepatosplenomegaly), heart and kidney injury, abnormal bone formation, muscle atrophy, and ocular disease.

Each LSD has a distinct clinical and pathological picture, which presumably relates to the nature of the accumulating substrate and the cell types in which it accumulates. For example, all forms of Pompe disease, which is caused by the accumulation of glycogen in the lysosome, are characterized by myopathy, which is expected given the role that glycogen has in the muscle (Katzin & Amato, 2008). And many of the sphingolipidoses are characterized by brain disease, which might be expected given the high levels at which glycosphingolipids exist in the brain. In some diseases, differing levels of residual enzyme activity can explain phenotypic variability, but in others, there is no obvious genotype–phenotype correlation. In fact, in some cases, patients with a similar genetic background, and sometimes even with the same genetic lesion, can present with remarkably disparate clinical symptoms or even be asymptomatic for the disease. Although these events are not properly understood, it has become increasingly evident that genetic pleiotropy plays a much larger role in disease than what had been previously thought.

3.1. Neuronal ceroid lipofuscinosis

Neuronal ceroid lipofuscinoses (NCLs) are a group of lysosomal storage diseases that are characterized by progressive neurodegeneration and accumulation of autofluorescent storage material ("lipofuscin") in the lysosomes of neurons and other cells. The NCLs comprise the most common group of progressive neurodegenerative diseases in children with an incidence as high as 1 in 12,500 live births (Scriver, 1995), and the most common NCL results from mutations in *CLN3*, causing juvenile NCL (Batten disease). So far, 14 loci have been described where mutations cause a form of NCL (CLN1-14).

Some of these proteins are known to be lysosomal-soluble hydrolases (CLN1 and 2), others are predicted transmembrane proteins (CLN3, 5, 6, and 8), while others like CLN11/GRN have no known direct lysosomal role. Mutations in any of the fourteen genes result in a different form of NCL and the onset can vary from infantile to adulthood.

The exact accumulated material in NCL varies in composition but is generally a combination of proteins, proteolipids, and metals (Palmer, Jolly, van Mil, Tyynela, & Westlake, 1997). A main component of the storage material that accumulates in the late-infantile variants and juvenile NCL is subunit C of the mitochondrial ATP synthase (Palmer et al., 1992); however, sphingolipid-activating proteins (saposins A and D) are enriched in the infantile forms of NCL (Seehafer & Pearce, 2006). Additionally, saposin D accumulates in congenital NCL (Siintola et al., 2006). The heterogeneity of storage material in different forms of NCL suggests that the disruption of various pathways is involved in this type of LSDs.

It is interesting to note that lack of any one of the NCL proteins does not result in severe mislocalization of other NCL proteins (Kyttala, Lahtinen, Braulke, & Hofmann, 2006). It is, thus, possible that deficiency of one NCL protein may either directly or indirectly lead to functional consequences upon another NCL protein by interfering with a phenomenon other than cellular localization. In fact, such an event seems to have already been described. Both the activity and the amount of mannose-6-phosphorylated TPP1 (CLN2) have been shown to be increased in the other forms of NCL (although this may not be specific to NCLs, since TPP1 activity has also been shown to be increased in Alzheimer's disease) (Junaid & Pullarkat, 1999). Furthermore, it has been suggested that the sequential action of TPP1 and cathepsin D is required for the lysosomal degradation of the subunit C of the mitochondrial ATP synthase in a pH-dependent manner (Kominami, 2002). Therefore, any change in cellular pH homeostasis, which has been reported in many NCLs (Holopainen, Saarikoski, Kinnunen, & Jarvela, 2001), would result in abnormal accumulation of subunit C. In addition, CLN8 can partially rescue the perturbed sphingolipid metabolism detected in CLN9-deficient cells (Schulz et al., 2006), indicating that CLN8 and the so far unidentified CLN9 protein may share functions in regulation of ceramide synthesis.

Lipofuscin, the aging pigment that characterizes NCLs, accumulates as part of senescence and, possibly because of that, it has been implicated in numerous age-related diseases such as Alzheimer's disease, Parkinson's disease, heart failure, and macular degeneration (Seehafer & Pearce, 2006). Its accumulation in postmitotic and aging cells has been observed in a variety of organisms, organs, and tissues and is considered a normal event in cellular senescence. In general, lipofuscin has been shown to be a mixture of oxidized proteins and lipids, carbohydrates, and trace amounts of metals. Specifically, lipofuscin consists of approximately two-thirds protein, one-third lipid, and the predominant metal is iron (Brunk & Terman, 2002a,b). The exact intracellular effects of lipofuscin in the cell are largely hypothetical and still under discussion. One main factor of lipofuscin cytotoxicity is believed to be its ability to incorporate transition metals, such as iron. Binding of iron results in a redox-active surface on lipofuscin, able to catalyze the Fenton reaction (Brunk & Terman, 2002b). This in turn leads to increased oxidative stress, which causes lysosomal membrane permeabilization, with ensuing relocation to the cytosol of iron and lysosomal hydrolytic enzymes, with resulting apoptosis or necrosis (Kurz, Eaton, & Brunk, 2010).

Despite the fact that the precise biological functions of all NCL proteins are still elusive, the fact that lipofuscin is a hallmark of the disease suggests that it plays an important role in the pathobiological process. It is plausible that much of this pathogenicity is driven by the adverse effects of iron, leading to increased cellular oxidative stress.

4. GENETICS OF LSDs

LSDs are typically monogenic, that is, they involve only a single gene, but for most forms, several mutations have been described in the same gene in different patients. Nearly all types of protein-changing mutations have been found to cause forms of NCL: missense, nonsense, splice site, partial deletions, and duplications. Some of these will lead to complete loss of enzyme activity, some to partial loss of function, while for others, the mechanism through which they exert their effect on the disease is not understood. Interestingly, no genotype–phenotype correlations have been found for the majority of LSDs, nor is the prediction of clinical course possible through the genetic findings alone. In fact, there have been descriptions of individuals who are asymptomatic for the disease even though they carry a mutation that is responsible for a severe form of the disease in another individual.

Gaucher disease is an autosomal recessively inherited LSD found in all ethnicities but with increased frequency among Ashkenazi Jews. Many aspects of Gaucher disease serve as an example of other LSDs because of the wide range of phenotypes encountered.

Classically, the disorder was classified into three distinct types (type 1: OMIM#230800; type 2: OMIM#230900; and type 3: OMIM#2301000). Type 1 Gaucher disease, by definition, was considered to be nonneuronopathic and is by far the most common form of the disease. Type 2 Gaucher disease is the acute neuronopathic form and was viewed as a rapidly progressive neurodegenerative disorder of late infancy, resulting in death within the first couple of years. Type 3, or chronic neuronopathic, Gaucher disease was a "catch all," encompassing patients who survived infancy but had some form of neurological involvement.

The gene for Gaucher disease, *GBA*, is located on chromosome 1 and encompasses 11 exons. There is a highly homologous pseudogene located nearby, sharing ~96% of the sequence in coding regions (Horowitz et al., 1989). A large number of DNA variants has been identified in *GBA*; most are point mutations, but others, including insertions and deletions, splice site alterations, and recombinant alleles, have also been described (Hruska, LaMarca, Scott, & Sidransky, 2008). The recombinant alleles can be due to rearrangements, duplications, or gene fusions of *GBA* with the nearby pseudogene (*GBAP*), which considerably complicates their detection. Additionally, some of the disease-causing mutations in *GBA* are present as normal sequence in the pseudogene, which clearly means that screening for mutations must be performed with absolute certainty of the variants' location.

Gaucher cells have been described to contain iron in the form of dense particles of ferritin that are free in the cytoplasm. Most of this iron resides within the same lysosomes that contain the storage deposits characteristic of Gaucher disease (Weisberger, Emmons, & Gorczyca, 2004). Exactly how these iron deposits relate to mutation-carrying GBA function is not understood, but it is plausible that iron levels modulate disease progression in Gaucher disease, by modifying levels of oxidative stress. Moreover, it has recently been shown that type 1 Gaucher disease is characterized by persistent hyperferritinemia and that monitoring of serum ferritin could be useful during patient follow-up (Mekinian et al., 2012).

As mentioned before, 14 loci (CLN1-14) are known to cause different forms of NCLs, with over 400 disease-causing mutations identified in over 1000 cases (for an up-to-date database of genetic variants in NCL, see http://www.ucl.ac.uk/ncl/mutation.shtml).

CLN1 is protein palmitoyl-protein thioesterase 1 (PPT1), a hydrolase that is targeted by M6P receptors to the lysosomes. In the lysosomes, it contributes to the degradation of palmitoylated proteins by deacylating cysteine thioesters, and those are found as the accumulated material in NCL1. Mutations in *PPT1* presumably act through a loss-of-function mechanism, since one of the most common mutations is the R151X substitution (Das et al., 1998).

CLN2 is tripeptidyl-peptidase 1 (TPP1), a serine protease that cleaves tripeptides from the amino terminus of small proteins before these tripeptides are degraded by other lysosomal proteases. Nonsense mutations have also been described in *TPP1* although the vast majority are missense variants.

The *CLN3* gene, which is mutated in Batten disease, contains 15 exons and spans over 15 kb. It has been shown that the CLN3 protein is an integral membrane protein that colocalizes with lysosomes (Jarvela, Lehtovirta, Tikkanen, Kyttala, & Jalanko, 1999). It has been proposed to be an arginine transporter in the lysosome (Ramirez-Montealegre & Pearce, 2005) and a palmitoyl-protein delta-9 desaturase (Narayan, Rakheja, Tan, Pastor, & Bennett, 2006). The most common mutation in CLN3 is a genomic deletion of 1.02 kb, which is responsible for 73% of Batten disease cases (Isolation of a novel gene underlying Batten disease & CLN3.The International Batten Disease Consortium, 1995).

Currently, the most recently identified gene for NCL is *KCTD7/CLN14* (Staropoli et al., 2012), where homozygous mutations give rise to disease with infantile onset. Interestingly, mutations in *KCTD7* had been previously identified as the cause of progressive myoclonic epilepsy without lysosomal storage (Van Bogaert et al., 2007). The mechanism through which mutations in this gene lead to NCL is far from understood, but an interplay with the ubiquitin–proteasome system seems evident, given the known interaction of KCTD7 with cullin-3, a scaffolding component of several E3 ubiquitin–ligase complexes that selectively tag proteins for degradation by the proteasome.

In NCLs, early reports have described some changes in metal levels, including iron and zinc, in the blood cells of patients (Johansson, Lindh, Westermarck, Heiskala, & Santavuori, 1990), but these have not been indisputably confirmed. Additionally, it has been shown that altered metal homeostasis may also occur in some NCL models. Recently, changes in expression of the metal transporter, ZnT6, in both CLN3 and CLN6 mouse cerebellar neuron precursor cell lines have been reported (Cao et al., 2011), further strengthening the role of metals in these diseases.

5. IRON DISORDERS OF GENETIC ORIGIN AND THE LYSOSOME

HFE-related iron overload disease (type 1 hemochromatosis) is, by far, the most frequent form of genetic iron overload disorders. Type 1 hemochromatosis is a recessive disease that is linked to mutations of the *HFE* gene located on chromosome 6. Homozygosity for the p.Cys282Tyr (C282Y) mutation explains more than 90% of type 1 hemochromatosis. This mutation inhibits the molecular cascade that results in decreased hepatic production of hepcidin. Hepcidin is a peptide hormone produced by the liver that appears to be the master regulator of iron homeostasis in humans and other mammals (Nemeth & Ganz, 2006). Its main function is to regulate iron transport across the gut mucosa, thereby preventing excess iron absorption and maintaining normal iron levels within the body. Hepcidin also inhibits transport of iron out of the macrophages. As a result of mutation in *HFE*, increased plasma iron leads to increased Tf saturation, which is associated with the appearance of nontransferrin-bound iron. This form of iron is avidly taken up by parenchymal cells in the liver, the pancreas, and the heart, leading to excess iron in these organs. The liver increases in volume (hepatomegaly), releases more transaminases into the plasma due to hepatocytic iron-related damage, and can develop scarring called fibrosis, which can lead to cirrhosis and, later on, to hepatocellular carcinoma. Excessive iron deposition in the pancreas, endocrine glands, and heart can lead to insulin-dependent diabetes, hypogonadism, and cardiac failure, respectively (Brissot, Bardou-Jacquet, Jouanolle, & Loreal, 2011).

In patients presenting with iron overload, it is common to see intracellular accumulation of iron in the lysosome (Iancu & Shiloh, 1994). It is thought that this event occurs to serve as a protective role in removing this redox-active metal from other cytoplasmic components and to provide a route for iron removal from the liver through lysosome-mediated biliary excretion (Barton, 2010). The excessive accumulation of iron in the lysosomes can also lead to lysosomal fragility, impaired lysosomal function, and eventual cellular injury through the release of hydrolytic enzymes and iron into the cytoplasm.

It has long been proposed that iron overload results in an acquired lysosomal storage disease through this mechanism (Peters & Seymour, 1976; Selden, Owen, Hopkins, & Peters, 1980; Seymour & Peters, 1978). Several studies have shown lysosomal changes in patients presenting with hemochromatosis; however, definitive evidence of fragile lysosomes releasing their contents into the cytoplasm and thus leading to cellular injury is yet to be shown.

6. LSDs AND COMMON NEURODEGENERATIVE DISEASES: GENETICS BRIDGING THE GAP

Over the last few years, special attention has been given to the role of the lysosome and lysosomal dysfunction as a primary event in the pathobiology of neurodegenerative disease. This is partially because there is increasing evidence that patients with lysosomal storage diseases also exhibit neurodegeneration, which suggests the importance of normal lysosomal function for neuronal survival. Although the potential mechanism for this involvement to occur is easy to imagine (most of these diseases are characterized by protein deposition in one or more areas of the brain, and the lysosome is part of the cell's clearing mechanism), it really was not until an association between Gaucher disease and Parkinson's disease was found that the link became increasingly evident.

The relationship between the two diseases, which have wildly disparate clinical presentations, came not from molecular data, but from astute clinical observation detailing an increase in prevalence of Parkinson's disease among families with history for Gaucher disease (Machaczka, Rucinska, Skotnicki, & Jurczak, 1999; Neudorfer et al., 1996). As mentioned previously, mutations in the gene encoding the lysosomal enzyme glucocerebrosidase cause Gaucher disease in a recessive fashion, but although a number of mutations in different genes had formerly been identified as causing Parkinson's disease, *GBA* was not one of them. In this way, the molecular association between these two diseases remained much of a mystery for quite some time. It was only in 2008 that the relationship between the two diseases became apparent through genetic results. It was found that the same mutations in *GBA* that, when homozygous, cause Gaucher disease increase an individual's risk to develop PD when they are heterozygous (Bras et al., 2009; Neumann et al., 2009; Sidransky et al., 2009). The mechanism through which this increase in risk occurs is still not properly understood, partially because it is still not clear if these mutations cause Gaucher through a loss- or gain-of-function mechanism (Hardy, Lewis, Revesz, Lees, & Paisan-Ruiz, 2009; Westbroek, Gustafson, & Sidransky, 2011). However, the fact that glucocerebrosidase catalyzes the breakdown of the glycolipid glucosylceramide to ceramide and glucose and Gaucher patients typically present enlarged macrophages resulting from the intracellular accumulation of glucosylceramide strongly supports the notion that these mutations act through a loss-of-function mechanism. With this hypothesis in mind, it has been suggested that the lysosome in general and the ceramide pathway in particular would play an important role in parkinsonism and Lewy body diseases (Bras, Singleton, Cookson, & Hardy, 2008).

The findings at *GBA* in PD have also brought to light a novel principle in which one disease is severe, has an early onset, and is caused by homozygous loss-of-function mutations, while the other is a late onset disease with increased susceptibility caused by heterozygous (probably with partial loss-of-function) variants in the same gene as the first.

Usually, mutations in a gene cause specific phenotypes either in the heterozygous or in the homozygous state, but not in both. In autosomal recessive diseases, heterozygous individuals are usually healthy. In autosomal dominant disorders, the allele frequency for the mutation is low; thus, homozygous individuals are very rare. When observed, these homozygous cases either are very similar to the heterozygous-affected family members (as is the case in Huntington's disease (Durr et al., 1999), Parkinson's disease (Nichols et al., 2005), and Creutzfeldt–Jakob disease (Gabizon et al., 1993)) or have a more severe form of the same phenotype (e.g., spinocerebellar ataxias 2, 3, and 6 (Lerer, Merims, Abeliovich, Zlotogora, & Gadoth, 1996; Matsumura et al., 1997; Sanpei et al., 1996; Sobue et al., 1996)). The low allele frequency for mutations in dominant diseases can be explained by natural selection: Homozygous loss-of-function mutations cause early-onset disorders and many individuals with these mutations die before reaching reproductive age, contributing to the rare frequency of the disease and of the mutations. On the other hand, heterozygous variants confer risk to a disorder with an onset usually occurring beyond reproductive age, making these variants and diseases more common in the population.

The relationship between PD and a lysosomal disorder is not reserved for Gaucher disease only; recently, a family presenting with a typical form of NCL was identified as having a mutation in the gene *ATP13A2* (Bras, Verloes, Schneider, Mole, & Guerreiro, 2012). Mutations in *ATP13A2* had been previously described as causing a form of parkinsonism called Kufor–Rakeb syndrome, a very rare form of autosomal recessive hereditary parkinsonism with dementia and juvenile onset (Ramirez et al., 2006). The detailed function of the ATP13A2 protein is not known, but it is believed to be a member of the P5 subfamily of ATPases, which transports inorganic cations and other substrates. Recent data have shown that Kufor–Rakeb-causing mutations in this gene lead to impaired lysosomal acidification, decreased proteolytic processing of lysosomal enzymes, reduced degradation of lysosomal substrates, and diminished lysosomal-mediated clearance of autophagosomes (Dehay et al., 2012). This cellular phenotype is clearly reminiscent of LSDs and further implies common mechanisms between these diseases and parkinsonism. The vast majority of mutations in *ATP13A2* associated with Kufor–Rakeb are truncating mutations arising by deletion, duplication, insertion, or splice site mutations. Although it is not yet clear if there is a relationship between mutation type and disease severity, it is plausible that truncating loss-of-function mutations lead to Kufor–Rakeb, while less severe missense mutations lead to a different phenotype, possibly due to the fact that the protein does not undergo nonsense-mediated decay.

Neurodegeneration with brain iron accumulation-2 (NBIA-2) is characterized by the disruption of cellular mechanisms leading to the accumulation of iron in the basal ganglia. Mutations in the gene *PLA2G6* were recently described as the cause of NBIA-2 (Morgan et al., 2006). PLA2G6 belongs to the family of A2 phospholipases, which catalyze the release of fatty acids from phospholipids and play a role in a wide range of physiological functions (Tang et al., 1997). Similarly to *ATP13A2*, mutations in *PLA2G6* have also been found to cause additional phenotypes, in this case parkinsonism with dystonia (Paisan-Ruiz et al., 2009). These data provide further evidence for a link between iron homeostasis and parkinsonism.

Another recent result linking lysosomal disorders with a common neurodegenerative disease came from studies in the gene encoding the progranulin protein. Progranulin (GRN) is a 593-amino acid, cysteine-rich protein that undergoes proteolytic cleavage by extracellular proteases, such as elastase, giving rise to smaller peptide fragments termed granulins or epithelins (He & Bateman, 2003). It is widely expressed in a variety of tissues, including the nervous system, where it has the highest expression in neurons and activated microglia (Ahmed et al., 2010; Ryan et al., 2009). At the subcellular level, progranulin colocalizes with the endoplasmic reticulum and Golgi markers in the secretory pathway and the lysosomal marker Lamp1 (Almeida, Zhou, & Gao, 2011; Hu et al., 2010). In 2006, heterozygous mutations in GRN were discovered to be a cause of frontotemporal lobar degeneration (FTD) (Baker et al., 2006; Cruts et al., 2006). FTD is the second most common presenile dementia disorder after Alzheimer's disease, and since the initial description of disease-causing mutations, more than 70 mutations in GRN, almost all of which result in null alleles, have been identified in FTD patients. Clinical manifestations of heterozygous loss-of-function GRN mutations include variants of the FTD spectrum, parkinsonism, and the corticobasal syndrome (van Swieten & Heutink, 2008). Recently, however, a homozygous nonsense mutation has been described as the cause of typical adult NCL in two individuals, presenting with progressive loss of vision, retinal dystrophy, cerebellar ataxia, and seizures (Smith et al., 2012). These results have shown, once again, the existence of pleiotropic effects from the same gene, where homozygous mutations give rise to a severe early-onset disease (NCL), while heterozygous mutations give rise to a later-onset, more common, neurodegenerative disorder; the parallels with GBA in Gaucher and PD are obvious and further strengthen the role of the lysosome in these disorders.

A further example of the principle that loss-of-function homozygosity and heterozygosity give rise to different but related disorders comes from the field of Alzheimer's disease. Here, heterozygous mutations in the gene encoding the triggering receptor expressed on myeloid cells 2 (*TREM2*), an innate immune receptor, have been shown to substantially increase an individual's risk to developing Alzheimer's disease (Guerreiro et al., 2013). Some of the same mutations, when in the homozygous state, cause a rare form of dementia with bone involvement, called Nasu–Hakola disease (Paloneva et al., 2002). A portion of the Nasu–Hakola-causing mutations are nonsense, thus suggesting that the mechanism through which these mutations exert their effect in disease is loss-of-function. Since some of the mutations are the same as the ones that exert risk for Alzheimer's disease, it is likely that the mechanism in AD is also driven by loss-of-function. This is interesting, particularly since it has been shown that TREM2 activation leads to increased lysosomal function (Takahashi, Prinz, Stagi, Chechneva, & Neumann, 2007). Further showing pleiotropy is the fact that the same mutations that cause Nasu–Hakola disease have also been found to cause typical FTD (Guerreiro et al., 2012) and no bone phenotype typical of Nasu–Hakola. Although the role of the lysosomal pathway has long been studied in Alzheimer's disease, these recent results clearly support the notion that the lysosomal pathway may be involved in the pathogenesis of dementia in general.

7. CONCLUSIONS AND FUTURE PERSPECTIVES

Iron is indispensable for vital biological functions such as oxygen transport, mitochondrial energy production, DNA synthesis, and DNA repair. Yet too much iron can be toxic, which means that a tightly controlled iron homeostasis mechanism must always be in place and the lysosome is known to play a pivotal role in this mechanism. Lysosomal dysfunction—such as occurs in aging and lysosomal storage diseases—can be a major factor limiting function and longevity of the organism. A growing body of evidence is showing that LSDs may share some of their pathobiological events with common neurodegenerative diseases. This, in turn, suggests that iron may play a more substantive role in these disorders than what has been appreciated. In addition, these data also showed that mutations' zygosity in the same gene is responsible for disparate clinical presentations, with homozygosity causing a rare early-onset severe phenotype and heterozygosity causing a milder later-onset disease.

As technologies continue to improve our ability to study the human genome, we will likely see more of this molecular interplay leading to different diseases. So far, it is clear that LSDs and neurodegenerative diseases share molecular events that lead to disease. It would certainly not be surprising if iron was found to play a major role in all of these disorders. A more profound appreciation of the role of the lysosome in aging and neurodegeneration, possibly driven by our knowledge from LSDs, could lead to an improved understanding of the molecular events in these diseases.

REFERENCES

Ahmed, Z., Sheng, H., Xu, Y. F., Lin, W. L., Innes, A. E., Gass, J., et al. (2010). Accelerated lipofuscinosis and ubiquitination in granulin knockout mice suggest a role for progranulin in successful aging. *The American Journal of Pathology*, *177*(1), 311–324 (Research Support, Non-U.S. Gov't).

Aisen, P., Leibman, A., & Zweier, J. (1978). Stoichiometric and site characteristics of the binding of iron to human transferrin. *The Journal of Biological Chemistry*, *253*(6), 1930–1937 (Research Support, U.S. Gov't, P.H.S.).

Almeida, S., Zhou, L., & Gao, F. B. (2011). Progranulin, a glycoprotein deficient in frontotemporal dementia, is a novel substrate of several protein disulfide isomerase family proteins. *PLoS One*, *6*(10), e26454 (Research Support, N.I.H., Extramural Research Support, Non-U.S. Gov't).

Anderson, G. J., & Vulpe, C. D. (2009). Mammalian iron transport. *Cellular and Molecular Life Sciences*, *66*(20), 3241–3261 (Review).

Arosio, P., Ingrassia, R., & Cavadini, P. (2009). Ferritins: A family of molecules for iron storage, antioxidation and more. *Biochimica et Biophysica Acta*, *1790*(7), 589–599 (Research Support, Non-U.S. Gov't Review).

Baker, M., Mackenzie, I. R., Pickering-Brown, S. M., Gass, J., Rademakers, R., Lindholm, C., et al. (2006). Mutations in progranulin cause tau-negative frontotemporal dementia linked to chromosome 17. *Nature*, *442*(7105), 916–919 (Research Support, N.I.H., Extramural Research Support, Non-U.S. Gov't).

Balla, J., Jacob, H. S., Balla, G., Nath, K., Eaton, J. W., & Vercellotti, G. M. (1993). Endothelial-cell heme uptake from heme proteins: Induction of sensitization and desensitization to oxidant damage. *Proceedings of the National Academy of Sciences of the United States of America*, *90*(20), 9285–9289 (In Vitro Research Support, U.S. Gov't, P.H.S.).

Balla, G., Jacob, H. S., Balla, J., Rosenberg, M., Nath, K., Apple, F., et al. (1992). Ferritin: A cytoprotective antioxidant strategem of endothelium. *The Journal of Biological Chemistry*, *267*(25), 18148–18153 (Research Support, U.S. Gov't, P.H.S.).

Barton, J. C. (2010). *Handbook of iron overload disorders*. Cambridge; New york: Cambridge University Press. ix, 376 p.

Beutler, E. (1988). Gaucher disease. *Blood Reviews*, *2*(1), 59–70 (Research Support, Non-U.S. Gov't Research Support, U.S. Gov't, P.H.S. Review).

Beutler, E. (2006). Gaucher disease: Multiple lessons from a single gene disorder. *Acta Paediatrica*, *95*(451), 103–109 (Research Support, N.I.H., Extramural Research Support, Non-U.S. Gov't Review).

Bras, J., Paisan-Ruiz, C., Guerreiro, R., Ribeiro, M. H., Morgadinho, A., Januario, C., et al. (2009). Complete screening for glucocerebrosidase mutations in Parkinson disease patients from Portugal. *Neurobiology of Aging*, *30*(9), 1515–1517.

Bras, J., Singleton, A., Cookson, M. R., & Hardy, J. (2008). Emerging pathways in genetic Parkinson's disease: Potential role of ceramide metabolism in Lewy body disease. *The FEBS Journal*, *275*(23), 5767–5773.

Bras, J., Verloes, A., Schneider, S. A., Mole, S. E., & Guerreiro, R. J. (2012). Mutation of the parkinsonism gene ATP13A2 causes neuronal ceroid-lipofuscinosis. *Human Molecular Genetics*, *21*(12), 2646–2650 (Research Support, Non-U.S. Gov't).

Bright, N. A., Gratian, M. J., & Luzio, J. P. (2005). Endocytic delivery to lysosomes mediated by concurrent fusion and kissing events in living cells. *Current Biology*, *15*(4), 360–365 (Comparative Study Research Support, Non-U.S. Gov't).

Brissot, P., Bardou-Jacquet, E., Jouanolle, A. M., & Loreal, O. (2011). Iron disorders of genetic origin: A changing world. *Trends in Molecular Medicine*, *17*(12), 707–713 (Research Support, Non-U.S. Gov't Review).

Brunk, U., & Ericsson, J. L. (1972). Electron microscopical studies on rat brain neurons. Localization of acid phosphatase and mode of formation of lipofuscin bodies. *Journal of Ultrastructure Research*, *38*(1), 1–15.

Brunk, U. T., Neuzil, J., & Eaton, J. W. (2001). Lysosomal involvement in apoptosis. *Redox Report: Communications in Free Radical Research*, *6*(2), 91–97 (Review).

Brunk, U. T., & Terman, A. (2002a). Lipofuscin: Mechanisms of age-related accumulation and influence on cell function. *Free Radical Biology & Medicine*, *33*(5), 611–619 (Research Support, Non-U.S. Gov't Review).

Brunk, U. T., & Terman, A. (2002b). The mitochondrial-lysosomal axis theory of aging: Accumulation of damaged mitochondria as a result of imperfect autophagocytosis. *European Journal of Biochemistry / FEBS*, *269*(8), 1996–2002 (Research Support, Non-U.S. Gov't Review).

Cao, Y., Staropoli, J. F., Biswas, S., Espinola, J. A., MacDonald, M. E., Lee, J. M., et al. (2011). Distinct early molecular responses to mutations causing vLINCL and JNCL presage ATP synthase subunit C accumulation in cerebellar cells. *PLoS One*, *6*(2), e17118 (Research Support, N.I.H., Extramural Research Support, Non-U.S. Gov't).

Cruts, M., Gijselinck, I., van der Zee, J., Engelborghs, S., Wils, H., Pirici, D., et al. (2006). Null mutations in progranulin cause ubiquitin-positive frontotemporal dementia linked to chromosome 17q21. *Nature*, *442*(7105), 920–924 (Research Support, Non-U.S. Gov't).

Das, A. K., Becerra, C. H., Yi, W., Lu, J. Y., Siakotos, A. N., Wisniewski, K. E., et al. (1998). Molecular genetics of palmitoyl-protein thioesterase deficiency in the U.S. *The Journal of Clinical Investigation*, *102*(2), 361–370 (Research Support, Non-U.S. Gov't Research Support, U.S. Gov't, P.H.S.).

De Domenico, I., McVey Ward, D., & Kaplan, J. (2008). Regulation of iron acquisition and storage: Consequences for iron-linked disorders. *Nature Reviews Molecular Cell Biology*, *9*(1), 72–81 (Research Support, N.I.H., Extramural Review).

de Duve, C. (2005). The lysosome turns fifty. *Nature Cell Biology*, 7(9), 847–849 (Historical Article).

Dehay, B., Ramirez, A., Martinez-Vicente, M., Perier, C., Canron, M. H., Doudnikoff, E., et al. (2012). Loss of P-type ATPase ATP13A2/PARK9 function induces general lysosomal deficiency and leads to Parkinson disease neurodegeneration. *Proceedings of the National Academy of Sciences of the United States of America*, *109*(24), 9611–9616 (Research Support, Non-U.S. Gov't).

Desjardins, M. (1995). Biogenesis of phagolysosomes: The 'kiss and run' hypothesis. *Trends in Cell Biology*, *5*(5), 183–186.

Dunn, L. L., Suryo Rahmanto, Y., & Richardson, D. R. (2007). Iron uptake and metabolism in the new millennium. *Trends in Cell Biology, 17*(2), 93–100 (Research Support, Non-U.S. Gov't Review).

Durr, A., Hahn-Barma, V., Brice, A., Pecheux, C., Dode, C., & Feingold, J. (1999). Homozygosity in Huntington's disease. *Journal of Medical Genetics, 36*(2), 172–173 (Case Reports Letter).

Eskelinen, E. L., Tanaka, Y., & Saftig, P. (2003). At the acidic edge: Emerging functions for lysosomal membrane proteins. *Trends in Cell Biology, 13*(3), 137–145 (Research Support, Non-U.S. Gov't Review).

Gabizon, R., Rosenmann, H., Meiner, Z., Kahana, I., Kahana, E., Shugart, Y., et al. (1993). Mutation and polymorphism of the prion protein gene in Libyan Jews with Creutzfeldt-Jakob disease (CJD). *American Journal of Human Genetics, 53*(4), 828–835 (Research Support, Non-U.S. Gov't Research Support, U.S. Gov't, Non-P.H.S. Research Support, U.S. Gov't, P.H.S.).

Galy, B., Ferring, D., Benesova, M., Benes, V., & Hentze, M. W. (2004). Targeted mutagenesis of the murine IRP1 and IRP2 genes reveals context-dependent RNA processing differences in vivo. *RNA, 10*(7), 1019–1025 (Research Support, Non-U.S. Gov't).

Germain, D. P. (2004). Gaucher's disease: A paradigm for interventional genetics. *Clinical Genetics, 65*(2), 77–86 (Review).

Guerreiro, R. J., Lohmann, E., Bras, J. M., Gibbs, J. R., Rohrer, J. D., Gurunlian, N., et al. (2012). Using exome sequencing to reveal mutations in TREM2 presenting as a frontotemporal dementia-like syndrome without bone involvement. *Archives of Neurology, 1–7.*

Guerreiro, R., Wojtas, A., Bras, J., Carrasquillo, M., Rogaeva, E., Majounie, E., et al. (2013). TREM2 variants in Alzheimer's disease. *The New England Journal of Medicine, 368*(2), 117–127 (Meta-Analysis Research Support, N.I.H., Extramural Research Support, N.I.H., Intramural Research Support, Non-U.S. Gov't).

Haglund, K., Sigismund, S., Polo, S., Szymkiewicz, I., Di Fiore, P. P., & Dikic, I. (2003). Multiple monoubiquitination of RTKs is sufficient for their endocytosis and degradation. *Nature Cell Biology, 5*(5), 461–466 (Research Support, Non-U.S. Gov't).

Hardy, J., Lewis, P., Revesz, T., Lees, A., & Paisan-Ruiz, C. (2009). The genetics of Parkinson's syndromes: A critical review. *Current Opinion in Genetics & Development, 19*(3), 254–265 (Review).

He, Z., & Bateman, A. (2003). Progranulin (granulin-epithelin precursor, PC-cell-derived growth factor, acrogranin) mediates tissue repair and tumorigenesis. *Journal of Molecular Medicine, 81*(10), 600–612 (Research Support, Non-U.S. Gov't Review).

Helenius, A., Mellman, I., Wall, D., & Hubbard, A. (1983). Endosomes. *Trends in Biochemical Sciences, 8*(7), 245–250.

Holopainen, J. M., Saarikoski, J., Kinnunen, P. K., & Jarvela, I. (2001). Elevated lysosomal pH in neuronal ceroid lipofuscinoses (NCLs). *European Journal of Biochemistry / FEBS, 268*(22), 5851–5856 (Research Support, Non-U.S. Gov't).

Horowitz, M., Wilder, S., Horowitz, Z., Reiner, O., Gelbart, T., & Beutler, E. (1989). The human glucocerebrosidase gene and pseudogene: Structure and evolution. *Genomics, 4*(1), 87–96 (Comparative Study Research Support, Non-U.S. Gov't Research Support, U.S. Gov't, P.H.S.).

Hruska, K. S., LaMarca, M. E., Scott, C. R., & Sidransky, E. (2008). Gaucher disease: Mutation and polymorphism spectrum in the glucocerebrosidase gene (GBA). *Human Mutation, 29*(5), 567–583 (Review).

Hu, F., Padukkavidana, T., Vaegter, C. B., Brady, O. A., Zheng, Y., Mackenzie, I. R., et al. (2010). Sortilin-mediated endocytosis determines levels of the frontotemporal dementia protein, progranulin. *Neuron, 68*(4), 654–667 (Comparative Study Research Support, N.I.H., Extramural Research Support, Non-U.S. Gov't).

Huang, F., Kirkpatrick, D., Jiang, X., Gygi, S., & Sorkin, A. (2006). Differential regulation of EGF receptor internalization and degradation by multiubiquitination within the kinase domain. *Molecular Cell, 21*(6), 737–748 (Research Support, N.I.H., Extramural Research Support, Non-U.S. Gov't).

Iancu, T., & Shiloh, H. (1994). Morphologic observations in iron overload: An update. In C. Hershko, A. Konijn, & P. Aisen (Eds.), *Progress in iron research, Vol. 356.* (pp. 255–265). US: Springer.

Isolation of a novel gene underlying Batten disease, CLN3. The International Batten disease consortium. (1995). *Cell, 82*(6), 949–957 (Research Support, Non-U.S. Gov't Research Support, U.S. Gov't, Non-P.H.S. Research Support, U.S. Gov't, P.H.S.).

Jarvela, I., Lehtovirta, M., Tikkanen, R., Kyttala, A., & Jalanko, A. (1999). Defective intracellular transport of CLN3 is the molecular basis of Batten disease (JNCL). *Human Molecular Genetics, 8*(6), 1091–1098 (Research Support, Non-U.S. Gov't).

Johansson, E., Lindh, U., Westermarck, T., Heiskala, H., & Santavuori, P. (1990). Altered elemental profiles in neuronal ceroid lipofuscinosis. *Journal of Trace Elements and Electrolytes in Health and Disease, 4*(3), 139–142 (Research Support, Non-U.S. Gov't).

Jolly, R. D., Douglas, B. V., Davey, P. M., & Roiri, J. E. (1995). Lipofuscin in bovine muscle and brain: A model for studying age pigment. *Gerontology, 41*(Suppl. 2), 283–295 (Research Support, U.S. Gov't, P.H.S.).

Journet, A., Chapel, A., Kieffer, S., Roux, F., & Garin, J. (2002). Proteomic analysis of human lysosomes: Application to monocytic and breast cancer cells. *Proteomics, 2*(8), 1026–1040 (Research Support, Non-U.S. Gov't).

Junaid, M. A., & Pullarkat, R. K. (1999). Increased brain lysosomal pepstatin-insensitive proteinase activity in patients with neurodegenerative diseases. *Neuroscience Letters, 264*(1–3), 157–160 (Research Support, U.S. Gov't, P.H.S.).

Katzin, L. W., & Amato, A. A. (2008). Pompe disease: A review of the current diagnosis and treatment recommendations in the era of enzyme replacement therapy. *Journal of Clinical Neuromuscular Disease, 9*(4), 421–431 (Review).

Klionsky, D. J. (2007). Autophagy: From phenomenology to molecular understanding in less than a decade. *Nature Reviews Molecular Cell Biology, 8*(11), 931–937 (Research Support, N.I.H., Extramural Review).

Kominami, A. E. (2002). What are the requirements for lysosomal degradation of subunit c of mitochondrial ATPase? *IUBMB Life, 54*(2), 89–90.

Kurz, T., Eaton, J. W., & Brunk, U. T. (2010). Redox activity within the lysosomal compartment: Implications for aging and apoptosis. *Antioxidants & Redox Signaling, 13*(4), 511–523 (Review).

Kurz, T., Terman, A., Gustafsson, B., & Brunk, U. T. (2008). Lysosomes and oxidative stress in aging and apoptosis. *Biochimica et Biophysica Acta, 1780*(11), 1291–1303 (Review).

Kyttala, A., Lahtinen, U., Braulke, T., & Hofmann, S. L. (2006). Functional biology of the neuronal ceroid lipofuscinoses (NCL) proteins. *Biochimica et Biophysica Acta, 1762*(10), 920–933 (Review).

Lerer, I., Merims, D., Abeliovich, D., Zlotogora, J., & Gadoth, N. (1996). Machado-Joseph disease: Correlation between the clinical features, the CAG repeat length and homozygosity for the mutation. *European Journal of Human Genetics, 4*(1), 3–7.

Levine, B., & Klionsky, D. J. (2004). Development by self-digestion: Molecular mechanisms and biological functions of autophagy. *Developmental Cell, 6*(4), 463–477 (Research Support, Non-U.S. Gov't Research Support, U.S. Gov't, P.H.S. Review).

Levy, J. E., Jin, O., Fujiwara, Y., Kuo, F., & Andrews, N. C. (1999). Transferrin receptor is necessary for development of erythrocytes and the nervous system. *Nature Genetics, 21*(4), 396–399 (Comparative Study Research Support, Non-U.S. Gov't Research Support, U.S. Gov't, P.H.S.).

Luzio, J. P., Pryor, P. R., & Bright, N. A. (2007). Lysosomes: Fusion and function. *Nature Reviews. Molecular Cell Biology, 8*(8), 622–632 (Research Support, Non-U.S. Gov't Review).

Machaczka, M., Rucinska, M., Skotnicki, A. B., & Jurczak, W. (1999). Parkinson's syndrome preceding clinical manifestation of Gaucher's disease. *American Journal of Hematology, 61*(3), 216–217 (Case Reports Letter).

Matsumura, R., Futamura, N., Fujimoto, Y., Yanagimoto, S., Horikawa, H., Suzumura, A., et al. (1997). Spinocerebellar ataxia type 6. Molecular and clinical features of 35 Japanese patients including one homozygous for the CAG repeat expansion. *Neurology, 49*(5), 1238–1243 (Research Support, Non-U.S. Gov't).

McNeil, P. L., & Kirchhausen, T. (2005). An emergency response team for membrane repair. *Nature Reviews Molecular Cell Biology, 6*(6), 499–505 (Research Support, N.I.H., Extramural Research Support, U.S. Gov't, Non-P.H.S. Research Support, U.S. Gov't, P.H.S. Review).

Mekinian, A., Stirnemann, J., Belmatoug, N., Heraoui, D., Fantin, B., Fain, O., et al. (2012). Ferritinemia during type 1 Gaucher disease: Mechanisms and progression under treatment. *Blood Cells, Molecules & Diseases, 49*(1), 53–57 (Research Support, Non-U.S. Gov't).

Mellman, I. (1996). Endocytosis and molecular sorting. *Annual Review of Cell and Developmental Biology, 12*, 575–625 (Review).

Mellman, I., Fuchs, R., & Helenius, A. (1986). Acidification of the endocytic and exocytic pathways. *Annual Review of Biochemistry, 55*, 663–700 (Review).

Morgan, N. V., Westaway, S. K., Morton, J. E., Gregory, A., Gissen, P., Sonek, S., et al. (2006). PLA2G6, encoding a phospholipase A2, is mutated in neurodegenerative disorders with high brain iron. *Nature Genetics, 38*(7), 752–754 (Research Support, N.I.H., Extramural Research Support, Non-U.S. Gov't).

Mukherjee, S., Ghosh, R. N., & Maxfield, F. R. (1997). Endocytosis. *Physiological Reviews, 77*(3), 759–803 (Research Support, Non-U.S. Gov't Research Support, U.S. Gov't, P.H.S. Review).

Munoz, M., Villar, I., & Garcia-Erce, J. A. (2009). An update on iron physiology. *World Journal of Gastroenterology, 15*(37), 4617–4626 (Review).

Nakagawa, I., Amano, A., Mizushima, N., Yamamoto, A., Yamaguchi, H., Kamimoto, T., et al. (2004). Autophagy defends cells against invading group a streptococcus. *Science, 306*(5698), 1037–1040 (Research Support, Non-U.S. Gov't).

Narayan, S. B., Rakheja, D., Tan, L., Pastor, J. V., & Bennett, M. J. (2006). CLN3P, the Batten's disease protein, is a novel palmitoyl-protein Delta-9 desaturase. *Annals of Neurology, 60*(5), 570–577.

Nemeth, E., & Ganz, T. (2006). Regulation of iron metabolism by hepcidin. *Annual Review of Nutrition, 26*, 323–342 (Research Support, N.I.H., Extramural Research Support, Non-U.S. Gov't Review).

Neudorfer, O., Giladi, N., Elstein, D., Abrahamov, A., Turezkite, T., Aghai, E., et al. (1996). Occurrence of Parkinson's syndrome in type I Gaucher disease. *QJM: Monthly Journal of the Association of Physicians, 89*(9), 691–694 (Research Support, Non-U.S. Gov't).

Neumann, J., Bras, J., Deas, E., O'Sullivan, S. S., Parkkinen, L., Lachmann, R. H., et al. (2009). Glucocerebrosidase mutations in clinical and pathologically proven Parkinson's disease. *Brain, 132*(Pt 7), 1783–1794.

Nichols, W. C., Pankratz, N., Hernandez, D., Paisan-Ruiz, C., Jain, S., Halter, C. A., et al. (2005). Genetic screening for a single common LRRK2 mutation in familial Parkinson's disease. *Lancet, 365*(9457), 410–412 (Multicenter Study Research Support, Non-U.S. Gov't Research Support, U.S. Gov't, P.H.S.).

Nixon, R. A., Yang, D. S., & Lee, J. H. (2008). Neurodegenerative lysosomal disorders: A continuum from development to late age. *Autophagy*, *4*(5), 590–599 (Research Support, N.I.H., Extramural Research Support, Non-U.S. Gov't Review).

Paisan-Ruiz, C., Bhatia, K. P., Li, A., Hernandez, D., Davis, M., Wood, N. W., et al. (2009). Characterization of PLA2G6 as a locus for dystonia-parkinsonism. *Annals of Neurology*, *65*(1), 19–23 (Research Support, N.I.H., Intramural Research Support, Non-U.S. Gov't).

Palmer, D. N., Fearnley, I. M., Walker, J. E., Hall, N. A., Lake, B. D., Wolfe, L. S., et al. (1992). Mitochondrial ATP synthase subunit c storage in the ceroid-lipofuscinoses (Batten disease). *American Journal of Medical Genetics*, *42*(4), 561–567 (Research Support, Non-U.S. Gov't Research Support, U.S. Gov't, P.H.S.).

Palmer, D. N., Jolly, R. D., van Mil, H. C., Tyynela, J., & Westlake, V. J. (1997). Different patterns of hydrophobic protein storage in different forms of neuronal ceroid lipofuscinosis (NCL, Batten disease). *Neuropediatrics*, *28*(1), 45–48 (Research Support, U.S. Gov't, P.H.S.).

Paloneva, J., Manninen, T., Christman, G., Hovanes, K., Mandelin, J., Adolfsson, R., et al. (2002). Mutations in two genes encoding different subunits of a receptor signaling complex result in an identical disease phenotype. *American Journal of Human Genetics*, *71*(3), 656–662 (Research Support, Non-U.S. Gov't).

Peters, T. J., & Seymour, C. A. (1976). Acid hydrolase activities and lysosomal integrity in liver biopsies from patients with iron overload. *Clinical Science and Molecular Medicine*, *50*(1), 75–78.

Pryor, P. R., Mullock, B. M., Bright, N. A., Gray, S. R., & Luzio, J. P. (2000). The role of intraorganellar Ca(2+) in late endosome-lysosome heterotypic fusion and in the reformation of lysosomes from hybrid organelles. *The Journal of Cell Biology*, *149*(5), 1053–1062 (Research Support, Non-U.S. Gov't).

Ramirez, A., Heimbach, A., Grundemann, J., Stiller, B., Hampshire, D., Cid, L. P., et al. (2006). Hereditary parkinsonism with dementia is caused by mutations in ATP13A2, encoding a lysosomal type 5 P-type ATPase. *Nature Genetics*, *38*(10), 1184–1191 (Research Support, Non-U.S. Gov't).

Ramirez-Montealegre, D., & Pearce, D. A. (2005). Defective lysosomal arginine transport in juvenile Batten disease. *Human Molecular Genetics*, *14*(23), 3759–3773 (Research Support, N.I.H., Extramural Research Support, Non-U.S. Gov't).

Richardson, D. R., & Ponka, P. (1997). The molecular mechanisms of the metabolism and transport of iron in normal and neoplastic cells. *Biochimica et Biophysica Acta*, *1331*(1), 1–40 (Research Support, Non-U.S. Gov't Review).

Rodriguez, A., Webster, P., Ortego, J., & Andrews, N. W. (1997). Lysosomes behave as Ca2+ – regulated exocytic vesicles in fibroblasts and epithelial cells. *The Journal of Cell Biology*, *137*(1), 93–104 (Research Support, U.S. Gov't, P.H.S.).

Rouault, T. A. (2006). The role of iron regulatory proteins in mammalian iron homeostasis and disease. *Nature Chemical Biology*, *2*(8), 406–414 (Research Support, N.I.H., Intramural Review).

Ryan, C. L., Baranowski, D. C., Chitramuthu, B. P., Malik, S., Li, Z., Cao, M., et al. (2009). Progranulin is expressed within motor neurons and promotes neuronal cell survival. *BMC Neuroscience*, *10*, 130 (Research Support, Non-U.S. Gov't).

Sanpei, K., Takano, H., Igarashi, S., Sato, T., Oyake, M., Sasaki, H., et al. (1996). Identification of the spinocerebellar ataxia type 2 gene using a direct identification of repeat expansion and cloning technique, DIRECT. *Nature Genetics*, *14*(3), 277–284 (Research Support, Non-U.S. Gov't).

Schulz, A., Mousallem, T., Venkataramani, M., Persaud-Sawin, D. A., Zucker, A., Luberto, C., et al. (2006). The CLN9 protein, a regulator of dihydroceramide synthase.

The Journal of Biological Chemistry, 281(5), 2784–2794 (Research Support, Non-U.S. Gov't).

Scriver, C. R. (1995). *The metabolic and molecular bases of inherited disease* (7th ed.). New York: McGraw-Hill, Health Professions Division.

Seehafer, S. S., & Pearce, D. A. (2006). You say lipofuscin, we say ceroid: Defining autofluorescent storage material. *Neurobiology of Aging, 27*(4), 576–588 (Research Support, N.I.H., Extramural Review).

Selden, C., Owen, M., Hopkins, J. M., & Peters, T. J. (1980). Studies on the concentration and intracellular localization of iron proteins in liver biopsy specimens from patients with iron overload with special reference to their role in lysosomal disruption. *British Journal of Haematology, 44*(4), 593–603.

Seymour, C. A., & Peters, T. J. (1978). Organelle pathology in primary and secondary haemochromatosis with special reference to lysosomal changes. *British Journal of Haematology, 40*(2), 239–253.

Shapiro, B. E., Logigian, E. L., Kolodny, E. H., & Pastores, G. M. (2008). Late-onset Tay-Sachs disease: The spectrum of peripheral neuropathy in 30 affected patients. *Muscle & Nerve, 38*(2), 1012–1015.

Sharp, P., & Srai, S. K. (2007). Molecular mechanisms involved in intestinal iron absorption. *World Journal of Gastroenterology, 13*(35), 4716–4724 (Research Support, Non-U.S. Gov't Review).

Sidransky, E., Nalls, M. A., Aasly, J. O., Aharon-Peretz, J., Annesi, G., Barbosa, E. R., et al. (2009). Multicenter analysis of glucocerebrosidase mutations in Parkinson's disease. *The New England Journal of Medicine, 361*(17), 1651–1661.

Siintola, E., Partanen, S., Stromme, P., Haapanen, A., Haltia, M., Maehlen, J., et al. (2006). Cathepsin D deficiency underlies congenital human neuronal ceroid-lipofuscinosis. *Brain: A Journal of Neurology, 129*(Pt 6), 1438–1445 (Case Reports Research Support, Non-U.S. Gov't).

Smith, K. R., Damiano, J., Franceschetti, S., Carpenter, S., Canafoglia, L., Morbin, M., et al. (2012). Strikingly different clinicopathological phenotypes determined by progranulin-mutation dosage. *American Journal of Human Genetics, 90*(6), 1102–1107 (Research Support, Non-U.S. Gov't).

Sobue, G., Doyu, M., Nakao, N., Shimada, N., Mitsuma, T., Maruyama, H., et al. (1996). Homozygosity for Machado-Joseph disease gene enhances phenotypic severity. *Journal of Neurology, Neurosurgery, and Psychiatry, 60*(3), 354–356 (Case Reports Letter).

Spada, M., Pagliardini, S., Yasuda, M., Tukel, T., Thiagarajan, G., Sakuraba, H., et al. (2006). High incidence of later-onset fabry disease revealed by newborn screening. *American Journal of Human Genetics, 79*(1), 31–40 (Research Support, N.I.H., Extramural Research Support, Non-U.S. Gov't).

Staropoli, J. F., Karaa, A., Lim, E. T., Kirby, A., Elbalalesy, N., Romansky, S. G., et al. (2012). A homozygous mutation in KCTD7 links neuronal ceroid lipofuscinosis to the ubiquitin-proteasome system. *American Journal of Human Genetics, 91*(1), 202–208 (Case Reports Research Support, N.I.H., Extramural Research Support, Non-U.S. Gov't).

Takahashi, K., Prinz, M., Stagi, M., Chechneva, O., & Neumann, H. (2007). TREM2-transduced myeloid precursors mediate nervous tissue debris clearance and facilitate recovery in an animal model of multiple sclerosis. *PLoS Medicine, 4*(4), e124 (Research Support, Non-U.S. Gov't).

Tang, J., Kriz, R. W., Wolfman, N., Shaffer, M., Seehra, J., & Jones, S. S. (1997). A novel cytosolic calcium-independent phospholipase A2 contains eight ankyrin motifs. *The Journal of Biological Chemistry, 272*(13), 8567–8575.

Terman, A., Gustafsson, B., & Brunk, U. T. (2007). Autophagy, organelles and ageing. *The Journal of Pathology, 211*(2), 134–143 (Review).

Van Bogaert, P., Azizieh, R., Desir, J., Aeby, A., De Meirleir, L., Laes, J. F., et al. (2007). Mutation of a potassium channel-related gene in progressive myoclonic epilepsy. *Annals of Neurology*, *61*(6), 579–586 (Research Support, Non-U.S. Gov't).

van Swieten, J. C., & Heutink, P. (2008). Mutations in progranulin (GRN) within the spectrum of clinical and pathological phenotypes of frontotemporal dementia. *Lancet Neurology*, 7(10), 965–974 (Review).

Weisberger, J., Emmons, F., & Gorczyca, W. (2004). Cytochemical diagnosis of Gaucher's disease by iron stain. *British Journal of Haematology*, *124*(6), 696.

Westbroek, W., Gustafson, A. M., & Sidransky, E. (2011). Exploring the link between glucocerebrosidase mutations and parkinsonism. *Trends in Molecular Medicine*, *17*(9), 485–493 (Research Support, N.I.H., Intramural Review).

Wilson, J., Huynh, C., Kennedy, K. A., Ward, D. M., Kaplan, J., Aderem, A., et al. (2008). Control of parasitophorous vacuole expansion by LYST/Beige restricts the intracellular growth of Leishmania amazonensis. *PLoS Pathogens*, *4*(10), e1000179 (Research Support, N.I.H., Extramural).

Wraith, J. E. (2002). Lysosomal disorders. *Seminars in Neonatology*, 7(1), 75–83 (Review).

Yu, Z., Persson, H. L., Eaton, J. W., & Brunk, U. T. (2003). Intralysosomal iron: A major determinant of oxidant-induced cell death. *Free Radical Biology & Medicine*, *34*(10), 1243–1252 (Comparative Study Research Support, Non-U.S. Gov't Research Support, U.S. Gov't, P.H.S.).

Zhang, Y., Mikhael, M., Xu, D., Li, Y., Soe-Lin, S., Ning, B., et al. (2010). Lysosomal proteolysis is the primary degradation pathway for cytosolic ferritin and cytosolic ferritin degradation is necessary for iron exit. *Antioxidants & Redox Signaling*, *13*(7), 999–1009 (Research Support, Non-U.S. Gov't).

CHAPTER TWELVE

Manganese and the Brain

Karin Tuschl[1], Philippa B. Mills, Peter T. Clayton
Clinical and Molecular Genetics Unit, UCL Institute of Child Health, London, United Kingdom
[1]Corresponding author: e-mail address: k.tuschl@ucl.ac.uk

Contents

Abstract

Manganese (Mn) is an essential trace metal that is pivotal for normal cell function and metabolism. Its homeostasis is tightly regulated; however, the mechanisms of Mn homeostasis are poorly characterized. While a number of proteins such as the divalent metal transporter 1, the transferrin/transferrin receptor complex, the ZIP family metal transporters ZIP-8 and ZIP-14, the secretory pathway calcium ATPases SPCA1 and SPCA2, ATP13A2, and ferroportin have been suggested to play a role in Mn transport, the degree that each of them contributes to Mn homeostasis has still to be determined. The recent discovery of SLC30A10 as a crucial Mn transporter in humans has shed further light on our understanding of Mn transport across the cell. Although essential, Mn is toxic at high concentrations. Mn neurotoxicity has been attributed to impaired dopaminergic (DAergic), glutamatergic and GABAergic transmission, mitochondrial dysfunction, oxidative stress, and neuroinflammation. As a result of preferential accumulation of Mn in the DAergic cells of the basal ganglia, particularly the globus pallidus, Mn toxicity causes extrapyramidal motor dysfunction. Firstly described as "manganism" in miners during the nineteenth century, this movement disorder resembles Parkinson's disease characterized by hypokinesia and postural instability. To date, a variety of acquired

International Review of Neurobiology, Volume 110
ISSN 0074-7742
http://dx.doi.org/10.1016/B978-0-12-410502-7.00013-2

causes of brain Mn accumulation can be distinguished from an autosomal recessively inherited disorder of Mn metabolism caused by mutations in the *SLC30A10* gene. Both, acquired and inherited hypermanganesemia, lead to Mn deposition in the basal ganglia associated with pathognomonic magnetic resonance imaging appearances of hyperintense basal ganglia on T1-weighted images. Current treatment strategies for Mn toxicity combine chelation therapy to reduce the body Mn load and iron (Fe) supplementation to reduce Mn binding to proteins that interact with both Mn and Fe. This chapter summarizes our current understanding of Mn homeostasis and the mechanisms of Mn toxicity and highlights the clinical disorders associated with Mn neurotoxicity.

1. MANGANESE: AN ESSENTIAL TRACE METAL AND NEUROTOXIN

Manganese (Mn) is an essential trace element that is crucial for the function of a variety of enzyme families including oxidoreductases, transferases, hydrolases, lyases, isomerases, and ligases. As such, Mn is involved in amino acid, lipid, protein, and carbohydrate metabolism as well as protein glycosylation, immune function, regulation of blood sugar, production of cell energy, reproduction, digestion, bone growth, and blood clotting (Aschner & Aschner, 2005; Aschner, Erikson, & Dorman, 2005; Roth, 2006). Additionally, Mn forms a constituent of metalloenzymes such as the Mn superoxide dismutase located at the mitochondria that facilitates the detoxification of superoxide free radicals (Martinez-Finley, Gavin, Aschner, & Gunter, 2013). Mn is required in all human tissues at concentrations ranging from 0.3 to 2.9 μg Mn/g wet tissue, and these tissue levels are under tight homeostatic regulation. The liver, kidney, pancreas, bone, and parts of the brain including basal ganglia and cerebellum are tissues particularly rich in Mn (Aschner et al., 2005). Our diet is the major source of Mn, which is present ubiquitously in a variety of foods. Especially Mn-rich are grains, green vegetables, fruits, nuts, spices, and tea (Aschner & Aschner, 2005; Roth, 2006). The gut tightly controls the body Mn load as only 1–5% of ingested Mn is absorbed from the intestine (Davidsson, Cederblad, Lonnerdal, & Sandstrom, 1991). Following Mn absorption, any excess Mn present in the portal circulation is promptly excreted into the bile by the liver, the key regulator of Mn elimination. A significant fraction of biliary excreted Mn undergoes enterohepatic circulation (Roth, 2006). Increased dietary exposure to Mn induces homeostatic adaptation and leads to reduced gastrointestinal Mn absorption and increased biliary excretion of

Mn (Abrams et al., 1976; Britton & Cotzias, 1966; Davis, Wolf, & Greger, 1992; Malecki, Radzanowski, Radzanowski, Gallaher, & Greger, 1996). Mn absorption is also affected by several dietary factors such as the presence of other trace elements (Davidsson et al., 1991). Particularly well studied is its interaction with iron (Fe). High intake of Fe reduces blood Mn levels, while Fe-deficiency anemia enhances intestinal Mn absorption (Davis et al., 1992; Fitsanakis et al., 2011).

Nutritional Mn deficiency has not been described in humans and has been attributed to its ubiquitous presence in the diet. However, experimental models of Mn deficiency have suggested that low levels of Mn cause poor bone growth, skeletal abnormalities, ataxia, and abnormal glucose tolerance (Keen et al., 1999). Furthermore, erythematous skin changes and altered mood have been observed in subjects on experimental low Mn diets (Friedman et al., 1987; Penland & Johnson, 1993). Mn overload from excess dietary or respiratory Mn exposure on the other hand is a well-recognized condition known as "manganism." Mn accumulates in the basal ganglia and causes distinct neurotoxic effects. The first description of Mn neurotoxicity occurred in 1837 by Couper who published a case series of five Mn ore crushers who suffered from parkinsonian symptoms including tremor in the extremities, gait disturbance, and whispering speech (Couper, 1837). Rodier further described Mn poisoning in 151 workers of Moroccan mines a century later (Rodier, 1955). Manganism has since been extensively characterized; this parkinsonian syndrome can be divided in three overlapping stages: An initial stage of emotional and cognitive disturbances with impaired fine motor coordination, followed by a second stage with worsening psychotic behavior, slurred speech, clumsiness, and masklike facies. During the final stage, patients develop disabling limb rigidity, dystonia, impairment of balance, and a characteristic cock-walk gait (Huang, Chu, Lu, & Calne, 1997; Roth, 2009). While some symptoms overlap with idiopathic Parkinson's disease (PD) such as bradykinesia, there are a number of dissimilarities including (a) less-frequent resting tremor, (b) more frequent dystonia, (c) a particular propensity to fall backward, (d) failure to achieve a sustained therapeutic response to levodopa, and (e) failure to show a reduction in fluorodopa uptake by positron emission tomography (PET) (Calne, Chu, Huang, Lu, & Olanow, 1994).

Excessive Mn accumulation in the brain can occur either through Mn overexposure or through impaired hepatobiliary excretion. The predominant route of exposure in occupational settings is the inhalation of Mn-laden dust during mining, welding, smelting, or battery manufacturing.

Individuals living near ferroalloy industries or in areas with high use of methylcyclopentadienyl Mn tricarbonyl (MMT), a gasoline fuel additive, are at risk of airborne Mn exposure and accumulate increased levels of Mn. Whether inhaled Mn is directly taken up via the olfactory or trigeminal nerve, transported across the pulmonary epithelium, or ingested following mucociliary clearance from the lungs has still to be elucidated (Heilig et al., 2006; Leavens, Rao, Andersen, & Dorman, 2007; Roth, 2006). Excessive dietary exposure to Mn can lead to increased absorption and accumulation of Mn in the brain and has been described in communities receiving drinking water with high Mn content (Bouchard et al., 2011; Khan et al., 2012). High doses of Mn in parenteral nutrition (PN) that bypass gut and liver control mechanisms can further lead to Mn neurotoxicity (Chalela, Bonillha, Neyens, & Hays, 2011). Similarly, cases of Mn poisoning have been observed in intravenous methcathinone drug addicts who use potassium permanganate as the oxidant in the synthesis of the drug (Stepens et al., 2008). Mn accumulation in end-stage liver disease occurs due to impaired hepatic function that leads to insufficient biliary excretion of Mn and subsequent accumulation of Mn in the basal ganglia, a condition described as acquired hepatocerebral degeneration (AHD) (Butterworth, 2012).

While acquired causes of hypermanganesemia have been known about and diagnosed for a long time, the first genetic Mn overload syndrome caused by mutations in the Mn transporter gene *SLC30A10* was only recently identified (Quadri et al., 2012; Tuschl et al., 2012). Loss-of-function mutations in *SLC30A10*, a Mn transporter forming part of the secretory pathway, lead to accumulation of Mn in the liver and brain associated with an extrapyramidal movement disorder and chronic liver disease that may progress to cirrhosis.

2. MANGANESE TRANSPORT AND HOMEOSTASIS

The mechanisms of Mn homeostasis and Mn transport have been the subject of many studies; however, the exact routes of Mn transfer into and out of the cell remain controversial. Uptake and efflux of both Mn and Fe are regulated by a number of shared transporters suggesting a strong interdependency of Mn and Fe transport.

2.1. Enteral Mn uptake

Absorption of Mn into the enterocyte is thought to occur in the proximal small bowel via the divalent metal transporter 1 (DMT1). DMT1 transports both Mn and Fe and also a range of other cations. DMT1 is a H^+ symporter that transports one H^+ and one divalent cation in the same direction (Sacher, Cohen, & Nelson, 2001). DMT1 is ubiquitously present in all tissues with high levels of DMT1 found in the duodenum. DMT1 expression is regulated by Fe status associated with increased DMT1 expression at the enterocyte membrane during Fe deficiency (Gunshin et al., 1997). Consequently, in Fe deficiency, Mn absorption across the intestine is increased with subsequent Mn accumulation in the brain. Conversely, diets high in Mn lead to decreased plasma Fe levels with an increase in transferrin (Tf) and total iron-binding capacity (TIBC) (Fitsanakis, Zhang, Garcia, & Aschner, 2010; Garcia, Gellein, Syversen, & Aschner, 2007). Mutations in *SLC11A2*, the gene encoding DMT1, in the microcytic (*mk*) mice and Belgrade (*b/b*) rats affect both Mn and Fe transport across the intestine and result in severe microcytic anemia (Burdo et al., 2001; Chua & Morgan, 1997; Knopfel, Zhao, & Garrick, 2005). Efflux from the enterocyte is believed to occur via Ferroportin (Fpn), which is discussed in more detail later. Once absorbed into the portal circulation, the main proportion of Mn, around 80%, is bound to β-globulin and albumin as bivalent Mn^{2+}. Some is oxidized to Mn^{3+} by ceruloplasmin and bound to Tf as trivalent Mn^{3+} (Aschner, Guilarte, Schneider, & Zheng, 2007).

2.2. Manganese transport into the brain

The exact mechanisms involved in Mn transport into the brain are not well understood. Three routes of uptake into the central nervous system have been proposed including entry across the cerebral capillaries at the blood brain barrier (BBB), crossing the choroid plexus into the cerebrospinal fluid (CSF), or direct intraaxonal uptake via olfactory or trigeminal presynaptic nerve endings (Aschner et al., 2005; Yokel, 2009).

The major contributors to Mn transfer into the brain involve DMT1 (via Tf-dependent and Tf-independent pathways), ZIP-8 and ZIP-14, members of the solute carrier-39 family, voltage-regulated, store-operated, and ionotropic glutamate receptor Ca^{2+} channels, and uptake as Mn citrate complex (Au, Benedetto, & Aschner, 2008; Crossgrove, Allen, Bukaveckas, Rhineheimer, & Yokel, 2003; Crossgrove & Yokel, 2004, 2005) (Fig. 12.1).

Mn^{3+} has a high affinity to Tf, and in the blood, around 20% of total Mn is bound to Tf as Mn^{3+}. Binding of the Tf–Mn^{3+} complex to the Tf receptor (TfR) at the plasma membrane causes internalization of the Tf–TfR complex into endosomal vesicles. The endosomal V-ATPase facilitates

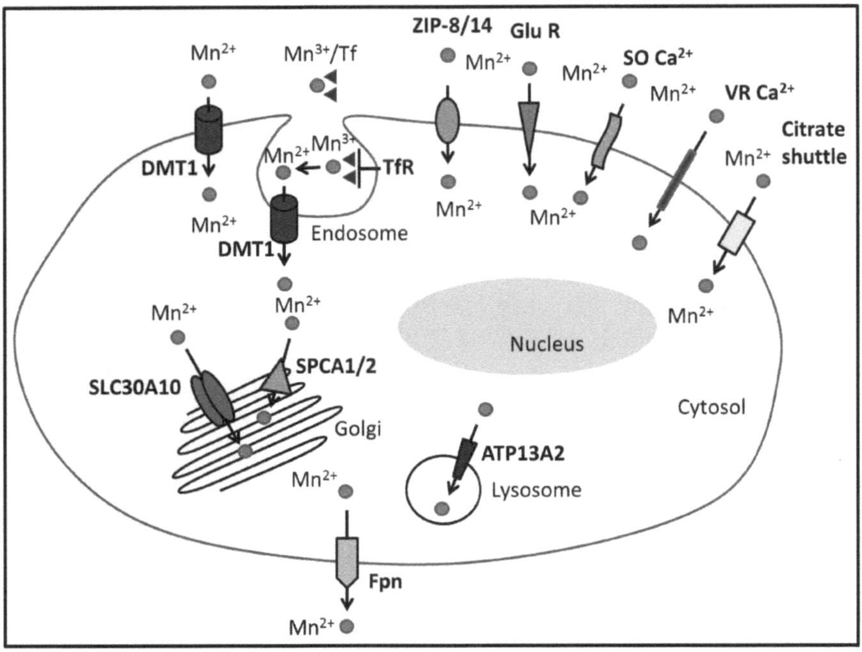

Figure 12.1 Transporters known to be involved in Mn trafficking across the cell. DMT1, divalent metal transporter 1; Tf, transferrin; TfR, transferrin receptor; Glu R, glutamate receptor; So Ca^{2+}, store-operated Ca^{2+} channel; VR Ca^{2+}, voltage-regulated Ca^{2+} channel; Fpn, ferroportin. (For color version of this figure, the reader is referred to the online version of this chapter.)

acidification of the vesicle and leads to release of the metal, followed by conversion of Mn^{3+} to Mn^{2+} and activation of endosomal DMT1 with subsequent uptake of Mn into the cytosol (Au et al., 2008; Gunter et al., 2012). Tf and DMT1 colocalize at the endosomal membrane at the BBB supporting the theory of Tf-dependent uptake of Mn (Gruenheid et al., 1999). Additionally, DMT1 has been suggested to transfer divalent Mn^{2+} at the BBB via a Tf-independent way. However, there are contradictory data as to whether DMT1 is actually expressed in capillary endothelial cells (Burdo et al., 2001; Wang, Ong, & Connor, 2001). Furthermore, it has been shown that Mn^{2+} uptake requires a neutral pH that is not consistent with the acidic pH optimum required for DMT1-facilitated uptake (Yokel & Crossgrove, 2004). On the other hand, highest expression levels of DMT1 in the brain are found in the dopamine (DA)-rich areas of the basal ganglia, the brain regions most affected by Mn toxicity (Huang, Ong, & Connor, 2004). In addition,

DMT1 expression is induced upon Mn exposure, which is associated with increased dopaminergic (DAergic) neurotoxicity (Sriram, Lin, Jefferson, Roberts, Chapman, et al., 2010).

ZIP-8 and ZIP-14, both members of the solute carrier-39 family, are expressed in tissues involved in the regulation of Mn homeostasis (including duodenum, liver, brain, lungs, and kidney) and transfer Mn, Fe, zinc (Zn), and cadmium (Cd) into the cell. They are localized on apical surfaces of brain capillaries and have been shown to transport Mn^{2+} with high affinity (Girijashanker et al., 2008; He et al., 2006; Wang et al., 2012). Maximal activity occurs at a pH of 7.5, which is consistent with the finding of Crossgrove and Yokel that Mn uptake positively correlates with pH (Crossgrove & Yokel, 2005). ZIP-8 and ZIP-14 are Mn^{2+}/HCO_3^- symporters that use a HCO_3^- gradient across the plasma membrane as the driving force for Mn uptake (Girijashanker et al., 2008; He et al., 2006). Studies on the function of ZIP-8 and ZIP-14 have shown that they also play a role in the absorption of Mn in the liver and the proximal tubule in the kidney. Since they are expressed at the BBB, it is likely that they contribute to Mn absorption into the brain (Fujishiro, Yano, Takada, Tanihara, & Himeno, 2012; Girijashanker et al., 2008).

The relative contribution of Ca^{2+} channels to Mn transport into the brain has still to be determined. Evidence suggests that both voltage-regulated and store-operated Ca^{2+} channels mediate brain Mn influx (Crossgrove & Yokel, 2005; Lucaciu, Dragu, Copaescu, & Morariu, 1997). The ionotropic glutamate receptor channel is yet another candidate proposed to facilitate Mn uptake into the brain (Kannurpatti, Joshi, & Joshi, 2000).

Mn citrate represents the major non-protein-bound species of Mn to enter the brain at the BBB. The influx transfer coefficient for Mn citrate was shown to be greater than that of Mn^{2+} alone and Tf–Mn^{3+} (Crossgrove et al., 2003).

Additional to Mn transfer at the BBB, Mn uptake appears to occur via the choroid plexus. Studies in rats have shown that Mn readily crosses the filtration barrier of the choroid plexus with subsequent absorption of Mn at the ventricular ependymal surfaces and transport into the subarachnoid space (London, Toney, Gabel, & Funk, 1989). In rats, intravenous injection of $^{54}MnCl_2$ leads to substantial accumulation of ^{54}Mn in the choroid plexus within 1 h of injection followed by a gradual increase of Mn concentrations in the surrounding brain tissues (Takeda, Sawashita, & Okada, 1998).

Airborne Mn can also enter the brain via direct intraaxonal transport along the olfactory nerve. Studies in rats administering intranasal $MnCl_2$ confirmed accumulation of Mn in the olfactory bulb and led to an increase in brain Mn concentration in a dose-, time-, and tissue-dependent manner. Repeated application resulted in deposition of Mn in the striatum, the target site of Mn toxicity (Gianutsos, Morrow, & Morris, 1997).

2.3. Mn detoxification and efflux from the cell

The cytoplasmic Fe exporter Fpn has been shown to act as an effective mediator of Mn efflux. Fpn is highly expressed at the plasma membrane of all relevant Fe-exporting cells such as enterocytes, hepatocytes, and macrophages but is also present in neurons and oligodendrocytes. Mn exposure significantly increases Fpn expression with subsequent reduction of total cell Mn and attenuated Mn toxicity (Madejczyk & Ballatori, 2012; Yin et al., 2010). Patients with mutations in the Fpn gene *SLC40A1* develop hemochromatosis type 4 with Fe overload within the reticuloendothelial system (Santos et al., 2012). To date, the effect of Fpn mutations on Mn metabolism have not been studied. Hepcidin, a peptide hormone produced by the liver, plays a pivotal role for systemic Fe homeostasis by regulating the activity of Fpn. High Fe levels and inflammation increase hepcidin expression followed by internalization and degradation of the hepcidin–Fpn complex (Tussing-Humphreys, Pusatcioglu, Nemeth, & Braunschweig, 2012). Recently, Mn has also been shown to induce hepcidin expression in hepatocytes providing a further link between Mn and Fe homeostasis (Bartnikas, 2012).

Recently, SLC30A10 has been identified as a key player in the regulation of Mn homeostasis. Recognized as the disease-causing gene in an inherited Mn overload syndrome, it is thought to mediate Mn sequestration into the Golgi (Quadri et al., 2012; Tuschl et al., 2012). SLC30A10 belongs to the cation diffusion facilitator family of divalent metal ion transporters (Sreedharan, Stephansson, Schioth, & Fredriksson, 2011). They share the same structure of six transmembrane domains (TMD) with cytosolic N- and C-termini and a conserved cation efflux domain at the C-terminus. Sequence homology with transporters from the same family has suggested a role of SLC30A10 in Zn transport (Seve, Chimienti, Devergnas, & Favier, 2004). However, recent work examining its function in a yeast model has confirmed its ability to transport Mn and protect cells from Mn toxicity. An alteration in the metal-binding affinity of the transporter has been attributed to evolutionary changes of the amino acid sequence of the protein (Tuschl et al., 2012). Structures required for Zn binding such as a histidine-rich region in a cytosolic loop and a conserved HxxxD motif in TMD II and V have been replaced by sequences rich in arginines, lysines, and serines. Together with a conserved cytosolic cysteine residue at TMD IV, these data suggest binding specificity for Mn (Quadri et al., 2012). Studies investigating the subcellular localization of SLC30A10 have confirmed its position in the Golgi. High extracellular Zn concentrations cause the trafficking of SLC30A10 to the plasma membrane (Bosomworth, Thornton, Coneyworth, Ford, & Valentine, 2012). The effect of Mn exposure on subcellular localization of SLC30A10 has not been studied. *SLC30A10* is highly expressed in the small intestine, liver, and brain tissues consistent with Mn accumulation seen in these tissues in humans with *SLC30A10* mutations (Bosomworth et al., 2012). High concentrations of Mn lead to increased *SLC30A10* gene and protein expression in HepG2 hepatocellular carcinoma cells consistent with a role of SLC30A10 in the detoxification of Mn from the cytoplasm (Quadri et al., 2012). Further research is required to study whether SLC30A10 can traffic to the plasma membrane during Mn exposure and facilitate transport of Mn out of the cell.

The secretory pathway Ca^{2+}/Mn^{2+} ATPases SPCA1 and SPCA2 localized at the Golgi present an alternative way of cytosolic Mn detoxification. Both pumps have high affinity for Ca^{2+} and Mn^{2+} ions and sequester these ions into the Golgi. Overexpression of *SPCA1* facilitates Mn accumulation in the Golgi and increases cell viability (Mukhopadhyay & Linstedt, 2011). While *SPCA1* is ubiquitously expressed, *SPCA2* has a more restricted expression pattern limited to the gastrointestinal tract, brain, and testes. Monoallelic mutations in *SPCA1* are known to cause Hailey–Hailey disease, a blistering skin disorder, while biallelic sequence changes with complete loss of function are thought to be incompatible with life (Vanoevelen et al., 2005; Xiang, Mohamalawari, & Rao, 2005). Mice exposed to high levels of Mn develop Mn accumulation in areas of the brain that show high expression of SPCA1 further supporting a role of SPCA1 in Mn detoxification (Sepulveda, Berrocal, Marcos, Wuytack, & Mata, 2007). Similarly, SPCA1 has been shown to facilitate Mn detoxification in the liver by sequestering Mn into the Golgi and early endosomes (Leitch et al., 2011). The sarco/endoplasmic reticulum Ca^{2+} ATPases (SERCA) have also been suggested to transport both Ca^{2+} and Mn^{2+}; however, Mn^{2+} is only a weak competitor of Ca^{2+} for the transport sites (Chiesi & Inesi, 1980; Gomes da & Madeira, 1986; Van et al., 2004).

The discovery of *ATP13A2* (*PARK9*) as the gene affected in a rare form of juvenile-onset PD, also known as Kufor–Rakeb syndrome, has provided an additional link between Mn toxicity and PD. *ATP13A2* encodes a P5-type cation-transporting ATPase located at the lysosome that protects cells from Mn-induced toxicity. Overexpression of *ATP13A2* leads to decreased levels of intracellular Mn. Expression of *ATP13A2* is found to be highest in the substantia nigra, the region of DAergic neuron loss in PD. Exposure to excess Mn further induces *ATP13A2* gene expression followed by sequestration of Mn into the lysosome (Gitler et al., 2009; Schmidt, Wolfe, Stiller, & Pearce, 2009; Tan et al., 2011).

3. MECHANISMS OF MANGANESE TOXICITY

In recent years, enormous progress has been made in our understanding of the biochemical mechanisms associated with Mn neurotoxicity.

3.1. Mn and its effect on dopamine transmission

Mn neurotoxicity causes an extrapyramidal motor disorder that resembles idiopathic PD. However, it remains debatable whether manganism and PD share common cellular mechanisms. The hallmark of PD is the degeneration of the nigrostriatal DAergic neurons in the substantia nigra pars compacta that innervate the caudate and putamen. This is consistent with a loss of dopamine transporter (DAT) levels and reduced dopa decarboxylase activity in the substantia nigra in single-photon emission computed tomography (SPECT) and PET studies of PD patients (Guilarte, 2011). In contrast, the most significant effects of Mn toxicity occur at the globus pallidus, while synthesis of DA and DAT levels in the substantia nigra is preserved in most cases (Yamada et al., 1986). Following excessive exposure to Mn the metal accumulates in the basal ganglia, particularly in the globus pallidus (Dorman et al., 2006; Guilarte et al., 2006). While some studies have described a loss of DAergic neurons within the substantia nigra upon Mn exposure, the majority of studies report intact presynaptic DAergic neurons (Guilarte, 2011). Given the overlapping clinical symptoms between manganism and PD, it is expected that DA neurotransmission is affected in both conditions. However, the disturbance is likely to occur at different points within the complex neuronal pathways in the basal ganglia (Roth, Li, Sridhar, & Khoshbouei, 2013). Indeed, recent studies have confirmed that chronic Mn exposure significantly impacts on amphetamine-induced DA release in the striatum, while the neurons in the substantia nigra remain intact (Guilarte et al., 2008).

Furthermore, a significant decrease in DAT levels in the striatum was observed in Mn-exposed rats (McDougall et al., 2008). Mn was also found to decrease DA uptake and DA efflux in DAT-transfected human embryonic kidney cells by promoting redistribution of DAT from the cell surface to the internal compartment of the cell (Roth et al., 2013). Other studies showed that expression of D2–DA receptors (D2R) is downregulated in the basal ganglia and the midbrain following Mn exposure (Butterworth, Spahr, Fontaine, & Layrargues, 1995; Mousseau, Perney, Layrargues, & Butterworth, 1993; Sriram, Lin, Jefferson, Roberts, Chapman, et al., 2010). In PD, on the other hand, the level of D2Rs is either unaffected or increased. Overall, there is increasing evidence that Mn neurotoxicity causes abnormal presynaptic DAergic signaling and decrease in DA release, while DA neuron integrity within the substantia nigra remains preserved with normal levels of DA (Guilarte et al., 2008).

On the other hand, Intrastriatal Mn injections in the rat brain have been found to cause DAergic neuron loss (Brouillet, Shinobu, McGarvey, Hochberg, & Beal, 1993). Studies in *Caenorhabditis elegans* have shown that Mn has the ability to induce DAergic-specific neurodegeneration. This neurotoxicity is dependent on the presence of extracellular DA and requires functional DA reuptake via DAT (Benedetto, Au, Avila, Milatovic, & Aschner, 2010). It was concluded that DA is required for Mn toxicity to occur and that Mn and DA work together synergistically, resulting in selective DAergic neurotoxicity. Mn was suggested to mediate oxidation of extracellular DA, which, after uptake via DAT, may cause oxidative injury and degeneration of DAergic neurons (Chakraborty & Aschner, 2012).

3.2. Oxidative stress and mitochondrial dysfunction

Exposure to high levels of Mn has been shown to be involved in the production of reactive oxygen species (ROS) and enhanced oxidative stress. The most common ROS include the superoxide radical ($O_2^{\bullet -}$), hydrogen peroxide (H_2O_2), and hydroxyl radical ($OH^{\bullet}$) leading to damage of nucleic acids, proteins, and phospholipids (Martinez-Finley et al., 2013). Within the cell, Mn preferentially concentrates in the mitochondria via the Ca^{2+} uniporter (Gavin, Gunter, & Gunter, 1992). Long-term Mn exposure increases oxidative stress markers including p38(MAPK) phosphorylation and caspase activity in the striatum (Cordova et al., 2013). Furthermore, intracellular levels and activation of antioxidants such as glutathione and catalase are augmented following Mn exposure (HaMai & Bondy, 2004;

Settivari, Levora, & Nass, 2009; Yoon et al., 2011). Consistent with a role of Mn in the production of oxidative stress, treatment of Mn-exposed cells with antioxidants attenuates Mn toxicity (Cordova et al., 2013; Prabhakaran, Ghosh, Chapman, & Gunasekar, 2008; Stephenson et al., 2013). The basal ganglia are areas in the brain with high oxidative activity promoting oxidation from Mn^{2+} to Mn^{3+}, the species with a greater pro-oxidant potential, augmenting the autoxidation of DA (Donaldson, McGregor, & LaBella, 1982). The main species of Mn in the mitochondria is Mn^{2+} bound to ATP (Farina, Avila, da Rocha, & Aschner, 2012). Mn^{2+} is thought to directly interfere with oxidative phosphorylation causing ROS production (Gavin et al., 1992). However, the exact pathways that lead to Mn-induced production of ROS are still unknown. It is suggested that the first step in ROS generation is the production of $O_2^{\bullet -}$ that can be converted to H_2O_2 by the Mn and Cu/Zn superoxide dismutase in the mitochondria and cytoplasm, respectively. H_2O_2 can be further converted to $OH^{\bullet}$ in the presence of Mn or other transition metals (Goldstein, Meyerstein, & Czapski, 1993; Martinez-Finley et al., 2013).

Mn^{2+} interferes with Ca^{2+} homeostasis within the mitochondria by occupying Ca^{2+} binding sites. Together with the generation of oxidative stress, this leads to the induction of a process named mitochondrial permeability transition. Opening of a permeability transition pore mediates increased solubility of the mitochondrial membrane for ions and protons causing rapid swelling and ultrastructural changes associated with loss of the mitochondrial inner membrane potential, impaired oxidative phosphorylation, and ATP synthesis, thereby inducing apoptosis and neurodegeneration (Farina et al., 2012; Martinez-Finley et al., 2013; Rao & Norenberg, 2004).

3.3. Glutamine, glutamate, and γ-aminobutyric acid (GABA)

Glutamine is an important precursor in the brain for the synthesis of both the excitatory neurotransmitter glutamate and the inhibitory neurotransmitter GABA and plays a critical role in neuron–glia interactions (Sidoryk-Wegrzynowicz, Lee, Albrecht, & Aschner, 2009). Glutamate and GABA neurons both project signals within the basal ganglia and are involved in the regulation of motor coordination. Astrocytes facilitate the regulation of the glutamate–glutamine cycling between astrocytes and neurons and have in fact been shown to accumulate Mn to a much higher extent than neurons (Aschner, Gannon, & Kimelberg, 1992; Farina et al., 2012;

Rao & Norenberg, 2004; Wedler et al., 1994). Mn exposure impairs astrocytic glutamine transport by reducing the expression of specific glutamine transporters with subsequent decrease in glutamine uptake (Sidoryk-Wegrzynowicz et al., 2009). Mn is thought to promote ubiquitination and degradation of the transporter protein by a mechanism involving the protein kinase C signaling pathway (Sidoryk-Wegrzynowicz, Lee, Mingwei, & Aschner, 2011). In addition, the uptake of glutamate via the astrocytic glutamate–aspartate transporter is affected by Mn, leading to increased extracellular glutamate levels and neuronal excitability (Erikson, Suber, & Aschner, 2002). This is consistent with the finding of N-methyl-D-aspartate excitotoxic lesions following the local intrastriatal injection of $MnCl_2$ into rat striatum (Brouillet et al., 1993). Glutamate further augments Mn toxicity as both Mn and glutamate toxicity affect Ca^{2+} accumulation in the mitochondria promoting apoptotic cell death (Roth, Sridhar, & Singleton, 2012).

Upon Mn exposure, extracellular GABA concentrations are increased due to altered expression levels of GABA transporters and receptors causing reduced uptake of GABA (Anderson et al., 2008; Fordahl et al., 2010). The authors suggest that the increased extracellular levels of GABA in the striatum affect the activity of the GABA projections to the substantia nigra, thus facilitating increased inhibition and impaired striatal DA release via the nigrostriatal pathway. Consistent with these findings, magnetic resonance spectroscopy studies in Mn-exposed workers confirm raised levels of GABA in the thalamus and adjacent brain regions (Racette et al., 2012).

In summary, there is strong evidence that Mn toxicity affects the balance of several neurotransmitters in the basal ganglia including DA, glutamate, and GABA, thus leading to changes in DAergic, glutamatergic, and GABAergic transmission. To what extent each of these pathways contributes to Mn neurotoxicity remains subject to debate; however, it is likely that a multifactorial interplay of several cellular mechanisms affects neurotransmitter signaling within the basal ganglia causing an extrapyramidal movement disorder.

3.4. Neuroinflammation

While Mn can affect neurons directly through oxidative damage and mitochondrial dysfunction, Mn has also been shown to enhance neurotoxicity by activation of glial cells and subsequent release of inflammatory cytokines and nonneuronal ROS. Mn has the ability to induce the release of interleukin

(IL)-1β, IL-6, and tumor necrosis factor-α from microglial cells (Filipov & Dodd, 2012). Mn-induced neuronal injury has been shown to require the presence of astrocytes, confirming the crucial role of astrocytes in Mn neurotoxicity. This is further supported by the significantly higher levels of Mn accumulating in astrocytes during Mn exposure compared with neurons (Wedler et al., 1994). In addition, Mn has been shown to alter astrocyte morphology and lead to Alzheimer type II astrocytosis in the globus pallidus (Hazell, Normandin, Norenberg, Kennedy, & Yi, 2006). This is associated with increased production of nitric oxide (NO) and increased expression of NO synthase in astrocytes (Spranger et al., 1998). Mn potentiates cytokine-induced expression of NO synthase and production of NO in astrocytes via activation of soluble guanylate cyclase, leading to ERK-dependent enhancement of NF-kappaB signaling (Moreno, Sullivan, Carbone, Hanneman, & Tjalkens, 2008). Overall, inflammation of glial cells within the basal ganglia with subsequent neurotoxic injury seems to be an important mechanism of Mn toxicity.

4. NEURODEGENERATIVE DISORDERS ASSOCIATED WITH Mn TOXICITY

4.1. Inherited hypermanganesemia caused by mutations in *SLC30A10*

Recently, the first inborn error of Mn metabolism with accumulation of Mn in the liver and brain has been identified. Recessive loss-of-function mutations in *SLC30A10* cause a movement disorder and chronic liver disease reminiscent of Wilson's disease (Quadri et al., 2012; Tuschl et al., 2012). As with acquired causes of hypermanganesemia, the MRI is pathognomonic for Mn accumulation with T1-weighted hyperintensity of the basal ganglia and white matter (Stamelou et al., 2012). Further characteristics include hematologic abnormalities such as polycythemia and abnormal Fe indices. While liver toxicity appears to be a major contributor to the mortality of this disease, liver disease has not been reported in cases of chronic environmental manganism. In animal models, however, Mn has been shown to promote intrahepatic cholestasis and induce hepatocellular necrosis (Huang et al., 2011; Witzleben, Pitlick, Bergmeyer, & Benoit, 1968). Furthermore, acute Mn poisoning has been reported to cause fatal hepatic failure in human cases (Sanchez et al., 2012). Depending on the time of disease onset, two distinct phenotypes can be distinguished: a childhood-onset form with predominant dystonia and an adult-onset form with predominant parkinsonism.

4.1.1 Clinical phenotype

The majority of affected individuals present with gait disturbance within the first 10 years of life. Motor symptoms include dystonia causing a characteristic high-stepping gait ("cock-walk"), painful limb spasms, and fine motor impairment as well as dysarthria, truncal ataxia, bradykinesia, and retropulsion (Quadri et al., 2012; Tuschl et al., 2012). Fewer cases develop spastic paraparesis and typical pyramidal tract signs (Delnooz et al., 2013; Gospe et al., 2000). Motor impairment is progressive and many patients become wheelchair-bound. Some individuals show signs of motor neuropathy with distal weakness and areflexia (Delnooz et al., 2013; Quadri et al., 2012). Intellect appears normal and affected individuals are able to attend mainstream school.

To date, a single sibship with adult-onset parkinsonism has been described. Two brothers presented at the age of 47 and 57 years with progressive gait disturbance and bradykinesia. Neurological examination revealed hypomimia, monotone speech, mild rigidity, global bradykinesia, wide-based gait with freezing and starting hesitation, and moderate postural instability without evidence of tremor, dystonia, or cerebellar and pyramidal disturbances. Treatment attempts with L-dopa and DA agonists were unsuccessful (Quadri et al., 2012).

In both childhood- and adult-onset forms, neurological symptoms are accompanied by polycythemia that precedes the manifestation of the movement disorder in many cases. Hemoglobin values can exceed 20 g/dL; however, individuals remain asymptomatic without evidence of ischemic complications. The exact mechanisms leading to polycythemia are not known. One possibility is the ability of Mn to induce erythropoietin gene expression (Ebert & Bunn, 1999; Tuschl et al., 2012).

Given the interdependency of Mn and Fe transport, it is not surprising that abnormal Fe indices are a feature of this disorder. Hallmarks of inherited hypermanganesemia are low serum ferritin and Fe levels and elevated TIBC (Quadri et al., 2012; Tuschl et al., 2012).

Hepatic involvement is variable in this disorder. While some patients die of cirrhosis-associated complications in adolescence to early adulthood, others survive into late adulthood without liver disease. Hence, liver disease is not pathognomonic for this disease but when present should further suggest the diagnosis. To date, the majority of identified individuals have evidence of hepatic involvement including hepatomegaly, raised liver transaminases, and unconjugated hyperbilirubinemia (Tuschl et al., 2012). However, pure neurological phenotypes presenting with dystonia alone can occur. Furthermore, marked intrafamilial phenotypic variability has been observed. Within one family, two siblings have survived into late adulthood with an isolated movement disorder, while their sister died of liver failure at the age of 46 years with minimal neurological symptoms (Delnooz et al., 2013; Quadri et al., 2012). Factors determining an individual's susceptibility for liver disease progression have yet to be identified and are likely to involve genetic and environmental aspects.

4.1.2 Diagnosis

Diagnosis of inherited hypermanganesemia is suggested by the typical clinical findings, biochemical characteristics, and MRI brain appearances. While whole blood Mn levels are not usually part of the diagnostic workup of dystonia or parkinsonism, the association of extrapyramidal movement disorder, chronic liver disease, and polycythemia together with T1 hyperintensity of the basal ganglia should prompt determination of whole blood Mn levels. In all affected individuals reported to date, whole blood Mn levels are elevated; hence, whole blood Mn seems to be a reliable disease marker. However, blood Mn levels are affected by chelation therapy and Fe treatment and should be interpreted together with other disease parameters (Tuschl, Clayton, Gospe, & Mills, 1993).

Brain MRI appearances are pathognomonic for the disease and are similar to those seen in acquired cases of hypermanganesemia. Mn accumulation in the basal ganglia causes T1 hyperintensities affecting the globus pallidus, putamen, caudate, subthalamic, and dentate nuclei, while the thalamus and ventral pons are spared. When the disease is extensive, white matter and anterior pituitary involvement can also be observed. T2-weighted images may show changes, however, to a much lesser extent and are often reported as normal (Quadri et al., 2012; Tuschl et al., 2012). Improvement of blood Mn levels during treatment can lead to normalization of the MRI brain appearances (Stamelou et al., 2012) (Fig. 12.2).

Liver ultrasound and MRI can be helpful to assess liver involvement. Liver biopsy performed in individuals with progressive liver disease has confirmed accumulation of Mn in the liver. Rhodanine staining for Mn is positive and the liver Mn level measured as μg/g wet weight is above the normal range of 1–2 μg/g. Histologically, various degrees of liver fibrosis, steatosis, and cirrhosis have been reported (Quadri et al., 2012; Tuschl et al., 2012).

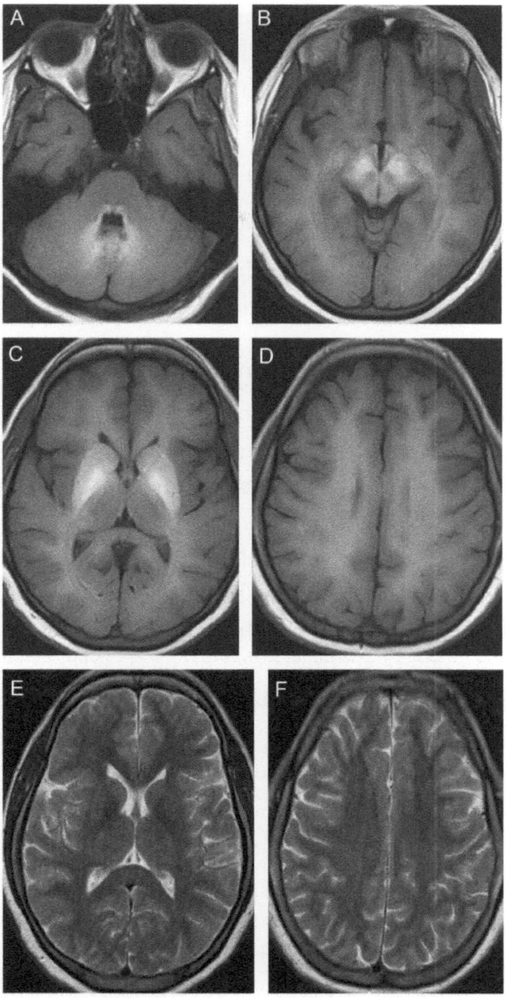

Figure 12.2 Characteristic MRI brain appearances of a patient with *SLC30A10* mutations. (A–D) Transaxial T1-weighted images showing abnormal high signal return from all white matter as well as more prominent signal return from the globus pallidus, putamen, caudate, subthalamic, and dentate nucleus bilaterally. (E and F) Transaxial T2-weighted images showing abnormal low signal return from the globus pallidus in the same distribution as the regions of highest signal return on T1-weighted images.

Once secondary causes of hypermanganesemia have been excluded, mutation analysis of *SLC30A10* confirms the diagnosis. To date, only 12 homozygous sequence alterations in *SLC30A10* have been found in 12 unrelated families. It is therefore not possible to say whether there is any firm genotype–phenotype correlation. However, the mutation found in the sibship with adult-onset parkinsonism affects the terminal 3' end of the coding sequence, resulting in a protein that is truncated by only the last 49 amino acids. It is possible that this mutant allele produces a protein with residual function causing a milder phenotype (Quadri et al., 2012; Tuschl et al., 2012).

4.1.3 Treatment

Chelation therapy with intravenous disodium calcium edetate has proven effective in two patients with inherited hypermanganesemia, one with childhood-onset dystonia and the other with adult-onset parkinsonism (Quadri et al., 2012; Tuschl et al., 2012). Response to an initial 5-day course of twice-daily disodium calcium edetate at 20 mg/kg/dose (maximum 2 g/day) markedly increased urinary Mn excretion, suggesting effective chelation of Mn. Disodium calcium edetate is given intravenously because of poor enteral absorption (Tuschl et al., 2008). Long-term treatment with monthly 5-day courses of disodium calcium edetate (20 mg/kg/dose twice daily) led to significant reduction of whole blood Mn levels, normalization of hemoglobin and Fe indices, and improvement of MRI brain appearances. Biochemical changes were accompanied by clinical recovery with normalization of gait, increased mobility, improved fine motor movements, and halt of liver disease progression (Quadri et al., 2012; Stamelou et al., 2012). Disodium calcium edetate has the potential to cause nephrotoxicity and can impair the homeostasis of other heavy metals such as copper and Zn (Bradberry & Vale, 2009). However, in cases of inherited hypermanganesemia, it has proved an effective and safe chelator with minimal adverse effects. Nephrotoxicity was not observed even during long-term treatment. A mild decrease in blood Zn levels was easily corrected using low-dose Zn supplementation. Other heavy-metal concentrations remained stable. While on treatment, monitoring of serum electrolytes including calcium, phosphate, and magnesium; renal and liver function; full blood count; and serum concentrations of trace metals such as zinc, copper, and selenium is recommended (Tuschl et al., 1993).

In addition to chelation therapy, Fe supplementation has further stabilized blood Mn levels and reduced the Mn load in some patients (Ribeiro, Dos Santos-Neto, Braga-Neto, & Barsottini, 2013; Tuschl et al., 2008). This is not surprising given the interdependency of Fe and Mn levels in states of deficiency as discussed earlier (Fitsanakis et al., 2010). In cases of inherited hypermanganesemia, Fe indices are suggestive of Fe depletion; hence, optimization of Fe intake was attempted in order to lower intestinal Mn uptake. Indeed, Fe supplementation alone and in addition to chelation therapy effectively lowered blood Mn and further reduced the body Mn load (Ribeiro et al., 2013; Tuschl et al., 2008).

Other chelation agents including D-penicillamine and para-aminosalicylic acid (PASA) have been used with little effect (Brna, Gordon, Dooley, & Price, 2011; Tuschl et al., 2008). A single individual has been treated with dimercaptosuccinic acid (DMSA) and Fe leading to a marked improvement of neurological symptoms (unpublished). Whether this effect is due to the chelating action of DMSA or Fe supplementation itself is debatable.

In cases of end-stage liver disease, liver transplantation should be considered. However, to date, transplantation has not been attempted in individuals with inherited hypermanganesemia; hence, no data on the outcome are available.

4.2. Secondary causes of hypermanganesemia

4.2.1 Overexposure to Mn

4.2.1.1 Environmental manganism

The majority of cases of manganism occur during occupational exposure to Mn-laden dust in mining, in welding and smelting industries, and in steel, aluminum, and dry battery manufacturing (Coelho et al., 2012; Myers et al., 2009; Rivera-Mancia, Rios, & Montes, 2011). Mn concentrations in air above 1 mg/m^3 can lead to Mn accumulation in the brain correlating with T1-weighted hyperintensities of the globus pallidus and other basal ganglia (Nelson, Golnick, Korn, & Angle, 1993). Occupational Mn exposure causes typical motor symptoms of manganism, a parkinsonian hypokinetic movement disorder. Removal from exposure generally leads to resolution of MRI brain abnormalities and improvement of neurological symptoms (Herrero et al., 2006; Nelson et al., 1993). However, irreversible cases have been reported particularly following a prolonged exposure and a longer symptomatic period (Herrero et al., 2006; Huang et al., 1993). Poor response to L-dopa is a hallmark of manganism and allows differentiation from idiopathic PD (Koller, Lyons, & Truly, 2004). Chelation therapy with disodium calcium edetate has been used to effectively lower blood Mn levels and improve neurological symptoms. However, not all patients with manganism show a favorable response (Herrero et al., 2006). Other chelation agents such as DMSA and PASA have been used in single cases. While DMSA has been reported to have a negligible effect on blood and urine Mn levels, PASA has successfully treated a single case of occupational manganism (Angle, 1995; Jiang et al., 2006).

Environmental Mn exposure has been linked to impaired cognitive development in children and adults. Raised concentrations of Mn in drinking water appear to have an adverse impact on cognitive performance (Bouchard et al., 2011; Khan et al., 2012). A further study reported an inverted U-shaped association between blood Mn levels and cognitive development at 12 months of age, suggesting a negative impact of both high and low Mn levels. This is consistent with Mn's role as an essential trace metal at low concentrations and a neurotoxin at higher concentrations (Roels et al., 2012). Neonates might be at particular risk of Mn toxicity since intestinal Mn absorption during the neonatal period is increased, while biliary excretion of Mn is poorly developed (Aschner & Aschner, 2005). Altered neonatal Mn homeostasis might reflect a higher requirement of Mn during development, which is further suggested by higher levels of blood Mn in pregnant women (Roels et al., 2012). Certain infant and soy formulas contain high levels of Mn and have been suggested to represent sources of excess Mn (Cockell, Bonacci, & Belonje, 2004; Krachler, Rossipal, & Irgolic, 1998). This is of particular interest since an association between high Mn levels and attention deficit hyperactivity disorder has been proposed stressing the importance of Mn as an environmental toxin (Scassellati, Bonvicini, Faraone, & Gennarelli, 2012).

Public health concerns regarding the use of MMT, a gasoline fuel additive, as a potential environmental air pollutant have been raised (Angle, 1995; Dorman, Andersen, Roper, & Taylor, 2012). MMT yields a mixture of phosphate, sulfate, and oxide forms of Mn (Angle, 1995; Dobson, Erikson, & Aschner, 2004). One study has suggested an association between raised airborne Mn levels and the risk of PD. However, it is possible that the predicted airborne concentrations of Mn in the neighborhood of subjects did not adequately represent personal exposure to Mn (Angle, 1995; Finkelstein & Jerrett, 2007). Also, air Mn levels in cities with use of MMT are below the current inhalation reference concentration for inhalable Mn of 0.05 $\mu g\ Mn/m^3$ set by the US Environmental Protection Agency (Dobson et al., 2004). Whether MMT poses a risk of Mn overexposure remains debatable.

4.2.1.2 Mn and PD

Individuals with Mn-induced parkinsonism resemble patients with idiopathic PD. However, these syndromes can be differentiated using clinical and neuroimaging studies. While idiopathic PD leads to nigrostriatal DA neuron degeneration, DA synthesis is preserved in manganism. In contrast, Mn neurotoxicity affects the DAergic system at the point of DA release (Guilarte, 2011). Clinically, parkinsonian symptoms in manganism tend to occur bilaterally with a higher incidence of dystonia, postural impairment and action tremor, and lack of response to L-dopa treatment and often display a more rapid disease progression (Cersosimo & Koller, 2006). Recent neuroimaging studies suggest significant neurobiochemical differences between manganism and PD consistent with normal DA synthesis in Mn neurotoxicity (Huang, 2007). Although manganism and idiopathic PD are clinically distinct, several studies have found an association between chronic Mn overexposure and idiopathic PD. In a cohort of individuals with chronic environmental Mn exposure, a higher incidence of PD has been reported. Furthermore, individuals with PD have collectively raised Mn levels in blood and CSF when compared with healthy controls, suggesting that Mn is indeed a risk factor for the development of PD (Ahmed & Santosh, 2010; Finkelstein & Jerrett, 2007; Fukushima, Tan, Luo, & Kanda, 2010; Gorell et al., 1997, 1999; Hozumi et al., 2011; Willis et al., 2010).

It is possible that individual genetic susceptibility to Mn neurotoxicity accelerates the onset of PD or manganism. This could explain the higher rate of PD in areas of environmental Mn exposure as well as the different susceptibilities displayed by Mn-exposed workers to the development of manganism (Roth, 2009). During the last decade, a vast number of genes involved in the pathogenesis of PD have been identified (Singleton, Farrer, & Bonifati, 2013). Of these, *PARK2* and *PARK9* have been found to play a significant role in the protection against Mn toxicity. *PARK2* encodes parkin, an ubiquitin E3 ligase that facilitates the conjugation of ubiquitin to a variety of proteins followed by their degradation in the proteasome. Mn increases the expression of parkin and overexpression of parkin selectively inhibits Mn and Mn/DA-induced cell death of DAergic SH-SY5Y cells (Higashi et al., 2004). As parkin is responsible for the ubiquitination of DMT1, mutations in *PARK2* lead to increased levels of DMT1 that could facilitate increased accumulation of Mn in the brain (Roth, 2009; Roth, Singleton, Feng, Garrick, & Paradkar, 2010). Furthermore, mutations in PARK2 have been shown to affect mitochondrial homeostasis and increase Mn-induced apoptotic signaling (Roth, Ganapathy, & Ghio, 2012). Mn exposure leads to downregulation of the expression of parkin in DAergic brain areas, further supporting a link between Mn and PD (Sriram, Lin, Jefferson, Roberts, Wirth, et al., 2010; Wang et al., 2009).

Mutations in *PARK9*, the gene encoding ATP13A2, cause Kufor–Rakeb syndrome characterized by rapid progressive juvenile-onset PD, pyramidal signs, vertical gaze palsy, and cognitive decline (Behrens et al., 2010). Expression of *ATP13A2*, a lysosomal P5-type ATPase, is upregulated following Mn exposure. Overexpression of *ATP13A2* reduces intracellular Mn concentrations, thereby protecting cells from Mn-induced cell death (Schmidt et al., 2009; Tan et al., 2011). Knockdown of *ATP13A2* in *C. elegans* enhances α-synuclein misfolding, a key mechanism in the pathogenesis of PD (Gitler et al., 2009). An additional link between *ATP13A2* mutations and Mn neurotoxicity has been suggested since polymorphisms in *PARK9* were found to be associated with impaired motor coordination in elderly individuals from an area with high environmental Mn exposure (Rentschler et al., 2012).

Besides genetic factors, biochemical susceptibility factors such as Fe-deficiency anemia and subclinical impairment of liver function are likely to affect an individual's risk to the development of PD and Mn toxicity (Lucchini, Martin, & Doney, 2009). Lifetime exposure to Mn might be significantly increased in cases of Fe-deficiency anemia that result in increased intestinal absorption of Mn (Fitsanakis et al., 2010). Similarly, individuals with mildly impaired liver function may accumulate higher levels of Mn due to impaired biliary excretion. Indeed, individuals with parkinsonism from an area with high environmental Mn exposure were found to have a significantly raised aspartate aminotransferase/alanine aminotransferase (*AST/ALT*) ratio, suggesting that subclinical liver dysfunction may pose an increased risk factor for Mn-related neurodegeneration (Squitti et al., 2009).

4.2.1.3 Excess Mn in total parenteral nutrition

During PN, Mn bypasses the gut, the enterohepatic circulation, and physiological biliary excretion by the liver. Hence, PN poses a risk of Mn overexposure (Slicker & Vermilyea, 2009). Several cases of Mn toxicity in individuals on PN have been described in the literature (Chalela et al., 2011; Fell et al., 1996; Hsieh, Liang, Peng, & Lee, 2007; Kikuchi, 2009; Klos, Chandler, Kumar, Ahlskog, & Josephs, 2006; Komaki, Maisawa, Sugai, Kobayashi, & Hashimoto, 1999; Masumoto et al., 2001; Nagatomo et al., 1999). Mn overexposure is of particular concern in individuals who develop PN-associated liver disease, a common complication of long-term PN, causing cholestasis and impaired biliary excretion (Alves et al., 1997; Sue, Chen, & Chen, 1996; Xu & Li, 2012). Since the recognition of PN-associated Mn toxicity, recommendations for the daily dose of parenteral Mn have been made that range from 0.01 to 2.2 mg. PN providing more than 0.1 mg Mn/day has been reported to lead to Mn accumulation and high intensity basal ganglia on T1-weighted MRI images (Erikson, Thompson, Aschner, & Aschner, 2007). In infants and neonates, the recommended daily dose is 1 μg/kg. Some protocols suggest stopping Mn supplementation in neonates when bilirubin levels reach more than 2 mg/dL since biliary excretion is poorly developed in the first weeks of life (Burjonrappa & Miller, 2012). Because Mn is often a contaminant in PN, some patients are likely to continue to receive excessive doses of Mn despite attempts at minimizing the amount of Mn in the PN (Slicker & Vermilyea, 2009).

Despite adhering to current recommendations of Mn dosing, Mn accumulation in the basal ganglia and raised blood Mn levels remain a concern in patients on long-term PN (Abdalian, Saqui, Fernandes, & Allard, 2012). Withdrawal from PN leads to normalization of blood Mn levels accompanied by resolution of brain MRI abnormalities over the following months. Neurodevelopment of children who receive PN appears not to be affected (Klos et al., 2006). A study of adult patients, however, has reported an increased risk of cognitive impairment (Kafritsa et al., 1998; Klos et al., 2006). Since Mn deficiency has not been an issue in patients on PN, some authors suggest that Mn should not be routinely prescribed for individuals on long-term PN (Hardy, 2009). Whole blood Mn levels seem to correlate with basal ganglia signal intensity on MRI and are recommended for monitoring of patients on long-term PN together with brain MRI in cases of suspected Mn neurotoxicity (Burjonrappa & Miller, 2012; Hardy, 2009).

4.2.1.4 Mn in methcathinone formulation

In recent years, several cases of Mn poisoning have occurred in intravenous methcathinone users (Levin, 2005; Sanotsky et al., 2007; Sikk et al., 2007; Stepens et al., 2008). In Eastern Europe and Russia, methcathinone, a psychostimulant otherwise known as ephedrone, is prepared from ephedrine and pseudoephedrine by adding the oxidant potassium permanganate in the presence of acetic acid. Regular use leads to Mn intoxication with typical neurological symptoms. Onset of symptoms occurs at an average of 5.8 years after the start of methcathinone use with some individuals becoming symptomatic within the first year of use (Stepens et al., 2008). Characteristic MRI brain appearances are accompanied by the typical cock-walk gait, particularly difficulties with backward motion, and speech impairment (Levin, 2005; Sikk et al., 2007; Stepens et al., 2008). Cognitive function appears preserved; however, some drug users show mild executive dysfunction (Selikhova et al., 2008). Following cessation of methcathinone use, blood Mn levels decrease and MRI brain appearances normalize. However, the neurological symptoms persist in the majority of cases, leaving many handicapped with an irreversible parkinsonian syndrome (Sikk et al., 2013; Stepens et al., 2008). Response to both L-dopa and chelation therapy with disodium calcium edetate is poor. It is likely that the oxidizing potential of permanganate together with the amphetamine-like effect of methcathinone enhances Mn neurotoxicity, causing irreversible neuronal damage. Neuroimaging studies again suggest intact presynaptic nigrostriatal neurons as observed in other cases of Mn toxicity (Sikk et al., 2010).

4.2.2 Impaired hepatic excretion of Mn

4.2.2.1 Acquired hepatocerebral degeneration

Following intestinal absorption, any excess Mn entering the portal circulation is readily excreted into the bile without reaching the systemic circulation. The liver therefore acts as a key player in the delicate regulation of Mn homeostasis (Aschner et al., 2007). In cases of advanced liver disease, hepatic Mn excretion is impaired. Portosystemic shunts allow Mn to bypass the liver, thereby escaping biliary elimination. Subsequently, Mn accumulates in the blood and brain leading to hypermanganesemia, extrapyramidal movement disorder, and Mn deposition in the basal ganglia on MRI brain images (Butterworth, 2000; Spahr et al., 1996). One percent of patients with liver cirrhosis are estimated to be affected by a movement disorder caused by Mn toxicity (Fernandez-Rodriguez, Contreras, De Villoria, & Grandas, 2010; Hauser, Zesiewicz, Martinez, Rosemurgy, & Olanow, 1996). Of patients with end-stage liver disease, this number is much higher; 20–90% of patients awaiting liver transplantation appear to be affected by AHD (Burkhard, Delavelle, Du, & Spahr, 2003; Ferrara & Jankovic, 2009). Blood Mn levels are significantly raised and correlate with T1-weighted signal abnormalities on MRI (Hauser et al., 1996; Rahelic, Kujundzic, Romic, Brkic, & Petrovecki, 2006). Pallidal Mn concentrations in patients with AHD are markedly increased and typically accompanied by Alzheimer type II astrocytosis similar to that seen in experimental models of Mn intoxication in rats (Lebda, El-Neweshy, & El-Sayed, 2012; Pomier-Layrargues, Spahr, & Butterworth, 1995; Weissenborn, Ehrenheim, Hori, Kubicka, & Manns, 1995). Motor symptoms include parkinsonism, dystonia, choreoathetosis, dysarthria, ataxia, pyramidal signs, and in rare cases myelopathy similar to transverse myelitis. As with other cases of hypermanganesemia, but in contrast to cases of PD, action tremor and postural instability are more frequently reported (Fernandez-Rodriguez et al., 2010; Ferrara & Jankovic, 2009; Meissner & Tison, 2011). Onset of neurological symptoms varies significantly between affected patients and can develop after weeks to decades of hepatic dysfunction (Ferrara & Jankovic, 2009). AHD seems to occur in any type of chronic liver disease; however, the presence of portosystemic shunting in addition to hepatocellular dysfunction seems to be a prerequisite for Mn accumulation (Burkhard et al., 2003; Butterworth, 2012; Rose et al., 1999). In rare cases, portosystemic shunts without hepatocellular dysfunction such as hereditary hemorrhagic telangiectasias or portal vein thrombosis have been reported to cause Mn toxicity and AHD (Kumar et al., 2008; Nolte et al., 1998). While the majority of cases of AHD have been described in the adult population, AHD can also be a complication of end-stage liver disease in children (Bekiesinska-Figatowska, Mierzewska, & Jurkiewicz, 2013; Papapetropoulos et al., 2008; Pinto et al., 2010). Motor symptoms are usually progressive and liver transplantation appears to be the only effective treatment. Liver transplantation leads to a marked improvement of neurological symptoms and normalization of blood Mn levels and MRI brain appearances. However, a certain degree of irreversibility of neurological symptoms has been reported. L-Dopa has proven beneficial in some cases, while other reports suggest negligible effect of L-dopa treatment (Butterworth, 2012). Chelation therapy has rarely been attempted in cases of AHD. A case report suggests a positive effect of trientene on neurological symptoms in AHD (Park, Kim, Choi, Lee, & Chung, 2008). However, without performing randomized controlled trials, treatment effects remain speculative.

5. PERSPECTIVE

Overall, there is increasing evidence that Mn plays an important role as a neurotoxin both in environmental exposure and in inherited dystonia/parkinsonism. With the discovery of *SLC30A10* as a crucial Mn transporter, we have made an important step towards a better understanding of Mn homeostasis. Future studies of the properties of this transporter will provide new answers about the regulation of *in vivo* Mn flux. Progress has also been made in elucidating the role of Mn in the development of PD. Studies of novel PD genes such as *PARK2* and *PARK9* have provided new insights into how Mn neurotoxicity may be involved in idiopathic PD although further research will be crucial to understand the mechanisms involved. It is becoming more evident that individual genetic susceptibility and environmental exposure (to Fe and to Mn) play an important role in the development of PD and Mn toxicity; however, these susceptibility factors remain poorly understood. Finally, while much has been learned about the mechanisms of Mn neurotoxicity, many more studies are required to translate these findings into potential treatments and biomarkers for Mn-related neurodegenerative disorders.

REFERENCES

Abdalian, R., Saqui, O., Fernandes, G., & Allard, J. P. (2012). Effects of manganese from a commercial multi-trace element supplement in a population sample of Canadian patients on long-term parenteral nutrition. *Journal of Parenteral and Enteral Nutrition*, *37*, 538–543.

Abrams, E., Lassiter, J. W., Miller, W. J., Neathery, M. W., Gentry, R. P., & Scarth, R. D. (1976). Absorption as a factor in manganese homeostasis. *Journal of Animal Science*, *42*, 630–636.

Ahmed, S. S., & Santosh, W. (2010). Metallomic profiling and linkage map analysis of early Parkinson's disease: A new insight to aluminum marker for the possible diagnosis. *PLoS One*, *5*, e11252.

Alves, G., Thiebot, J., Tracqui, A., Delangre, T., Guedon, C., & Lerebours, E. (1997). Neurologic disorders due to brain manganese deposition in a jaundiced patient receiving long-term parenteral nutrition. *Journal of Parenteral and Enteral Nutrition*, *21*, 41–45.

Anderson, J. G., Fordahl, S. C., Cooney, P. T., Weaver, T. L., Colyer, C. L., & Erikson, K. M. (2008). Manganese exposure alters extracellular GABA, GABA receptor and transporter protein and mRNA levels in the developing rat brain. *Neurotoxicology*, *29*, 1044–1053.

Angle, C. R. (1995). Dimercaptosuccinic acid (DMSA): Negligible effect on manganese in urine and blood. *Occupational and Environmental Medicine*, *52*, 846.

Aschner, J. L., & Aschner, M. (2005). Nutritional aspects of manganese homeostasis. *Molecular Aspects of Medicine*, *26*, 353–362.

Aschner, M., Erikson, K. M., & Dorman, D. C. (2005). Manganese dosimetry: Species differences and implications for neurotoxicity. *Critical Reviews in Toxicology*, *35*, 1–32.

Aschner, M., Gannon, M., & Kimelberg, H. K. (1992). Manganese uptake and efflux in cultured rat astrocytes. *Journal of Neurochemistry*, *58*, 730–735.

Aschner, M., Guilarte, T. R., Schneider, J. S., & Zheng, W. (2007). Manganese: Recent advances in understanding its transport and neurotoxicity. *Toxicology and Applied Pharmacology*, *221*, 131–147.

Au, C., Benedetto, A., & Aschner, M. (2008). Manganese transport in eukaryotes: The role of DMT1. *Neurotoxicology*, *29*, 569–576.

Bartnikas, T. B. (2012). Known and potential roles of transferrin in iron biology. *Biometals*, *25*, 677–686.

Behrens, M. I., Bruggemann, N., Chana, P., Venegas, P., Kagi, M., Parrao, T., et al. (2010). Clinical spectrum of Kufor–Rakeb syndrome in the Chilean kindred with ATP13A2 mutations. *Movement Disorders*, *25*, 1929–1937.

Bekiesinska-Figatowska, M., Mierzewska, H., & Jurkiewicz, E. (2013). Basal ganglia lesions in children and adults. *European Journal of Radiology*, *82*, 837–849.

Benedetto, A., Au, C., Avila, D. S., Milatovic, D., & Aschner, M. (2010). Extracellular dopamine potentiates mn-induced oxidative stress, lifespan reduction, and dopaminergic neurodegeneration in a BLI-3-dependent manner in *Caenorhabditis elegans*. *PLoS Genetics*, *6*, 1–18.

Bosomworth, H. J., Thornton, J. K., Coneyworth, L. J., Ford, D., & Valentine, R. A. (2012). Efflux function, tissue-specific expression and intracellular trafficking of the Zn transporter ZnT10 indicate roles in adult Zn homeostasis. *Metallomics*, *4*, 771–779.

Bouchard, M. F., Sauve, S., Barbeau, B., Legrand, M., Brodeur, M. E., Bouffard, T., et al. (2011). Intellectual impairment in school-age children exposed to manganese from drinking water. *Environmental Health Perspectives*, *119*, 138–143.

Bradberry, S., & Vale, A. (2009). A comparison of sodium calcium edetate (edetate calcium disodium) and succimer (DMSA) in the treatment of inorganic lead poisoning. *Clinical Toxicology (Philadelphia, PA)*, *47*, 841–858.

Britton, A. A., & Cotzias, G. C. (1966). Dependence of manganese turnover on intake. *American Journal of Physiology*, *211*, 203–206.

Brna, P., Gordon, K., Dooley, J. M., & Price, V. (2011). Manganese toxicity in a child with iron deficiency and polycythemia. *Journal of Child Neurology*, *26*, 891–894.

Brouillet, E. P., Shinobu, L., McGarvey, U., Hochberg, F., & Beal, M. F. (1993). Manganese injection into the rat striatum produces excitotoxic lesions by impairing energy metabolism. *Experimental Neurology*, *120*, 89–94.

Burdo, J. R., Menzies, S. L., Simpson, I. A., Garrick, L. M., Garrick, M. D., Dolan, K. G., et al. (2001). Distribution of divalent metal transporter 1 and metal transport protein 1 in the normal and Belgrade rat. *Journal of Neuroscience Research*, *66*, 1198–1207.

Burjonrappa, S. C., & Miller, M. (2012). Role of trace elements in parenteral nutrition support of the surgical neonate. *Journal of Pediatric Surgery*, *47*, 760–771.

Burkhard, P. R., Delavelle, J., Du, P. R., & Spahr, L. (2003). Chronic parkinsonism associated with cirrhosis: A distinct subset of acquired hepatocerebral degeneration. *Archives of Neurology*, *60*, 521–528.

Butterworth, R. F. (2000). Complications of cirrhosis III. Hepatic encephalopathy. *Journal of Hepatology*, *32*, 171–180.

Butterworth, R. F. (2012). Parkinsonism in cirrhosis: Pathogenesis and current therapeutic options. *Metabolic Brain Disease*, *28*, 261–267.

Butterworth, R. F., Spahr, L., Fontaine, S., & Layrargues, G. P. (1995). Manganese toxicity, dopaminergic dysfunction and hepatic encephalopathy. *Metabolic Brain Disease*, *10*, 259–267.

Calne, D. B., Chu, N. S., Huang, C. C., Lu, C. S., & Olanow, W. (1994). Manganism and idiopathic parkinsonism: Similarities and differences. *Neurology*, *44*, 1583–1586.

Cersosimo, M. G., & Koller, W. C. (2006). The diagnosis of manganese-induced parkinsonism. *Neurotoxicology*, *27*, 340–346.

Chakraborty, S., & Aschner, M. (2012). Altered manganese homeostasis: Implications for BLI-3-dependent dopaminergic neurodegeneration and SKN-1 protection in *C. elegans*. *Journal of Trace Elements in Medicine and Biology*, *26*, 183–187.

Chalela, J. A., Bonillha, L., Neyens, R., & Hays, A. (2011). Manganese encephalopathy: An under-recognized condition in the intensive care unit. *Neurocritical Care*, *14*, 456–458.

Chiesi, M., & Inesi, G. (1980). Adenosine 5'-triphosphate dependent fluxes of manganese and and hydrogen ions in sarcoplasmic reticulum vesicles. *Biochemistry*, *19*, 2912–2918.

Chua, A. C., & Morgan, E. H. (1997). Manganese metabolism is impaired in the Belgrade laboratory rat. *Journal of Comparative Physiology B*, *167*, 361–369.

Cockell, K. A., Bonacci, G., & Belonje, B. (2004). Manganese content of soy or rice beverages is high in comparison to infant formulas. *Journal of the American College of Nutrition*, *23*, 124–130.

Coelho, P., Costa, S., Silva, S., Walter, A., Ranville, J., Sousa, A. C., et al. (2012). Metal(loid) levels in biological matrices from human populations exposed to mining contamination–Panasqueira mine (Portugal). *Journal of Toxicology and Environmental Health, Part A, 75*, 893–908.

Cordova, F. M., Aguiar, A. S., Jr., Peres, T. V., Lopes, M. W., Goncalves, F. M., Pedro, D. Z., et al. (2013). Manganese-exposed developing rats display motor deficits and striatal oxidative stress that are reversed by Trolox. *Archives of Toxicology, 87*, 1231–1244.

Couper, J. (1837). On the effects of black oxide of manganese when inhaled into the lungs. *British Annals of Medicine Pharmacy, Vital Statistics, and General Science, 1*, 41–42.

Crossgrove, J. S., Allen, D. D., Bukaveckas, B. L., Rhineheimer, S. S., & Yokel, R. A. (2003). Manganese distribution across the blood–brain barrier. I. Evidence for carrier-mediated influx of managanese citrate as well as manganese and manganese transferrin. *Neurotoxicology, 24*, 3–13.

Crossgrove, J. S., & Yokel, R. A. (2004). Manganese distribution across the blood–brain barrier III. The divalent metal transporter-1 is not the major mechanism mediating brain manganese uptake. *Neurotoxicology, 25*, 451–460.

Crossgrove, J. S., & Yokel, R. A. (2005). Manganese distribution across the blood–brain barrier. IV. Evidence for brain influx through store-operated calcium channels. *Neurotoxicology, 26*, 297–307.

Davidsson, L., Cederblad, A., Lonnerdal, B., & Sandstrom, B. (1991). The effect of individual dietary components on manganese absorption in humans. *American Journal of Clinical Nutrition, 54*, 1065–1070.

Davis, C. D., Wolf, T. L., & Greger, J. L. (1992). Varying levels of manganese and iron affect absorption and gut endogenous losses of manganese by rats. *Journal of Nutrition, 122*, 1300–1308.

Delnooz, C. C., Wevers, R. A., Quadri, M., Clayton, P. T., Mills, P. B., Tuschl, K., et al. (2013). Phenotypic variability in a dystonia family with mutations in the manganese transporter gene. *Movement Disorders, 28*, 685–686.

Dobson, A. W., Erikson, K. M., & Aschner, M. (2004). Manganese neurotoxicity. *Annals of the New York Academy of Sciences, 1012*, 115–128.

Donaldson, J., McGregor, D., & LaBella, F. (1982). Manganese neurotoxicity: A model for free radical mediated neurodegeneration? *Canadian Journal of Physiology and Pharmacology, 60*, 1398–1405.

Dorman, D. C., Andersen, M. E., Roper, J. M., & Taylor, M. D. (2012). Update on a pharmacokinetic-centric alternative tier II program for MMT-part I: Program implementation and lessons learned. *Journal of Toxicology, 2012*, 946742.

Dorman, D. C., Struve, M. F., Marshall, M. W., Parkinson, C. U., James, R. A., & Wong, B. A. (2006). Tissue manganese concentrations in young male rhesus monkeys following subchronic manganese sulfate inhalation. *Toxicological Sciences, 92*, 201–210.

Ebert, B. L., & Bunn, H. F. (1999). Regulation of the erythropoietin gene. *Blood, 94*, 1864–1877.

Erikson, K. M., Suber, R. L., & Aschner, M. (2002). Glutamate/aspartate transporter (GLAST), taurine transporter and metallothionein mRNA levels are differentially altered in astrocytes exposed to manganese chloride, manganese phosphate or manganese sulfate. *Neurotoxicology, 23*, 281–288.

Erikson, K. M., Thompson, K., Aschner, J., & Aschner, M. (2007). Manganese neurotoxicity: A focus on the neonate. *Pharmacology and Therapeutics, 113*, 369–377.

Farina, M., Avila, D. S., da Rocha, J. B., & Aschner, M. (2012). Metals, oxidative stress and neurodegeneration: A focus on iron, manganese and mercury. *Neurochemistry International, 62*, 575–594.

Fell, J. M., Reynolds, A. P., Meadows, N., Khan, K., Long, S. G., Quaghebeur, G., et al. (1996). Manganese toxicity in children receiving long-term parenteral nutrition. *Lancet, 347*, 1218–1221.

Fernandez-Rodriguez, R., Contreras, A., De Villoria, J. G., & Grandas, F. (2010). Acquired hepatocerebral degeneration: Clinical characteristics and MRI findings. *European Journal of Neurology*, *17*, 1463–1470.

Ferrara, J., & Jankovic, J. (2009). Acquired hepatocerebral degeneration. *Journal of Neurology*, *256*, 320–332.

Filipov, N. M., & Dodd, C. A. (2012). Role of glial cells in manganese neurotoxicity. *Journal of Applied Toxicology*, *32*, 310–317.

Finkelstein, M. M., & Jerrett, M. (2007). A study of the relationships between Parkinson's disease and markers of traffic-derived and environmental manganese air pollution in two Canadian cities. *Environmental Research*, *104*, 420–432.

Fitsanakis, V. A., Zhang, N., Avison, M. J., Erikson, K. M., Gore, J. C., & Aschner, M. (2011). Changes in dietary iron exacerbate regional brain manganese accumulation as determined by magnetic resonance imaging. *Toxicological Sciences*, *120*, 146–153.

Fitsanakis, V. A., Zhang, N., Garcia, S., & Aschner, M. (2010). Manganese (Mn) and iron (Fe): Interdependency of transport and regulation. *Neurotoxicity Research*, *18*, 124–131.

Fordahl, S. C., Anderson, J. G., Cooney, P. T., Weaver, T. L., Colyer, C. L., & Erikson, K. M. (2010). Manganese exposure inhibits the clearance of extracellular GABA and influences taurine homeostasis in the striatum of developing rats. *Neurotoxicology*, *31*, 639–646.

Friedman, B. J., Freeland-Graves, J. H., Bales, C. W., Behmardi, F., Shorey-Kutschke, R. L., Willis, R. A., et al. (1987). Manganese balance and clinical observations in young men fed a manganese-deficient diet. *Journal of Nutrition*, *117*, 133–143.

Fujishiro, H., Yano, Y., Takada, Y., Tanihara, M., & Himeno, S. (2012). Roles of ZIP8, ZIP14, and DMT1 in transport of cadmium and manganese in mouse kidney proximal tubule cells. *Metallomics*, *4*, 700–708.

Fukushima, T., Tan, X., Luo, Y., & Kanda, H. (2010). Relationship between blood levels of heavy metals and Parkinson's disease in China. *Neuroepidemiology*, *34*, 18–24.

Garcia, S. J., Gellein, K., Syversen, T., & Aschner, M. (2007). Iron deficient and manganese supplemented diets alter metals and transporters in the developing rat brain. *Toxicological Sciences*, *95*, 205–214.

Gavin, C. E., Gunter, K. K., & Gunter, T. E. (1992). Mn^{2+} sequestration by mitochondria and inhibition of oxidative phosphorylation. *Toxicology and Applied Pharmacology*, *115*, 1–5.

Gianutsos, G., Morrow, G. R., & Morris, J. B. (1997). Accumulation of manganese in rat brain following intranasal administration. *Fundamental and Applied Toxicology*, *37*, 102–105.

Girijashanker, K., He, L., Soleimani, M., Reed, J. M., Li, H., Liu, Z., et al. (2008). Slc39a14 gene encodes ZIP14, a metal/bicarbonate symporter: Similarities to the ZIP8 transporter. *Molecular Pharmacology*, *73*, 1413–1423.

Gitler, A. D., Chesi, A., Geddie, M. L., Strathearn, K. E., Hamamichi, S., Hill, K. J., et al. (2009). Alpha-synuclein is part of a diverse and highly conserved interaction network that includes PARK9 and manganese toxicity. *Nature Genetics*, *41*, 308–315.

Goldstein, S., Meyerstein, D., & Czapski, G. (1993). The Fenton reagents. *Free Radical Biology & Medicine*, *15*, 435–445.

Gomes da, C. A., & Madeira, V. M. (1986). Magnesium and manganese ions modulate Ca2+ uptake and its energetic coupling in sarcoplasmic reticulum. *Archives of Biochemistry and Biophysics*, *249*, 199–206.

Gorell, J. M., Johnson, C. C., Rybicki, B. A., Peterson, E. L., Kortsha, G. X., Brown, G. G., et al. (1997). Occupational exposures to metals as risk factors for Parkinson's disease. *Neurology*, *48*, 650–658.

Gorell, J. M., Johnson, C. C., Rybicki, B. A., Peterson, E. L., Kortsha, G. X., Brown, G. G., et al. (1999). Occupational exposure to manganese, copper, lead, iron, mercury and zinc and the risk of Parkinson's disease. *Neurotoxicology*, *20*, 239–247.

Gospe, S. M., Jr., Caruso, R. D., Clegg, M. S., Keen, C. L., Pimstone, N. R., Ducore, J. M., et al. (2000). Paraparesis, hypermanganesaemia, and polycythaemia: A novel presentation of cirrhosis. *Archives of Disease in Childhood, 83*, 439–442.

Gruenheid, S., Canonne-Hergaux, F., Gauthier, S., Hackam, D. J., Grinstein, S., & Gros, P. (1999). The iron transport protein NRAMP2 is an integral membrane glycoprotein that colocalizes with transferrin in recycling endosomes. *Journal of Experimental Medicine, 189*, 831–841.

Guilarte, T. R. (2011). Manganese and Parkinson's disease: A critical review and new findings. *Ciência & Saúde Coletiva, 16*, 4549–4566.

Guilarte, T. R., Burton, N. C., McGlothan, J. L., Verina, T., Zhou, Y., Alexander, M., et al. (2008). Impairment of nigrostriatal dopamine neurotransmission by manganese is mediated by pre-synaptic mechanism(s): Implications to manganese-induced parkinsonism. *Journal of Neurochemistry, 107*, 1236–1247.

Guilarte, T. R., McGlothan, J. L., Degaonkar, M., Chen, M. K., Barker, P. B., Syversen, T., et al. (2006). Evidence for cortical dysfunction and widespread manganese accumulation in the nonhuman primate brain following chronic manganese exposure: A 1H-MRS and MRI study. *Toxicological Sciences, 94*, 351–358.

Gunshin, H., Mackenzie, B., Berger, U. V., Gunshin, Y., Romero, M. F., Boron, W. F., et al. (1997). Cloning and characterization of a mammalian proton-coupled metal-ion transporter. *Nature, 388*, 482–488.

Gunter, T. E., Gerstner, B., Gunter, K. K., Malecki, J., Gelein, R., Valentine, W. M., et al. (2012). Manganese transport via the transferrin mechanism. *Neurotoxicology, 34C*, 118–127.

HaMai, D., & Bondy, S. C. (2004). Oxidative basis of manganese neurotoxicity. *Annals of the New York Academy of Sciences, 1012*, 129–141.

Hardy, G. (2009). Manganese in parenteral nutrition: Who, when, and why should we supplement? *Gastroenterology, 137*, S29–S35.

Hauser, R. A., Zesiewicz, T. A., Martinez, C., Rosemurgy, A. S., & Olanow, C. W. (1996). Blood manganese correlates with brain magnetic resonance imaging changes in patients with liver disease. *Canadian Journal of Neurological Sciences, 23*, 95–98.

Hazell, A. S., Normandin, L., Norenberg, M. D., Kennedy, G., & Yi, J. H. (2006). Alzheimer type II astrocytic changes following sub-acute exposure to manganese and its prevention by antioxidant treatment. *Neuroscience Letters, 396*, 167–171.

He, L., Girijashanker, K., Dalton, T. P., Reed, J., Li, H., Soleimani, M., et al. (2006). ZIP8, member of the solute-carrier-39 (SLC39) metal-transporter family: Characterization of transporter properties. *Molecular Pharmacology, 70*, 171–180.

Heilig, E. A., Thompson, K. J., Molina, R. M., Ivanov, A. R., Brain, J. D., & Wessling-Resnick, M. (2006). Manganese and iron transport across pulmonary epithelium. *American Journal of Physiology Lung Cellular and Molecular Physiology, 290*, L1247–L1259.

Herrero, H. E., Discalzi, G., Valentini, C., Venturi, F., Chio, A., Carmellino, C., et al. (2006). Follow-up of patients affected by manganese-induced parkinsonism after treatment with CaNa2EDTA. *Neurotoxicology, 27*, 333–339.

Higashi, Y., Asanuma, M., Miyazaki, I., Hattori, N., Mizuno, Y., & Ogawa, N. (2004). Parkin attenuates manganese-induced dopaminergic cell death. *Journal of Neurochemistry, 89*, 1490–1497.

Hozumi, I., Hasegawa, T., Honda, A., Ozawa, K., Hayashi, Y., Hashimoto, K., et al. (2011). Patterns of levels of biological metals in CSF differ among neurodegenerative diseases. *Journal of Neurological Sciences, 303*, 95–99.

Hsieh, C. T., Liang, J. S., Peng, S. S., & Lee, W. T. (2007). Seizure associated with total parenteral nutrition-related hypermanganesemia. *Pediatric Neurology, 36*, 181–183.

Huang, C. C. (2007). Parkinsonism induced by chronic manganese intoxication—An experience in Taiwan. *Chang Gung Medical Journal, 30*, 385–395.

Huang, C. C., Chu, N. S., Lu, C. S., & Calne, D. B. (1997). Cock gait in manganese intoxication. *Movement Disorders, 12*, 807–808.
Huang, C. C., Lu, C. S., Chu, N. S., Hochberg, F., Lilienfeld, D., Olanow, W., et al. (1993). Progression after chronic manganese exposure. *Neurology, 43*, 1479–1483.
Huang, E., Ong, W. Y., & Connor, J. R. (2004). Distribution of divalent metal transporter-1 in the monkey basal ganglia. *Neuroscience, 128*, 487–496.
Huang, P., Chen, C., Wang, H., Li, G., Jing, H., Han, Y., et al. (2011). Manganese effects in the liver following subacute or subchronic manganese chloride exposure in rats. *Ecotoxicology and Environmental Safety, 74*, 615–622.
Jiang, Y. M., Mo, X. A., Du, F. Q., Fu, X., Zhu, X. Y., Gao, H. Y., et al. (2006). Effective treatment of manganese-induced occupational parkinsonism with p-aminosalicylic acid: A case of 17-year follow-up study. *Journal of Occupational and Environmental Medicine, 48*, 644–649.
Kafritsa, Y., Fell, J., Long, S., Bynevelt, M., Taylor, W., & Milla, P. (1998). Long-term outcome of brain manganese deposition in patients on home parenteral nutrition. *Archives of Disease in Childhood, 79*, 263–265.
Kannurpatti, S. S., Joshi, P. G., & Joshi, N. B. (2000). Calcium sequestering ability of mitochondria modulates influx of calcium through glutamate receptor channel. *Neurochemical Research, 25*, 1527–1536.
Keen, C. L., Ensunsa, J. L., Watson, M. H., Baly, D. L., Donovan, S. M., Monaco, M. H., et al. (1999). Nutritional aspects of manganese from experimental studies. *Neurotoxicology, 20*, 213–223.
Khan, K., Wasserman, G. A., Liu, X., Ahmed, E., Parvez, F., Slavkovich, V., et al. (2012). Manganese exposure from drinking water and children's academic achievement. *Neurotoxicology, 33*, 91–97.
Kikuchi, A. (2009). An 8-year-old boy presenting with progressive extrapyramidal symptoms during total parenteral nutrition. *No to Hattatsu, 41*, 325–326.
Klos, K. J., Chandler, M., Kumar, N., Ahlskog, J. E., & Josephs, K. A. (2006). Neuropsychological profiles of manganese neurotoxicity. *European Journal of Neurology, 13*, 1139–1141.
Knopfel, M., Zhao, L., & Garrick, M. D. (2005). Transport of divalent transition-metal ions is lost in small-intestinal tissue of b/b Belgrade rats. *Biochemistry, 44*, 3454–3465.
Koller, W. C., Lyons, K. E., & Truly, W. (2004). Effect of levodopa treatment for parkinsonism in welders: A double-blind study. *Neurology, 62*, 730–733.
Komaki, H., Maisawa, S., Sugai, K., Kobayashi, Y., & Hashimoto, T. (1999). Tremor and seizures associated with chronic manganese intoxication. *Brain & Development, 21*, 122–124.
Krachler, M., Rossipal, E., & Irgolic, K. J. (1998). Trace elements in formulas based on cow and soy milk and in Austrian cow milk determined by inductively coupled plasma mass spectrometry. *Biological Trace Element Research, 65*, 53–74.
Kumar, N., Boeve, B. F., Cowl, C. T., Ellison, J. W., Kamath, P. S., & Swanson, K. L. (2008). Hypermanganesemia, hereditary hemorrhagic telangiectasia, brain abscess: The hepatic connection. *Neurology, 71*, 1118–1119.
Leavens, T. L., Rao, D., Andersen, M. E., & Dorman, D. C. (2007). Evaluating transport of manganese from olfactory mucosa to striatum by pharmacokinetic modeling. *Toxicological Sciences, 97*, 265–278.
Lebda, M. A., El-Neweshy, M. S., & El-Sayed, Y. S. (2012). Neurohepatic toxicity of subacute manganese chloride exposure and potential chemoprotective effects of lycopene. *Neurotoxicology, 33*, 98–104.
Leitch, S., Feng, M., Muend, S., Braiterman, L. T., Hubbard, A. L., & Rao, R. (2011). Vesicular distribution of secretory pathway Ca(2)+–ATPase isoform 1 and a role in manganese detoxification in liver-derived polarized cells. *Biometals, 24*, 159–170.

Levin, O. S. (2005). "Ephedron" encephalopathy. *Zhurnal Nevrologii i Psikhiatrii Imeni S.S. Korsakova, 105*, 12–20.

London, R. E., Toney, G., Gabel, S. A., & Funk, A. (1989). Magnetic resonance imaging studies of the brains of anesthetized rats treated with manganese chloride. *Brain Research Bulletin, 23*, 229–235.

Lucaciu, C. M., Dragu, C., Copaescu, L., & Morariu, V. V. (1997). Manganese transport through human erythrocyte membranes. An EPR study. *Biochimica et Biophysica Acta, 1328*, 90–98.

Lucchini, R. G., Martin, C. J., & Doney, B. C. (2009). From manganism to manganese-induced parkinsonism: A conceptual model based on the evolution of exposure. *Neuromolecular Medicine, 11*, 311–321.

Madejczyk, M. S., & Ballatori, N. (2012). The iron transporter ferroportin can also function as a manganese exporter. *Biochimica et Biophysica Acta, 1818*, 651–657.

Malecki, E. A., Radzanowski, G. M., Radzanowski, T. J., Gallaher, D. D., & Greger, J. L. (1996). Biliary manganese excretion in conscious rats is affected by acute and chronic manganese intake but not by dietary fat. *Journal of Nutrition, 126*, 489–498.

Martinez-Finley, E. J., Gavin, C. E., Aschner, M., & Gunter, T. E. (2013). Manganese neurotoxicity and the role of reactive oxygen species. *Free Radical Biology & Medicine, 62*, 65–75.

Masumoto, K., Suita, S., Taguchi, T., Yamanouchi, T., Nagano, M., Ogita, K., et al. (2001). Manganese intoxication during intermittent parenteral nutrition: Report of two cases. *Journal of Parenteral and Enteral Nutrition, 25*, 95–99.

McDougall, S. A., Reichel, C. M., Farley, C. M., Flesher, M. M., Der-Ghazarian, T., Cortez, A. M., et al. (2008). Postnatal manganese exposure alters dopamine transporter function in adult rats: Potential impact on nonassociative and associative processes. *Neuroscience, 154*, 848–860.

Meissner, W., & Tison, F. (2011). Acquired hepatocerebral degeneration. *Handbook of Clinical Neurology, 100*, 193–197.

Moreno, J. A., Sullivan, K. A., Carbone, D. L., Hanneman, W. H., & Tjalkens, R. B. (2008). Manganese potentiates nuclear factor-kappaB-dependent expression of nitric oxide synthase 2 in astrocytes by activating soluble guanylate cyclase and extracellular responsive kinase signaling pathways. *Journal of Neuroscience Research, 86*, 2028–2038.

Mousseau, D. D., Perney, P., Layrargues, G. P., & Butterworth, R. F. (1993). Selective loss of pallidal dopamine D2 receptor density in hepatic encephalopathy. *Neuroscience Letters, 162*, 192–196.

Mukhopadhyay, S., & Linstedt, A. D. (2011). Identification of a gain-of-function mutation in a Golgi P-type ATPase that enhances Mn^{2+} efflux and protects against toxicity. *Heilkunde-Heilwege, 108*, 858–863.

Myers, J. E., Fine, J., Ormond-Brown, D., Fry, J., Thomson, A., & Thompson, M. L. (2009). Estimating the prevalence of clinical manganism using a cascaded screening process in a South African manganese smelter. *Neurotoxicology, 30*, 934–940.

Nagatomo, S., Umehara, F., Hanada, K., Nobuhara, Y., Takenaga, S., Arimura, K., et al. (1999). Manganese intoxication during total parenteral nutrition: Report of two cases and review of the literature. *Journal of Neurological Sciences, 162*, 102–105.

Nelson, K., Golnick, J., Korn, T., & Angle, C. (1993). Manganese encephalopathy: Utility of early magnetic resonance imaging. *British Journal of Industrial Medicine, 50*, 510–513.

Nolte, W., Wiltfang, J., Schindler, C. G., Unterberg, K., Finkenstaedt, M., Niedmann, P. D., et al. (1998). Bright basal ganglia in T1-weighted magnetic resonance images are frequent in patients with portal vein thrombosis without liver cirrhosis and not suggestive of hepatic encephalopathy. *Journal of Hepatology, 29*, 443–449.

Papapetropoulos, S., Tzakis, A., Sengun, C., Reddy, C., Boukas, K., Zitser, J., et al. (2008). Case of pediatric acquired chronic hepatocerebral degeneration. *Pediatric Neurology*, *38*, 67–70.

Park, H. K., Kim, S. M., Choi, C. G., Lee, M. C., & Chung, S. J. (2008). Effect of trientine on manganese intoxication in a patient with acquired hepatocerebral degeneration. *Movement Disorders*, *23*, 768–770.

Penland, J. G., & Johnson, P. E. (1993). Dietary calcium and manganese effects on menstrual cycle symptoms. *American Journal of Obstetrics and Gynecology*, *168*, 1417–1423.

Pinto, R. B., Froehlich, P. E., Pitrez, E. H., Bragatti, J. A., Becker, J., Cornely, A. F., et al. (2010). MR findings of the brain in children and adolescents with portal hypertension and the relationship with blood manganese levels. *Neuropediatrics*, *41*, 12–17.

Pomier-Layrargues, G., Spahr, L., & Butterworth, R. F. (1995). Increased manganese concentrations in pallidum of cirrhotic patients. *Lancet*, *345*, 735.

Prabhakaran, K., Ghosh, D., Chapman, G. D., & Gunasekar, P. G. (2008). Molecular mechanism of manganese exposure-induced dopaminergic toxicity. *Brain Research Bulletin*, *76*, 361–367.

Quadri, M., Federico, A., Zhao, T., Breedveld, G. J., Battisti, C., Delnooz, C., et al. (2012). Mutations in SLC30A10 cause parkinsonism and dystonia with hypermanganesemia, polycythemia, and chronic liver disease. *American Journal of Human Genetics*, *90*, 467–477.

Racette, B. A., Aschner, M., Guilarte, T. R., Dydak, U., Criswell, S. R., & Zheng, W. (2012). Pathophysiology of manganese-associated neurotoxicity. *Neurotoxicology*, *33*, 881–886.

Rahelic, D., Kujundzic, M., Romic, Z., Brkic, K., & Petrovecki, M. (2006). Serum concentration of zinc, copper, manganese and magnesium in patients with liver cirrhosis. *Collegium Antropologicum*, *30*, 523–528.

Rao, K. V., & Norenberg, M. D. (2004). Manganese induces the mitochondrial permeability transition in cultured astrocytes. *Journal of Biological Chemistry*, *279*, 32333–32338.

Rentschler, G., Covolo, L., Haddad, A. A., Lucchini, R. G., Zoni, S., & Broberg, K. (2012). ATP13A2 (PARK9) polymorphisms influence the neurotoxic effects of manganese. *Neurotoxicology*, *33*, 697–702.

Ribeiro, R. T., Dos Santos-Neto, D., Braga-Neto, P., & Barsottini, O. G. (2013). Inherited manganism. *Clinical Neurology and Neurosurgery*, *115*, 1536–1538.

Rivera-Mancia, S., Rios, C., & Montes, S. (2011). Manganese accumulation in the CNS and associated pathologies. *Biometals*, *24*, 811–825.

Rodier, J. (1955). Manganese poisoning in Moroccan miners. *British Journal of Industrial Medicine*, *12*, 21–35.

Roels, H. A., Bowler, R. M., Kim, Y., Claus, H. B., Mergler, D., Hoet, P., et al. (2012). Manganese exposure and cognitive deficits: A growing concern for manganese neurotoxicity. *Neurotoxicology*, *33*, 872–880.

Rose, C., Butterworth, R. F., Zayed, J., Normandin, L., Todd, K., Michalak, A., et al. (1999). Manganese deposition in basal ganglia structures results from both portal-systemic shunting and liver dysfunction. *Gastroenterology*, *117*, 640–644.

Roth, J. A. (2006). Homeostatic and toxic mechanisms regulating manganese uptake, retention, and elimination. *Biological Research*, *39*, 45–57.

Roth, J. A. (2009). Are there common biochemical and molecular mechanisms controlling manganism and parkinsonism. *Neuromolecular Medicine*, *11*, 281–296.

Roth, J. A., Ganapathy, B., & Ghio, A. J. (2012). Manganese-induced toxicity in normal and human B lymphocyte cell lines containing a homozygous mutation in parkin. *Toxicology In Vitro*, *26*, 1143–1149.

Roth, J. A., Li, Z., Sridhar, S., & Khoshbouei, H. (2013). The effect of manganese on dopamine toxicity and dopamine transporter (DAT) in control and DAT transfected HEK cells. *Neurotoxicology*, *35*, 121–128.

Roth, J. A., Singleton, S., Feng, J., Garrick, M., & Paradkar, P. N. (2010). Parkin regulates metal transport via proteasomal degradation of the 1B isoforms of divalent metal transporter 1. *Journal of Neurochemistry*, *113*, 454–464.

Roth, J. A., Sridhar, S., & Singleton, S. T. (2012). Effect of glutamate and riluzole on manganese-induced apoptotic cell signaling in neuronally differentiated mouse P19 cells. *Neurochemistry International*, *61*, 25–33.

Sacher, A., Cohen, A., & Nelson, N. (2001). Properties of the mammalian and yeast metal-ion transporters DCT1 and Smf1p expressed in *Xenopus laevis* oocytes. *Journal of Experimental Biology*, *204*, 1053–1061.

Sanchez, B., Casalots-Casado, J., Quintana, S., Arroyo, A., Martin-Fumado, C., & Galtes, I. (2012). Fatal manganese intoxication due to an error in the elaboration of Epsom salts for a liver cleansing diet. *Forensic Science International*, *223*, e1–e4.

Sanotsky, Y., Lesyk, R., Fedoryshyn, L., Komnatska, I., Matviyenko, Y., & Fahn, S. (2007). Manganic encephalopathy due to "ephedrone" abuse. *Movement Disorders*, *22*, 1337–1343.

Santos, P. C., Dinardo, C. L., Cancado, R. D., Schettert, I. T., Krieger, J. E., & Pereira, A. C. (2012). Non-HFE hemochromatosis. *Revista Brasileira de Hematologia e Hemoterapia*, *34*, 311–316.

Scassellati, C., Bonvicini, C., Faraone, S. V., & Gennarelli, M. (2012). Biomarkers and attention-deficit/hyperactivity disorder: A systematic review and meta-analyses. *Journal of the American Academy of Child and Adolescent Psychiatry*, *51*, 1003–1019.

Schmidt, K., Wolfe, D. M., Stiller, B., & Pearce, D. A. (2009). Cd^{2+}, Mn^{2+}, Ni^{2+} and Se^{2+} toxicity to *Saccharomyces cerevisiae* lacking YPK9p the orthologue of human ATP13A2. *Biochemical and Biophysical Research Communications*, *383*, 198–202.

Selikhova, M., Fedoryshyn, L., Matviyenko, Y., Komnatska, I., Kyrylchuk, M., Krolicki, L., et al. (2008). Parkinsonism and dystonia caused by the illicit use of ephedrone—A longitudinal study. *Movement Disorders*, *23*, 2224–2231.

Sepulveda, M. R., Berrocal, M., Marcos, D., Wuytack, F., & Mata, A. M. (2007). Functional and immunocytochemical evidence for the expression and localization of the secretory pathway Ca^{2+}–ATPase isoform 1 (SPCA1) in cerebellum relative to other Ca^{2+} pumps. *Journal of Neurochemistry*, *103*, 1009–1018.

Settivari, R., Levora, J., & Nass, R. (2009). The divalent metal transporter homologues SMF-1/2 mediate dopamine neuron sensitivity in *Caenorhabditis elegans* models of manganism and Parkinson disease. *Journal of Biological Chemistry*, *284*, 35758–35768.

Seve, M., Chimienti, F., Devergnas, S., & Favier, A. (2004). In silico identification and expression of SLC30 family genes: An expressed sequence tag data mining strategy for the characterization of zinc transporters' tissue expression. *BMC Genomics*, *5*, 32.

Sidoryk-Wegrzynowicz, M., Lee, E., Albrecht, J., & Aschner, M. (2009). Manganese disrupts astrocyte glutamine transporter expression and function. *Journal of Neurochemistry*, *110*, 822–830.

Sidoryk-Wegrzynowicz, M., Lee, E., Mingwei, N., & Aschner, M. (2011). Disruption of astrocytic glutamine turnover by manganese is mediated by the protein kinase C pathway. *Glia*, *59*, 1732–1743.

Sikk, K., Haldre, S., Aquilonius, S. M., Asser, A., Paris, M., Roose, A., et al. (2013). Manganese-induced parkinsonism in methcathinone abusers: Bio-markers of exposure and follow-up. *European Journal of Neurology*, *20*, 915–920.

Sikk, K., Taba, P., Haldre, S., Bergquist, J., Nyholm, D., Askmark, H., et al. (2010). Clinical, neuroimaging and neurophysiological features in addicts with manganese-ephedrone exposure. *Acta Neurologica Scandinavica*, *121*, 237–243.

Sikk, K., Taba, P., Haldre, S., Bergquist, J., Nyholm, D., Zjablov, G., et al. (2007). Irreversible motor impairment in young addicts–ephedrone, manganism or both? *Acta Neurologica Scandinavica*, *115*, 385–389.

Singleton, A. B., Farrer, M. J., & Bonifati, V. (2013). The genetics of Parkinson's disease: Progress and therapeutic implications. *Movement Disorders, 28*, 14–23.
Slicker, J., & Vermilyea, S. (2009). Pediatric parenteral nutrition: Putting the microscope on macronutrients and micronutrients. *Nutrition in Clinical Practice, 24*, 481–486.
Spahr, L., Butterworth, R. F., Fontaine, S., Bui, L., Therrien, G., Milette, P. C., et al. (1996). Increased blood manganese in cirrhotic patients: Relationship to pallidal magnetic resonance signal hyperintensity and neurological symptoms. *Hepatology, 24*, 1116–1120.
Spranger, M., Schwab, S., Desiderato, S., Bonmann, E., Krieger, D., & Fandrey, J. (1998). Manganese augments nitric oxide synthesis in murine astrocytes: A new pathogenetic mechanism in manganism? *Experimental Neurology, 149*, 277–283.
Squitti, R., Gorgone, G., Panetta, V., Lucchini, R., Bucossi, S., Albini, E., et al. (2009). Implications of metal exposure and liver function in Parkinsonian patients resident in the vicinities of ferroalloy plants. *Journal of Neural Transmission, 116*, 1281–1287.
Sreedharan, S., Stephansson, O., Schioth, H. B., & Fredriksson, R. (2011). Long evolutionary conservation and considerable tissue specificity of several atypical solute carrier transporters. *Gene, 478*, 11–18.
Sriram, K., Lin, G. X., Jefferson, A. M., Roberts, J. R., Chapman, R. S., Chen, B. T., et al. (2010). Dopaminergic neurotoxicity following pulmonary exposure to manganese-containing welding fumes. *Archives of Toxicology, 84*, 521–540.
Sriram, K., Lin, G. X., Jefferson, A. M., Roberts, J. R., Wirth, O., Hayashi, Y., et al. (2010). Mitochondrial dysfunction and loss of Parkinson's disease-linked proteins contribute to neurotoxicity of manganese-containing welding fumes. *FASEB Journal, 24*, 4989–5002.
Stamelou, M., Tuschl, K., Chong, W. K., Burroughs, A. K., Mills, P. B., Bhatia, K. P., et al. (2012). Dystonia with brain manganese accumulation resulting from SLC30A10 mutations: A new treatable disorder. *Movement Disorders, 27*, 1317–1322.
Stepens, A., Logina, I., Liguts, V., Aldins, P., Eksteina, I., Platkajis, A., et al. (2008). A Parkinsonian syndrome in methcathinone users and the role of manganese. *New England Journal of Medicine, 358*, 1009–1017.
Stephenson, A. P., Schneider, J. A., Nelson, B. C., Atha, D. H., Jain, A., Soliman, K. F., et al. (2013). Manganese-induced oxidative DNA damage in neuronal SH-SY5Y cells: Attenuation of thymine base lesions by glutathione and N-acetylcysteine. *Toxicology Letters, 218*, 299–307.
Sue, W. C., Chen, C. Y., & Chen, C. C. (1996). Dyskinesia from manganism in a hepatic dysfunction patient. *Zhonghua Minguo Guo Xiao Er Ke Yi Xue Hui Za Zhi, 37*, 59–64.
Takeda, A., Sawashita, J., & Okada, S. (1998). Manganese concentration in rat brain: Manganese transport from the peripheral tissues. *Neuroscience Letters, 242*, 45–48.
Tan, J., Zhang, T., Jiang, L., Chi, J., Hu, D., Pan, Q., et al. (2011). Regulation of intracellular manganese homeostasis by Kufor-Rakeb syndrome-associated ATP13A2 protein. *Journal of Biological Chemistry, 286*, 29654–29662.
Tuschl, K., Clayton, P. T., Gospe, S. M., Jr., Gulab, S., Ibrahim, S., Singhi, P., et al. (2012). Syndrome of hepatic cirrhosis, dystonia, polycythemia, and hypermanganesemia caused by mutations in SLC30A10, a manganese transporter in man. *American Journal of Human Genetics, 90*, 457–466.
Tuschl, K., Clayton, P. T., Gospe, S. M., & Mills, P. B. (1993). Dystonia/parkinsonism, hypermanganesemia, polycythemia, and chronic liver disease. *GeneReviews™ [Internet]*. Seattle, WA: University of Washington.
Tuschl, K., Mills, P. B., Parsons, H., Malone, M., Fowler, D., Bitner-Glindzicz, M., et al. (2008). Hepatic cirrhosis, dystonia, polycythaemia and hypermanganesaemia—A new metabolic disorder. *Journal of Inherited Metabolic Disease, 31*, 151–163.
Tussing-Humphreys, L., Pusatcioglu, C., Nemeth, E., & Braunschweig, C. (2012). Rethinking iron regulation and assessment in iron deficiency, anemia of chronic disease,

and obesity: Introducing hepcidin. *Journal of the Academy of Nutrition and Dietetics, 112*, 391–400.

Van, B. K., Dode, L., Vanoevelen, J., Callewaert, G., De, S. H., Missiaen, L., et al. (2004). The Ca2+/Mn2+ pumps in the Golgi apparatus. *Biochimica et Biophysica Acta, 1742*, 103–112.

Vanoevelen, J., Dode, L., Van, B. K., Fairclough, R. J., Missiaen, L., Raeymaekers, L., et al. (2005). The secretory pathway Ca^{2+}/Mn^{2+}–ATPase 2 is a Golgi-localized pump with high affinity for Ca^{2+} ions. *Journal of Biological Chemistry, 280*, 22800–22808.

Wang, C. Y., Jenkitkasemwong, S., Duarte, S., Sparkman, B. K., Shawki, A., Mackenzie, B., et al. (2012). ZIP8 is an iron and zinc transporter whose cell-surface expression is up-regulated by cellular iron loading. *Journal of Biological Chemistry, 287*, 34032–34043.

Wang, J., Rahman, M. F., Duhart, H. M., Newport, G. D., Patterson, T. A., Murdock, R. C., et al. (2009). Expression changes of dopaminergic system-related genes in PC12 cells induced by manganese, silver, or copper nanoparticles. *Neurotoxicology, 30*, 926–933.

Wang, X. S., Ong, W. Y., & Connor, J. R. (2001). A light and electron microscopic study of the iron transporter protein DMT-1 in the monkey cerebral neocortex and hippocampus. *Journal of Neurocytology, 30*, 353–360.

Wedler, F. C., Vichnin, M. C., Ley, B. W., Tholey, G., Ledig, M., & Copin, J. C. (1994). Effects of Ca(II) ions on Mn(II) dynamics in chick glia and rat astrocytes: Potential regulation of glutamine synthetase. *Neurochemical Research, 19*, 145–151.

Weissenborn, K., Ehrenheim, C., Hori, A., Kubicka, S., & Manns, M. P. (1995). Pallidal lesions in patients with liver cirrhosis: Clinical and MRI evaluation. *Metabolic Brain Disease, 10*, 219–231.

Willis, A. W., Evanoff, B. A., Lian, M., Galarza, A., Wegrzyn, A., Schootman, M., et al. (2010). Metal emissions and urban incident Parkinson disease: A community health study of Medicare beneficiaries by using geographic information systems. *American Journal of Epidemiology, 172*, 1357–1363.

Witzleben, C. L., Pitlick, P., Bergmeyer, J., & Benoit, R. (1968). Acute manganese overload. A new experimental model of intrahepatic cholestasis. *American Journal of Pathology, 53*, 409–422.

Xiang, M., Mohamalawari, D., & Rao, R. (2005). A novel isoform of the secretory pathway Ca^{2+}, Mn(2+)-ATPase, hSPCA2, has unusual properties and is expressed in the brain. *Journal of Biological Chemistry, 280*, 11608–11614.

Xu, Z. W., & Li, Y. S. (2012). Pathogenesis and treatment of parenteral nutrition-associated liver disease. *Hepatobiliary & Pancreatic Diseases International, 11*, 586–593.

Yamada, M., Ohno, S., Okayasu, I., Okeda, R., Hatakeyama, S., Watanabe, H., et al. (1986). Chronic manganese poisoning: A neuropathological study with determination of manganese distribution in the brain. *Acta Neuropathologica, 70*, 273–278.

Yin, Z., Jiang, H., Lee, E. S., Ni, M., Erikson, K. M., Milatovic, D., et al. (2010). Ferroportin is a manganese-responsive protein that decreases manganese cytotoxicity and accumulation. *Journal of Neurochemistry, 112*, 1190–1198.

Yokel, R. A. (2009). Manganese flux across the blood–brain barrier. *Neuromolecular Medicine, 11*, 297–310.

Yokel, R. A., & Crossgrove, J. S. (2004). Manganese toxicokinetics at the blood–brain barrier. *Research Report Health Effects Institute, 119*, 7–58.

Yoon, H., Kim, D. S., Lee, G. H., Kim, K. W., Kim, H. R., & Chae, H. J. (2011). Apoptosis induced by manganese on neuronal SK-N-MC cell line: Endoplasmic reticulum (ER) stress and mitochondria dysfunction. *Environmental Health and Toxicology, 26*, e2011017.

CHAPTER THIRTEEN

Update on Wilson Disease

Annu Aggarwal[1], Mohit Bhatt
Wilson Disease Clinic, Kokilaben Dhirubhai Ambani Hospital and Medical Research Institute, Mumbai, India
[1]Corresponding author: e-mail address: annu.aggarwal@gmail.com

Contents

Abstract

Wilson disease (WD) is an inherited disorder of chronic copper toxicosis characterized by excessive copper deposition in the body, primarily in the liver and the brain. It is a progressive disease and fatal if untreated. Excessive copper accumulation results from the inability of liver to excrete copper in bile. Copper is an essential trace metal and has a crucial role in many metabolic processes. Almost all of the body copper is protein bound. In WD, the slow but relentless copper accumulation overwhelms the copper chaperones (copper-binding proteins), resulting in high levels of free copper and copper-induced tissue injury. Liver is the central organ for copper metabolism, and copper is initially accumulated in the liver but over time spills to other tissues.

WD has protean clinical manifestations mainly attributable to liver, brain, and osseomuscular impairment. Diagnosis of WD is challenging and based on combination of clinical features and laboratory tests. Identification of various high-frequency mutations identified in different population studies across the world has revived interest in developing DNA chips for rapid genetic diagnosis of WD.

All symptomatic and all presymptomatic patients require lifelong decoppering with careful clinical tracking. Decoppering ensures that presymptomatic individuals remain symptom free. With judicious decoppering, given time, even patients

International Review of Neurobiology, Volume 110
ISSN 0074-7742
http://dx.doi.org/10.1016/B978-0-12-410502-7.00014-4

with severe neurological disability improve and can return to normal life and resume school or work at par with their peers. Treatment regimens and tracking patients using the WD-specific Global Assessment Scale for WD (GAS for WD) are discussed.

1. INTRODUCTION

Wilson disease (WD) is an inherited disorder of chronic copper toxicosis characterized by excessive copper deposition in the body, primarily in the liver and the brain (Walshe, 2009). Excessive copper accumulation results from the inability of liver to excrete copper in bile. Copper toxicosis in WD differs from chronic poisoning from other heavy metals like mercury or lead, because unlike these metals, copper has an essential role in many metabolic processes. Deficiency of copper leads to well-defined clinical and pathological defects (Scheinberg & Sternlieb, 1984). The necessity of copper is exemplified by Menkes disease, an inherited disorder of copper transport that results in relative copper deficiency and failure to incorporate copper in various enzymes. Menkes disease manifests as a progressive multisystemic disease and death in infancy (Tümer and Møller, 2010). Free copper is injurious to cells, and almost all of the body copper is protein bound. In WD, the ever-increasing positive copper balance overwhelms the copper chaperones (copper-binding proteins), resulting in high levels of free copper and copper-induced tissue injury.

The story of WD began with Samuel Kinnier Wilson, while still a registrar at the National Hospital, Queens Square, London, describing a new rare familial disease characterized by progressive lenticular degeneration and liver cirrhosis. The report included exhaustive clinical and pathological description of 12 patients published as a single issue of Brain. Four of the 12 patients were observed by Wilson, while details of two patients were obtained from medical records at the National Hospital. Six of the 12 patients' clinical histories were published earlier; of these, one brother and sister were reported by Gower in 1988 as "Tetanoid Chorea, associated with Cirrhosis of the Liver," one patient was described by Ormerod in 1890, and a further three siblings were described by Homen, of Helsingfors, also in 1890. The disease manifested as progressive involuntary movements, spasticity, dysarthria, and transitory mental symptoms and was fatal in all. All 12 patients had advanced liver cirrhosis (Wilson, 1912).

In contrast to dystonia-dominant syndrome of Samuel Wilson, Westphal and Strumpell independently described cases with pseudosclerosis or a tremor-dominant phenotype of WD (Pfeiffer, 2011).

Wilson believed that WD was a toxin-induced neurodegenerative disorder. Though curious about the presence of unexplained advanced cirrhosis in young people with WD, Wilson surmised that the liver involvement did not contribute to the clinical syndrome or the disease evolution. This is surprising as one of his patients died of hematemesis (Wilson, 1912). Bramwell (1916) drew attention to the role of the liver in the natural history of the disease in a family where 4 of 7 siblings died between 9 and 14 years of liver failure. The name hepatolenticular degeneration was first applied to WD by Hall (1921).

Excess of copper was reported in the liver of a patient dying of WD a year after Wilson's original publication (Rumpel, 1913). In the 1920s, there were other reports of high copper in the brain or liver of patients dying of WD (Glazebrook, 1945; Haurowitz, 1930; Vogt, 1929) and similarity of Kayser–Fleischer rings (KF rings) and sunflower cataracts with copper-containing foreign body (Siemerling & Oloff, 1922). However, the central role of copper was finally acknowledged only in 1948 when Cumings (1948) demonstrated excess copper in the brain and liver of patients with WD. In fact, Cumings not only irrefutably establish that copper played a central role in WD but also suggested that that treatment with the newly developed chelating agent British antilewisite (BAL, dimercaprol) might arrest the progress of the disease (Cumings, 1948). Use of BAL (Cumings, 1951, Denny Brown & Porter, 1951) was soon followed by development and introduction of effective oral chelators as orphan drugs for treating WD (Walshe, 2009).

In what remains a classic monograph, Samuel Wilson described a new invariably fatal disease afflicting the young. Wilson's quest of finding a cure was realized half a century later. Treatment for WD has now been available for over 50 years (Walshe, 1956). Today, with early diagnosis and judicious decoppering, disease progression can be halted and disability reversed. The challenge is to diagnose WD early, preferably before symptom onset, and titrate decoppering effectively.

2. COPPER HOMEOSTASIS

Copper is the third most abundant trace element in humans (after iron and zinc) and serves as a cofactor for many important enzymes including cytochrome c oxidase, copper–zinc superoxide dismutase (SOD), dopamine beta-hydroxylase, and ceruloplasmin (Bull & Cox, 1994). Though an essential metal, copper can be toxic. Copper exists in two forms, cuprous and cupric ions, which can interchange rapidly. This redox property is essential for the role of copper in its enzymatic activity. However, unbound or free copper can rapidly generate reactive oxygen species and have a deleterious effect on the cell (Rosencrantz & Schilsky, 2011). Therefore, copper homeostasis is tightly controlled, and there is virtually no free copper in the body (Tümer & Møller, 2010).

Daily intake of copper is 1–2 mg, and 98% of the absorbed copper is excreted via the liver into the bile and thereon lost in feces (Lutsenko, Barnes, Bartee, & Dmitriev, 2007). Dietary copper is absorbed in the small intestine by various nonspecific metal transporters and is exported into the portal circulation by Cu-transporting ATPase ATP7A (Menkes protein). ATP7B is present in the enterocytes, but its role is uncertain. In Menkes disease, inactivation of ATP7A in the enterocytes results in overall copper deficiency (Kaler, 1998; Kodama & Murata, 1999). The copper in the portal blood binds to amino acids like histidine or albumin and is then transported to the liver.

Liver is the central organ for copper storage and homeostasis. The liver avidly takes up most of the copper in portal circulation through the CTR1 transporter. Copper is stored in the liver, bound to metallothioneins, and delivered to various intracellular organelles via copper chaperones. Copper in the cytosol is delivered to copper–zinc SOD via copper chaperone of superoxide dismutase (CCS) (Culotta, Yang, & O'Halloran, 2006). Copper enters the mitochondria via COX17 chaperone for incorporation into cytochrome c oxidase, while Atox1 chaperone delivers copper to ATP7B located at the trans golgi complex (TGN) (Lutsenko et al., 2007; Prohaska & Gybina, 2004). ATP7B is a major copper transporter in the liver and plays a vital role in intracellular copper homeostasis. Under normal copper conditions, ATP7B resides at the TGN and facilitates copper incorporation into cuproproteins like ceruloplasmin, which is then secreted into the blood and transports copper to various tissues. Under high copper conditions, ATP7B is transferred to bile canaliculi and facilitates excretion of excess copper into bile. A defect in ATP7B transporter leads to impaired efflux of copper and excessive copper deposition in the liver.

Ceruloplasmin is an alpha-1-acid glycoprotein and is a major copper transporter in blood. Ceruloplasmin is low in 80–95% of the patients with WD. However, low ceruloplasmin is not responsible for WD. Decreased

ATP7B activity leads to impaired incorporation of copper into apoceruloplasmin and low ceruloplasmin levels. Ceruloplasmin has a role in iron metabolism. Mutation in the ceruloplasmin (CP) gene on chromosome 3 leads to aceruloplasminemia, a multisystemic chronic iron toxicosis characterized by absence of ceruloplasmin, and excessive iron deposition in the brain, liver, and pancreas (Harris et al., 1995; Miyajima et al., 1987). CP gene is normal in WD. Ceruloplasmin levels can be low in up to 10–20% of the people heterozygous for WD mutations and in patients with other chronic liver diseases, Menkes disease, and protein malnutrition (Scheinberg & Sternlieb, 1984).

Minute quantities of copper are secreted in saliva and gastric secretion and lost through skin in healthy individuals. Urinary copper excretion is negligible. In WD biliary copper excretion is impaired leading to copper deposition in the liver. Over years, copper gradually accumulates leading to progressive liver dysfunction. Ultimately, the liver can no longer hold more copper, and the unbound toxic metal spills into the blood and other tissues. Urinary copper excretion increases in patients with WD but cannot compensate for impaired biliary excretion of copper or halt the slow relentless increase in positive copper balance in the body (Pfeiffer, 2011).

3. GENETICS OF WD

Wilson described WD as a familial disease but did not consider it hereditary (Wilson, 1912). Genetic origin was proposed by Hall in 1928, and autosomal recessive inheritance was reported in 1953 (Bearn, 1953). The disease was mapped to chromosome 13, and the disease-causing gene *ATP7B* was identified and cloned in 1993 simultaneously by several groups (Bull, Thomas, Rommens, Forbes, & Cox, 1993; Tanzi et al., 1993).

ATP7B is a large gene consisting of 21 exons and is located on chromosome 13 (13q14-q21). The gene encodes for a 1465-amino acid protein consisting of six copper-binding domains, a transduction domain, a cation channel and phosphorylation domain, a nucleotide-binding domain, and eight hydrophobic transmembrane domains. *ATP7B* mutations result in loss of function. Till date, no gain-of-function *ATP7B* mutation has been described (Schmidt, 2009).

The discovery of the *ATP7B* gene was met with great enthusiasm and hope that this would allow early and accurate diagnosis of WD. However, it was soon realized that the disease can be caused by several (>600) mutations and most patients are compound-heterozygous (WD mutation database—http://www.wilsondisease.med.ualberta.ca/database). Therefore, identification of WD mutation in a given patient is a time-consuming process and offered only on research basis by selected laboratories in the world. While, identification of WD mutation is the gold standard to diagnose WD, previous studies have reported patients with unequivocal WD bearing only one or no WD mutation (Kenney & Cox, 2007; Nicastro, Ranucci, Vajro, Vegnente, & Iorio, 2010). Failure to identify WD mutation in a patient does not conclusively exclude WD.

Majority of the *ATP7B* mutations identified are missense while, minority are deletions/insertions, nonsense, and intronic splice site. Mutations are scattered along the entire length of the *ATP7B* gene. The possibility of other causative genes contributing to WD has been explored, but not realized (Coronado, Damaraju, Kohijoki, & Cox, 2003). With precise clinical diagnosis and rapid whole-gene sequencing, the *ATP7B* mutation detection rate in WD patients is near perfect (~98%) (Aggarwal et al., 2013; Coffey et al., 2013).

In the recent years, there has been resurgence in interest in using genetic analysis to diagnose WD in patients and screen their siblings. DNA chips incorporating few frequent mutations are being developed to allow for rapid diagnosis (Schmidt, 2009). This enthusiasm stems from various high-frequency mutations identified in different population studies. For instance, the p.H1069Q (exon 14) is a common WD mutation accounting for WD in 30–60% of the Caucasian population (Ferenci, 2006). Screening of single-mutation p.H1069Q accounts for two-thirds of the population in northern, central, and eastern Europe and the United States. While, p.R778L and p.R778G mutations (exon 8) among the Chinese and Taiwanese populations enable rapid genetic analysis of WD in these populations (Chuang et al., 1996; Thomas, Forbes, Roberts, Walshe, & Cox, 1995). It is interesting that in countries with varied ethnicity like India, a few mutations likely account for WD in a large proportion of the population. The regional distribution of some of the common WD mutations is summarized in Table 13.1.

Genotype–phenotype correlation in WD is challenging since patients present with varied clinical symptoms. Monozygotic twins with varied clinical phenotypes have been reported (Czlonkowska, Gromadzka, & Chabik, 2009). Compared to a relatively late onset (second to third decade of life) and a predominantly neurological phenotype of mild to moderate disability that could be associated with the most common European mutation p.H1069Q, the two common western Indian mutations (p.C271* and p.E122fs) seem to lead to earlier onset (first and second decade), and more severe disease. The age of onset of WD in Indian patients is earlier than that reported for patients

Table 13.1 High-frequency *ATP7B* mutation spectrum in various world populations

Country	Sample size (*n*)	Nucleotide/ amino acid change	Exon	Domain	Allele frequency (%)	Reference
Austria	125	H1069Q	14	SEPHL	34.1	Ferenci, 2006
		G710S	8		6.4	
		2299insC	8		3.6	
		R969Q	13		3.6	
Bulgaria	89	H1069Q	14	SEPHL	58.75	Todorov et al., 2005
		2304–2305 ins C			11.25	
		3400delC			3.75	
Hungary	42	H1069Q	14	SEPHL	42.8	Firneisz et al., 2002
Germany	82	H1069Q	14	SEPHL	63	Caca et al., 2001
		3400delC	15		9	
		2299ins C	8		4	
Poland	142	H1069Q	14	SEPHL	72	Gromadzka et al., 2005
		3402delC	15		8	
United Kingdom	42	H1069Q	14	SEPHL	17	Curtis et al., 1999
		M769V	8	Tm4	8	
Japan	47	2871delC	13		15.9	Okada et al., 2000
		R778L	8		13.4	
China	40	R778L	8			Gu et al., 2003
	44	R778L	8			Wu, Wang, Murong, & Lin, 2000
India	52	C271*	2	Cu3	20	Aggarwal et al., 2013
		E122fs	2	Cu1	11	
		L795F		Tm4/Td	6	
		T977M		Tm6	6	

Continued

Table 13.1 High-frequency *ATP7B* mutation spectrum in various world populations—cont'd

Country	Sample size (*n*)	Nucleotide/ amino acid change	Exon	Domain	Allele frequency (%)	Reference
	27	C271*	2	Cu3	~9	Santhosh et al., 2006
		G1061E	14	ATP N-binding	~9	
		C271*	2		18.5	Gupta et al., 2005
		G1708-1C	4	Cu6	9.6	
		448_452del5	2	Cu1	5.6	
	43	I1102T	15	ATP loop	6.1	Kumar et al., 2005
		P922H	13	Tm6	5.8	
		P922*	13	Tm6	5.8	
		G1010A-fs	13	Tm6	5.8	

in Europe, Korea, and South America (Deguti et al., 2004; Lee et al., 2011; Stapelbroek et al., 2004; Taly, Prashanth, & Sinha, 2009). Further, the clinical phenotype of Indian patients with WD, is possibly more severe than that observed in western populations (Aggarwal et al., 2013).

Identification of common mutations and exonic hotspots in a given population makes genetic studies promising and a practical diagnostic tool. Genetic diagnosis also plays an indispensable role in diagnosis of asymptomatic siblings of patient wiith WD.

4. CLINICAL MANIFESTATIONS

4.1. Presymptomatic WD

WD is a multisystemic disease characterized by liver, neurological, and osseomuscular involvement. Copper accumulation commences after birth and slowly progresses. The liver and the basal ganglion bears the brunt of copper toxicosis. Patients with WD can remain presymptomatic for decades. It is crucial to be aware of and identify the red flags of WD (Table 13.2) to allow early diagnosis and prompt treatment.

The initial symptoms are often innocuous, for example, asymptomatic elevation of liver enzymes, fleeting jaundice, change in handwriting, drop

Table 13.2 Red flags for Wilson disease

System	Warning symptoms/findings
Neurological	Dysarthria
	Change in handwriting
	Clumsiness
	Difficulty in walking
	Dropping grades at school
	Depression
	Emotional liability
	Aggressive/unruly behavior in school, home, or work
	Hypersexuality
	Running amok
Liver	Incidental unexplained abnormalities on liver function tests
	Unexplained fleeting jaundice
	Repeated jaundice
	Easy bruising
	Unexplained pruritus
	Unexplained anemia
	Unexplained thrombocytopenia
	Liver cirrhosis on ultrasound during episode of acute jaundice
Osseomuscular symptoms	Proximal lower-limb muscle weakness
	Bone pains (dismissed as "growing pains")
	Fleeting arthralgia (ankle, knee, wrist, and elbow)
	Unexplained monoarthritis
	Unexplained limp
	Unexplained fracture
	Back pain
	Joint pain and swelling

Continued

Table 13.2 Red flags for Wilson disease—cont'd

Eye	Incidental diagnosis of Kayser–Fleischer rings (astute optometrist!)
Family history	Wilson disease
	Unexplained liver or neurological disease in family
	Unexplained death in family
General complaints	Unexplained weight loss
	Weight gain (from ascites)
	Failure to thrive
	Menstrual abnormalities in girls

in grades at school, irritability, or recurrent joint pain and swelling. Occasionally, abrupt catastrophic hemolysis and acute liver failure herald the disease, with little time for therapeutic intervention (Scheinberg & Sternlieb, 1984; Taly, Meenakshi-Sundaram, Sinha, Swamy, & Arunodaya, 2007).

4.2. Neurological (extrapyramidal) manifestations

Neurological (motor) manifestations are the most frequent initial symptoms of WD and seen in 40–60% of the patients (Dastur, Manghani, & Wadia, 1968; Scheinberg & Sternlieb, 1984; Taly et al., 2007). Involuntary movements are a common manifestation of WD and are often associated with early onset midline symptoms of dysarthria, dysphagia, and poor axial motor control. It is uncommon for a child with WD to develop marked limb dystonia but have a normal speech and gait. Dysarthria is the most common neurological feature of WD. Of the 100 consecutive patients with WD presenting with neurological symptoms, and seen between 2009 and 2013 at the Wilson Disease Clinic at KDAH, Mumbai, India, 90% had speech involvement. Dysarthria is multifactorial and in most patients it is difficult to delineate the exact neurological mechanism(s) contributing to the dysarthria. Dystonia involving tongue and facial muscles can produce profound dysarthria, while cerebellar involvement can lead to a scanning and explosive quality to speech. Muscular rigidity leads to low-volume speech with inadequate tongue movement and imprecise articulation. In patients with severe neurological disabilty, mutism is common. Decoppering improves dysarthria, though in severely affected patients, the recovery may not be complete (Aggarwal, Nagral, Jankharia, Aggarwal, & Bhatt, 2009). Another frequent midline symptom seen in patients with

WD is dysphagia. Patients with severe neurological disability often require gastric tube feeding for weeks to months.

There are two signs characteristically associated with WD—Wilson facies and KF rings. Wilson facies were described by Wilson in his original publication (Wilson, 1912). They are characterized by a facetious (false) smile, pseudo-laughter, open mouth and drooling saliva, reduced eye blinking, exploratory eye movements, and a dull look (Aggarwal et al., 2009) (Fig. 13.1). Wilson facies give patients with WD a characteristic facial feature, so much so, that they start resembling each other (Fig. 13.1). The "rire spasmodique" of Boudin & Pepin (1959) was described as a fixed smile with open mouth and with a high-pitched cry. To an observant clinician, limb dystonia or a movement disorder with characteristic Wilson facies can be an important diagnostic due to WD. Improvement in WD

Figure 13.1 Collage of Wilson disease patients' facies demonstrating the typical Wilson facies—a diagnostic clinical sign of Wilson disease. As seen in the picture, Wilson facies are characterized by factitious smile, pseudo-laughter, open mouth, dull look, and staring expression in variable combination. The figure also demonstrates hand, nuchal, and truncal dystonia in some of the patients. The patients shown in the picture resemble each other though they carry different Wilson disease-related mutations. With copper chelation, the facies normalize. (For color version of this figure, the reader is referred to the online version of this chapter.)

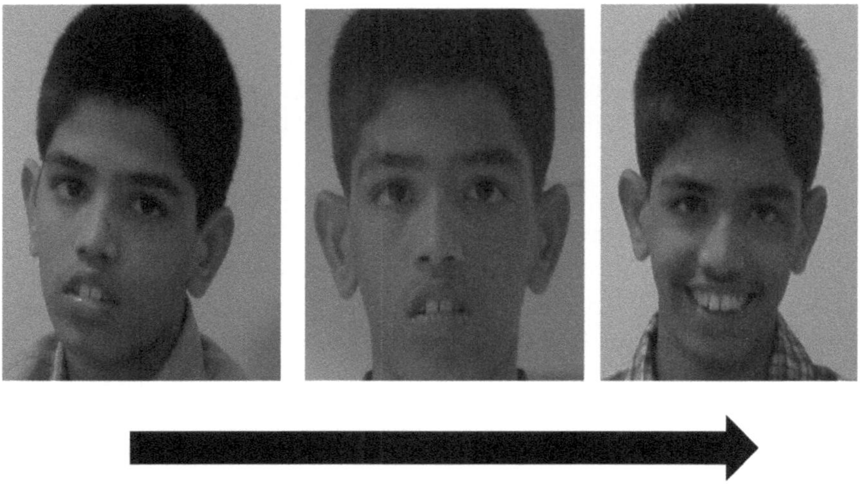

Figure 13.2 Improvement in Wilson facies over 18 months of decoppering. (For color version of this figure, the reader is referred to the online version of this chapter.)

facies is the earliest clinical sign of treatment (decoppering) response. Wilson facies can be graded as part of the Global Assessment Scale for WD and tracked to monitor decoppering response (Aggarwal et al., 2009) (Fig. 13.2).

In early 1900s, two ophthalmologists, Kayser and Fleischer, independently described the presence of corneal pigmentation in patients with pseudosclerosis (Walshe, 2006). Later, the rings were reported to represent copper deposition. KF rings are copper deposits in the corneal Descemet's membrane, visible as greenish discoloration at the outer corneal circumference. They first appear in the upper corneal limbus, followed by the lower limbus, and then form a complete ring that expands centripetally. KF rings can be seen using a torchlight directed tangentially at the cornea; however, early rings require slit lamp examination. The rings do not impair vision and with decoppering they clear in a sequence opposite to their deposition. Upto 95% of patients with neurologic symptoms and 44–62% of those with liver involvement have KF rings. (Scheinberg & Sternlieb, 1984). Extraocular movement abnormalities are rare and possibly secondary to copper deposition in the midbrain (Scheinberg & Sternlieb, 1984).

A whole range of movement disorders are seen in patients with WD. Patients usually have a mixed movemnt disorder. Four overlapping movement disorder syndromes are commonly recognized (see Pfeiffer, 2011):

1. Dystonic syndrome
2. Postural and action tremor with ataxia and titubation—pseudosclerosis form

3. Akinetic rigid syndrome (parkinsonism)
4. Choriform syndrome

The dystonic and the pseudosclerotic (cerebellar) phenotypes are the most commonly observed. Walsh and Yealland (1992) recorded the onset of WD in 136 patients between 1955 and 1987. Forty-five percent presented with parkinsonism, 32% with pseudosclerotic form, 15% with dystonic type, and 11% with chorea. The profile of neurological impairment observed in a patient cohort seen at the Wilson Disease Clinic, KDAH, Mumbai, India, is detailed in Table 13.3. Dystonia was the most common movement disorder and it was often associated with dystonic tremor and parkinsonism.

Early in the disease, dystonia may be limited to a limb, but if untreated, it invariably becomes generalized. As indicated earlier, speech involvement is an early feature. Interestingly, unlike neurodegeneration with brain iron accumulation, WD is not associated with opisthotonus posturing (Stamelou et al., 2013). Tremor in WD may be resting, postural, or intentional. The pseudosclerotic form is characterized by a typical proximal

Table 13.3 Profile of neurological impairment at diagnosis in 100 symptomatic patients with WD

Neurological feature	% of patients affected
Wilson facies	90.16
Scholastic backwardness	72.13
Depression	22.95
Psychosis	68.85
Dystonia	93.44
Tremor	45.90
Chorea	11.47
Parkinsonism	57.37
Speech impairment	91.80
Swallowing impairment	42.62
Salivation	55.73
Axial	68.85
Kayser–Fleischer rings	85.24

The initial symptom in all patients was neurological. Patients were seen at the Wilson Disease Clinic at KDAH, Mumbai, India, 2009–2013.

wing-beating tremor. Head tremor and isolated tongue tremor are described. Other movement disorders like chorea, myoclonus, and ataxia are infrequent (Prashanth, Taly, Sinha, Arunodaya, & Swamy, 2004; Walsh and Yealland, 1992). There may be regional differences in the WD phenotype. For instance, WD mutations frequent in western India were associated with earlier-onset disease and more severe phenotype than that described with the p.H1069Q mutation, the mutation most frequent in Europe and North America (Aggarwal et al., 2013).

Untreated WD patients continue to deteriorate, and subtle symptoms of dysarthria, behavior abnormalities, and facetious smiles give way to disabling movement disorders. Patients become progressively mute, have difficulty with axial balance, are prone to falls, and soon become bedbound. Without treatment, patients can deteriorate further and die from complications of being bedridden. If, however, decoppering is initiated, the outcomes can be very heartening. Bedbound patients can recover and resume normal life though it may take 1-3 years of judicious decoppering to reverse severe neurological disability. (Aggarwal & Bhatt, 2012).

4.3. Behavioral and cognitive problems

Neuropsychiatric manifestations are described in 20–70% of patients of WD. These may be innocuous or overlooked in the presence of severe extrapyramidal affection. Neuropsychiatric problems maybe the initial symptoms of WD or develop over time. Behavioral or cognitive problems are often ignored in children or attributed to schizophrenia or other primary psychiatric disorders. Neuroleptics are often prescribed, and if an extrapyramidal syndrome develops, it is attributed to neuroleptic related adverse effect (Scheinberg & Sternlieb, 1984; Walshe & Yealland, 1992).

Parents and school teachers are the first to observe subtle changes, like irritability, distractibility, and dropping scholastic performance. Wilson facies correlate with cognitive impairment and improve with decoppering (Fig. 13.2). Cognitive problems can continue for 6–12 months before other symptoms emerge. Aggression, reckless behavior, motor restlessness, running amok, emotional labiality, childlike behavior, impulse control disorders, and hypersexuality are common and different from positive symptoms observed in psychosis from schizophrenia. Antisocial behaviors can cause conflicts with law and lead to great anguish and unhappiness among family members. Expulsion from school or work is not uncommon.

Some of these children need hospitalization and neuroleptics for symptomatic relief. Suicidal attempts are reported (Aggarwal & Bhatt, 2008, Scheinberg & Sternlieb, 1984).

Aggarwal et al. (2009) reported patients with WD who were mute and bedbound at the start of treatment and developed severe psychosis in tandem with recovery of their motor function. The authors hypothesized that in these patients, psychosis was masked by mutism and severe motor disability and manifested once motor function improved. They called this phenomenon as emergent psychosis. Unlike neurological deterioration precipitated by sudden copper chelation, patients with emergent psychosis benefited from continued decoppering. On the contrary, WD patients with serious psychosis, if untreated, eventually developed severe motor disability and mutism that masked their psychosis, a phenomenon termed as concealed psychosis (Aggarwal & Bhatt, 2008; Walshe, 1989; Walshe & Dixon, 1986).

4.4. Hepatic manifestations

Liver failure is the most frequent presenting symptom of WD, and 40–50% of the patients present with liver dysfunction (Scheinberg & Sternlieb, 1984; Walshe, 1962). Symptoms of liver failure generally develop in first two decades but there are wide variations with the youngest reported patient being 2 years old, while the oldest was 74 years old (Pfeiffer, 2011, Czlonkowska, Rodo, & Gromadzka, 2008). Patients with initial liver symptoms present almost a decade earlier than those with initial neurological manifestations. Interestingly, patients with initial neurological disability often have a silent or overt liver disease (Aggarwal et al., 2013; Scheinberg & Sternlieb, 1984). This is not surprising as the liver is the primary target organ for copper deposition, and neurological manifestations likely occur once the liver can no longer accommodate any more copper and this excessive copper spills into the systemic circulation (Pfeiffer, 2011).

The most common mode of hepatic presentation is chronic silent cirrhosis with gradual decompensation from progressive portal hypertension and its complication. Acute viral hepatitis-like symptoms may occur, but there is usually evidence of chronic liver disease (thrombocytopenia or hemolytic anemia from hypersplenism, or cirrhosis on ultrasound) (Scheinberg & Sternlieb, 1984). Hepatitis is often associated with acute rise in serum ceruloplasmin, so a normal ceruloplasmin should be interpreted carefully. Autoimmune-like hepatic presentation is seen in a third of patients presenting with WD-related liver manifestations, and some respond to immunosuppressants leading to diagnostic confusion (Schilsky, Scheinberg, & Sternlieb, 1991; Sternlieb & Scheinberg, 1972).

The most fearsome manifestation of WD is fulminant hepatic failure and is seen in ~5% of the patients. This is rapidly progressive and invariably fatal unless liver transplant can be performed (Catana & Medici, 2012). Pathologically, most patients have evidence of silent cirrhosis though often not recognized antemortem (Korman et al., 2008). Diagnosis of WD in patients with fulminant hepatic failure is difficult. Urine collection for copper estimation is difficult due to hepatorenal syndrome, while liver biopsy is risky in face of coagulopathy. Further, liver copper and serum ceruloplasmin may be falsely elevated. Clues to diagnosis of WD are evidence of underlying cirrhosis, Coombs-negative hemolysis, and serum copper levels greater than 200 mg/dl (due to massive release of copper from the liver into the circulation) (Roberts & Schilsky, 2008). Unlike in acute fulminant viral hepatitis, in WD, the alkaline phosphates and aminotransferase levels are disproportionately low, while total bilirubin is disproportionately high due to concomitant rise of indirect bilirubin from copper induced hemolysis (Korman et al., 2008; Roberts & Schilsky, 2008).

It is recommended that WD be considered in all children with autoimmune hepatitis and in all adults with atypical autoimmune hepatitis or nonalcoholic steatohepatitis. Further, WD should be suspected in any patient presenting with acute hepatic failure associated with Coombs-negative intravascular hemolysis, modest elevations in serum aminotransferases, or low serum alkaline phosphatase with an alkaline phosphatase to bilirubin ratio of <2 (Roberts & Schilsky, 2008).

4.5. Hematological manifestations

Coombs-negative hemolytic anemia from copper toxicosis and unexplained thrombocytopenia, leucopenia, or pancytopenia from hypersplenism should alert the physician to the possibility of WD (Scheinberg & Sternlieb, 1984).

4.6. Osseomuscular manifestations

Bone and joint involvement (unrelated to penicillamine use) is a common though underrecognized feature of WD (Dastur et al., 1968; Golding & Walshe, 1977). Usually, osseomuscular symptoms are not volunteered, ignored as growing bone pains, or attributed to inadvertent injury. Osseomuscular abnormalities can be the presenting features or evolve later in the course of WD and are reported in up to 88% of the patients. The spectrum of involvement includes asymptomatic radiological abnormalities, fleeting joint pains, painful mono- or oligoarthropathy or spondylopathy, and nontraumatic fractures. Usually, large joints, like elbow, wrist, knees, and ankles are affected, but small joint pains can also be involved. The mechanism of joint involvement in WD is not clear (Aggarwal, Jankharia, & Bhatt, 2008; Golding & Walshe, 1977). Presence of osseomuscular symptoms in a person with liver or neurological impairment should alert the clinician to the diagnosis of WD.

5. DIAGNOSIS

WD is a multisystemic disease affecting the liver, brain, and osseomuscular system in varying combination with protean clinical manifestations (Dastur et al., 1968; Gill, Shankaran, & Desai, 1994; Taly et al., 2007, Walshe, 1962). A clinician is unlikely to see more than one patient in a lifetime of practice and no two WD patients are alike (Weiner & Lang, 1989). In a study cohort, patients consulted on an average of six doctors including specialists (range 2–11) before WD was diagnosed, contributing to a mean diagnostic delay of 2 years (Aggarwal et al., 2009). It is not unusual for families to have lost one or more child before WD is diagnosed in a younger sibling. In the posttreatment era, diagnostic failure has been the principal cause of death in WD (Członkowska, Tarnacka, Litwin, Gajda, & Rodo, 2005; Prashanth et al., 2004; Walshe, 2007). Maintaining a high degree of suspicion is the key to early diagnosis. Unexplained jaundice, alterations in liver function tests, neurological symptoms and osseomuscular problems, in a child or young adult should prompt consideration of WD (Table 13.2). Siblings of all patients with WD should be screened for the disease and kept under surveillance till adulthood unless WD is definitely ruled out by genetic analysis.

Genetic analysis is the gold standard for diagnosis of WD but has remained an impractical diagnostic tool as there are over 600 WD mutations described and many new continue to be reported. One or more disease-causing genes cannot be identified in 20% of the patients with clinically definite WD (Schmidt, 2009). This number of mutation-negative patients is likely to decrease with the ability to rapidly sequence and screen the whole *ATP7B* gene for WD mutations (Aggarwal et al., 2013; Coffey et al., 2013). While identification of WD mutations confirms WD, inability to find mutations does not conclusively exclude the disease. Recent studies from various parts of the world suggest that despite there being considerable genetic heterogeneity in various populations, a single or few *ATP7B* mutations may account for WD in a majority of patients in a given cohort (Table 13.1). Various research groups are therefore working on developing DNA chips targeting population-specific mutations to allow for rapid diagnosis of WD in that particular population. If WD mutations are identified in a patient, all of the patient's siblings should be screened for them.

Presently, diagnosis of WD is based on the combination of clinical features and laboratory tests. The presence of family history of WD, Wilson facies, and KF rings are important clinical clues for diagnosing WD. Constellation of neurological symptoms (extrapyramidal and behavioral problems), liver disease, and osseomuscular affection in young adults is uncommon and if present, it serves as a clinical marker of WD. Neurological symptoms are usually volunteered, but parents should be quizzed about subtle and often forgotten episodes of fleeting jaundice, incidental finding of raised liver enzymes, or innocuous features of osseomuscular involvement (Table 13.2). Diagnosis of WD in patients presenting with extrapyramidal syndrome or behavioral problems is usually straightforward and aided by the presence of KF rings, Wilson facies, abnormalities on brain scan, and clinical or subclinical liver disease. However, presymptomatic patients and those with pure liver disease often pose a diagnostic dilemma and necessitate careful consideration.

Serum ceruloplasmin, 24-h urinary copper excretion, liver function tests, brain imaging, and in select cases liver biopsy help establish the diagnosis of WD. Each of these tests has confounding features and test results need to be interpreted with caution (Table 13.4). Serum (total) copper or indices like serum non-ceruloplasmin-bound (free) copper (derived by simultaneous measurement of serum copper and serum ceruloplasmin levels) (Roberts & Schilsky, 2008) are difficult to measure reliably, may not reflect total body copper content, and may be confusing to interpret (Mak & Lam, 2008). Urine copper exertion measurement after penicillamine challenge has been standardized only for the pediatric population (Martins da Costa et al., 1992). Moreover, it is contentious if the penicillamine challenge adds to the diagnostic yield of basal urinary copper excretion (Gaffney, Fell, & O'Reilly, 2000; Scheinberg & Sternlieb, 1984). Measurement of incorporation of radioactive copper in ceruloplasmin is a research tool. Diagnostic algorithms proposed by American Association for Study of Liver Diseases (AASLD) (Roberts & Schilsky, 2008) and Ferenci's scoring (Ferenci et al., 2003) are useful in guiding investigations and interpreting test results.

Table 13.4 Diagnostic tests for WD and clues in interpretation of test results (Roberts and Schilsky, 2008; Scheinberg & Sternlieb, 1984)

Investigation	Value/finding supporting diagnosis of WD	Confounders		Precautions and test interpretation	Screening siblings
		False-positive results	False-negative results		
Kayser–Fleischer rings KF rings	Present	• Chronic cholestatic liver • Copper foreign body in eye	• 38–56% patients with WD-related liver disease • 5% patients with WD-related neurological disease	• Examination best done by experienced ophthalmologist/clinician • Can be viewed by torchlight shinned tangentially on the cornea • Early rings require slit lamp examination • Taking a picture of KF rings helps reconfirm their presence and track treatment response	If present in a normal sibling, confirms WD
Wilson facies	Present		• Absent in patients with pure liver involvement from WD • 5% patients with WD-related neurological disease		Cannot use as sole screening test to diagnose WD

Continued

Table 13.4 Diagnostic tests for WD and clues in interpretation of test results (Roberts and Schilsky, 2008; Scheinberg & Sternlieb, 1984)—cont'd

Investigation	Value/finding supporting diagnosis of WD	Confounders		Precautions and test interpretation	Screening siblings
		False-positive results	False-negative results		
Serum ceruloplasmin	<50 mg/L	• 20% of heterozygotes (people with 1 *ATP7B* mutation and 1 normal *ATP7B* allele) • Values are normally low till up to 6 months of age • Conditions of protein loss (renal, enteric, nutritional, etc.) • Menkes disease • Absent in aceruloplasminemia	• 5–15% of WD • Acute inflammation • Hyperestrogenemia like pregnancy, estrogen supplements, or treatment	• Can be evaluated enzymatically by their copper-dependent oxidase activity toward these substrates, or immunologic assays. Immunologic assays use may overestimate ceruloplasmin concentrations because they do not discriminate between apoceruloplasmin and holoceruloplasmin. Modestly low values need further investigation • Normal values do not exclude WD	Cannot use as sole screening test to diagnose WD

Basal 24-h urinary copper	>100 μg of copper/24 h		• Autoimmune hepatitis • Primary sclerosing cholangitis • Acute liver failure • Chronic active liver disease	• Ensure 24-h urinary collection • Avoid contamination with tap water • Measurement of 24-h urine volume and total creatinine excretion in the sample help assess completeness of urine sample collection	Cannot use as sole screening test to diagnose WD
Hepatic parenchymal copper	Copper content of 250 μg/g dry weight of liver. This is the best biochemical evidence of WD		Distribution of copper in liver is uneven	• Liver biopsy has a small but definite risk of complication, therefore it is reserved for symptomatic patients in whom simpler & safer tests have not helped establish definite diagnosis of WD. • Use disposable suction or cutting needle for liver biopsy and place	Not recommended

Continued

Table 13.4 Diagnostic tests for WD and clues in interpretation of test results (Roberts and Schilsky, 2008; Scheinberg & Sternlieb, 1984)—cont'd

Investigation	Value/finding supporting diagnosis of WD	Confounders: False-positive results	Confounders: False-negative results	Precautions and test interpretation	Screening siblings
				sample in copper-free container • Ensure adequate specimen (1–2 cm core recommended) • Copper content of <40–50 μg/g dry weight in untreated patients rules out WD • Transjugular liver biopsy is preferred if coagulopathy is a concern • Values (copper content of <70–250 μg/g dry weight of the liver): further testing is necessitated especially if there is active liver disease or other features of WD	

MRI brain in patients with neurological symptoms	Symmetrical hypersensitivities in the basal ganglion and brain stem on MRI T2W and FLAIR sequences	Leigh's disease, hypoxic ischemic encephalopathy, Japanese B encephalitis, methyl alcohol poisoning, osmotic disequilibrium syndrome, glutaric academia	MRI may be normal in patients with WD with neurological symptoms	MRI brain is preferred brain imaging MRI abnormalities may be seen before onset of neurological symptoms	Not recommended for screening

6. TREATMENT

WD is an inherited disorder of chronic copper toxicosis and is invariably fatal if not treated (Scheinberg & Sternlieb, 1984). The goal of WD treatment is decoppering with clinical tracking and this needs to be continued lifelong.

Copper starts accumulating in excessive amounts after birth, being sequestered initially in the liver and later spilling into brain and other body tissues. Symptoms from brain or liver failure usually occur with in the first two decades of life, and the disease progresses rapidly unless decoppering is initiated. Presymptomatic patients, (those who have WD mutations but as yet have not developed symptoms), become ill over time, unless copper accumulation is halted. All symptomatic and all presymptomatic patients require lifelong decoppering (Sternlieb & Scheinberg, 1968). Individuals heterozygous for WD mutation (one mutated and one normal *ATP7B* genes) have normal copper metabolism and do not necessitate treatment or monitoring.

Decoppering ensures that presymptomatic individuals remain symptom free. With judicious decoppering, given time, even patients with severe neurological disability improve and can return to normal life and resume school or work at par with their peers (Aggarwal & Bhatt, 2012; Scheinberg & Sternlieb, 1984, 1995; Walshe, 1989; Walshe & Dixon, 1986). Improvement and reversal of brain imaging abnormalities follow adequate decoppering (Aggarwal & Bhatt, 2012) (Fig. 13.3). KF rings and sunflower cataracts disappear. Liver failure stabilizes and is associated with clinical, biochemical, and histological improvement (Scheinberg & Sternlieb, 1984).

BAL was the first drug demonstrated to chelate copper in patients with WD (Cumings, 1951, Denny Brown & Porter, 1951) This was a major breakthrough and earned WD the distinction of being the first neurometabolic disease that could be treated. A few years later, Walshe conceived that penicillamine, the metabolite of penicillin, excreted in urine, had the requisite molecular structure that could chelate copper. He swallowed 1 g of penicillamine and, having survived 24 h without any ill effects, deemed the drug safe. Subsequent administration of penicillamine to patients with WD led to expected cupriuresis and clinical improvement. Introduction of penicillamine and later trientine as oral decoppering drugs, by Walshe, forever changed the natural history of the disease (Walshe, 2009).

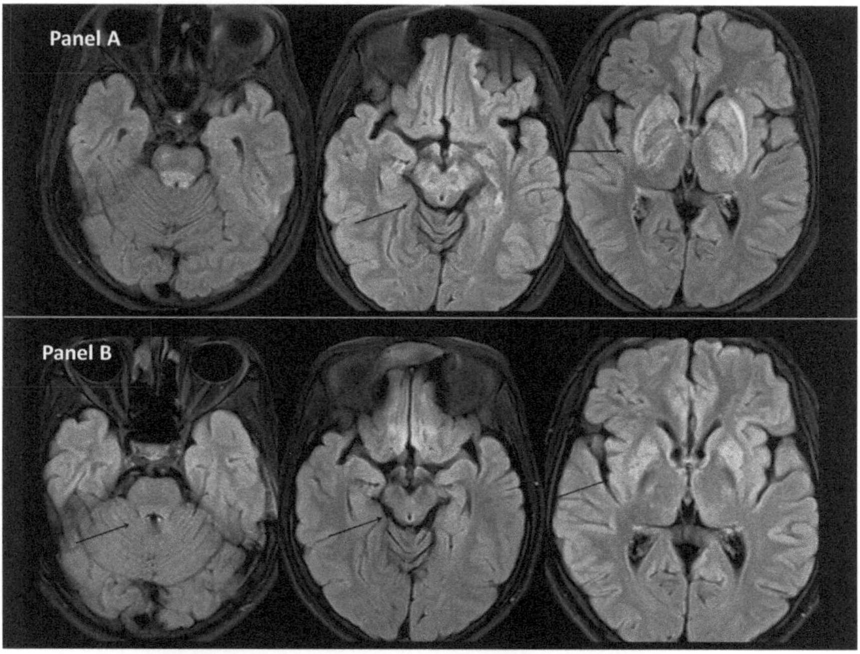

Figure 13.3 (A) Brain MRI of patient with Wilson disease shown in Fig. 13.2. The MRI demonstrates symmetrical hyperintensities on FLAIR-weighted images in the pons, midbrain, and basal ganglia (caudate, putamen, and globus pallidus). (B) Serial scan after 18 months of decoppering shows decrease in FLAIR hyperintensities, associated with atrophy of the caudate and putamen nuclei. (For color version of this figure, the reader is referred to the online version of this chapter.)

Copper is ubiquitous in water and food, and dietary restrictions are neither essential nor sufficient to prevent copper overload. In symptomatic patients, foods with high concentrations of copper, like shellfish, nuts, chocolate, mushrooms, and organ meats, should be avoided, for the first few years of initiation of decoppering. Once symptoms regress (indicating adequate decoppering), these dietary restrictions can be relaxed. Copper content of most municipal drinking water supplies is low and need not be tested (Pfeiffer, 2011; Roberts & Schilsky, 2008). In general, copper levels of 0.1 ppm in drinking water should prompt use of bottled or boiled water (Brewer et al., 2001). Domestic water softeners increase copper content of water (Yarze, Martin, Muñoz, & Friedman, 1992). Copper-containing vitamins or supplements should be avoided.

Globally, the largest experience of WD treatment is with penicillamine. It continues to be the most extensively used decoppering drug for WD, due to its wide availability and low cost. In much of the developing world, penicillamine is the only copper chelator available. Presently, trientine is available only in the United States and the United Kingdom; other countries need to import the drug. Trientine's relatively short shelf life and the need to store the drug in cool temperatures pose a problem in warm regions and add to the cost of therapy. Ammonium tetrathiomolybdate is an experimental drug and is not easily available (Pfeiffer, 2011; Roberts & Schilsky, 2008).

Penicillamine (dimethyl cysteine) is an avid copper chelator and the gold standard for treatment of WD. It is recommended for treatment of symptomatic and presymptomatic patients. Penicillamine should be administered on an empty stomach as food decreases the bioavailability of penicillamine by 50%. The drug is usually given in two or three divided doses to ensure compliance with fasting. It is best to supervise drug intake in children till they are older and symptom free and can be trusted to take the medication unsupervised. The aim of treatment with decoppering is to start slow and go slow. Penicillamine is initiated at doses of 125–250 mg per day. Dosage is gradually increased every 2–3 weeks by 125–250 mg, up to 1–3 g/day, under close clinical monitoring (see succeeding text). Clinical improvement is seen after few months of initiation of treatment. Patients with severe disability may require 2–3 years of decoppering to normalize. Once the positive copper balance in the body has been chelated, doses of the decoppering drugs can be reduced and patients can be maintained on a maintenance phase. Daily doses of penicillamine may be reduced to 250–750 mg in the maintenance phase. Penicillamine is a pyridoxine antagonist and 20 mg of pyridoxine should be supplemented. Other metal supplements should be avoided while patients are on penicillamine. Normal pregnancies are reported while on penicillamine (Scheinberg & Sternlieb, 1984).

Penicillamine-induced neurological worsening is seen in up to 10% of the patients and can resolve with downtitration of the drug followed by slower uptitration with close clinical monitoring (see succeeding text).

Hypersensitivity reactions occur 10–20% patients, within the first few weeks to months of commencing penicillamine. However, the incidence is much lower, if penicillamine is initiated gradually. The reactions include fever, rash, and lymphadenopathy, and usually. Resolve with short-course steroids. Penicillamine should be preferably discontinued. Once the hypersensitivity reaction has resolved, penicillamine may be reintroduced or substituted by trientine. Most patients with WD have thrombocytopenia (or uncommonly) concomitant leucopenia and anemia, from hypersplenism (liver cirrhosis). Rarely, platelet and leucocyte count may drop further as a hypersensitivity reaction to penicillamine, but these recover by downtitrating the penicillamine dose and a short-course of steroids. Late reactions are observed after years of penicillamine therapy. These include drug-induced systemic lupus erythematosus, nephrotic syndrome, Goodpasture's nephritis, hemolytic anemia, and thrombocytopenia. Symptoms of lupus and proteinuria from nephropathy resolve with steroids and penicillamine withdrawal. An alternative decoppering drug should be substituted and rechallenge with penicillamine is best avoided. There are case reports of myasthenia gravis and optic neuritis following penicillamine use. Long-term use of penicillamine may also lead to easy bruising or recurrent subcutaneous bleeding due to penicillamine-induced inhibition of collagen and elastin cross-linking. Elastosis perforans serpiginosa, pemphigus, and aphthous stomatitis are rare (Scheinberg & Sternlieb, 1984; Walshe, 2009).

Trientine (triethylene teramine dihydrochloride) was introduced by Walshe in 1968 as an alternative to penicillamine. It is designated as an orphan drug and approved by FDA for treatment of penicillamine-intolerant patients with WD. Trientine likely leads to less cupriuresis compared to penicillamine, but its continued administration is therapeutically effective (Walshe, 1973, 1982). Being a gentler chelator, trientine is less likely to lead to neurological deterioration following initiation of treatment.

Trientine has a polyamine-like structure, and copper is chelated by forming a stable complex with the four constituent nitrogen atoms in a planar ring. It lacks a sulfhydryl group and is therefore, possibly less immunogenic. Trientine has a poor gastric abortion; most of the absorbed drug is metabolized and only a small percentage is excreted in urine (Roberts & Schilsky, 2008; Walshe, 1973).

Trientine is useful in patients intolerant to penicillamine and those with renal or autoimmune disease. It is also an attractive initial therapy for the initial treatment for symptomatic WD, given its gentler chelation and less likelihood to cause neurological deterioration. Trientine has also been used as initial therapy in patients with decompensated liver disease.

Trientine has lesser adverse effects than penicillamine, and long-term use in penicillamine-intolerant patients is safe. Penicillamine-induced systemic lupus erythematosus and elastosis perforans serpiginosa can be reactivated following substitution of penicillamine with trientine (Walshe, 1982). *De novo* hypersensitivity reactions and lupus without prior penicillamine use have not been described. Neurological worsening is described with trientine but is likely less frequent than with penicillamine. Trientine-induced copper deficiency has led to reversible sideroblastic anemia and iron overload in the liver (Roberts & Schilsky, 2008). Normal pregnancies are reported in patients on treatment with trientine. Ingestion of 30 g of trientine, in a suicidal attempt, did not lead to apparent ill effects (Scheinberg & Sternlieb, 1984).

Doses of 750–2000 mg/day of trientine in 2–3 divided doses, are typical and may be reduced to 750–1500 mg/day in the maintenance phase. Doses are titrated as for penicillamine. The drug is best administered fasting. Coadministration of iron supplements should be avoided as trientine chelates iron and forms toxic complexes with iron.

BAL (dimercaprol) is no longer used for treatment of WD but may have application in ill patients who are intolerant to oral decoppering drugs. BAL is administered admixed peanut oil as deep intramuscular injections. The injections are painful, associated with waning decoppering effect and high incidence of adverse effects (Walshe, 2009; Walshe & Munro, 1995).

Zinc inhibits intestinal absorption of copper, by inducing metallothionein, a cysteine-rich protein present in many tissues including intestine, liver, and brain. Oral zinc is absorbed by the enterocytes and bound to the induced metallothionein, trapping the metal within the enterocytes. The metallothionein has a higher affinity for copper than for zinc and sequesters dietary copper, preventing its absorption into the portal circulation. The bound zinc and copper are stored in the enterocytes and excreted in feces when the cells are sloughed. Reabsorption of copper secreted in the salivary and gastric circulation is similarly impaired resulting in a mild negative copper balance. Zinc-induced metallothionein in hepatocytes may be beneficial in liver disease (Lee, Northup, & Berg, 2006; Marcellini et al., 2005).

Zinc does not chelate copper stored in the body. The negative cooper balance induced by zinc-mediated impairment of copper absorption is too small for zinc to be an effective as monotherapy in symptomatic patients (Scheinberg & Sternlieb, 1984). Neurological and hepatic deterioration, with fatal outcome, has been reported following use of zinc in symptomatic individuals (Lang, Rabas-Kolominsky, Engelhardt, Kobras, & Konig, 1993; Walshe & Munro, 1995). Zinc has been used as monotherapy in presymptomatic individuals, who have been diagnosed with WD but do not yet have any symptoms (Marcellini et al., 2005). It has also been used as "maintenance therapy" in patients who have been decoppered with penicillamine or trientine (Brewer et al., 1998, Sinha & Taly, 2008). Hepatitis and emergence of symptoms have occurred following institution of zinc in presymptomatic patients (Castilla-Higuero, Romero-Gomez, Suarez, &

Castro, 2000; Mishra, Kalra, & Seth, 2008). Caution should be exercised in using zinc, and any evidence of clinical deterioration should prompt replacement of zinc with a decoppering drug (penicillamine/trientine) (Subramanian, Vanek, & Bronstein, 2002).

Zinc is administered as acetate, sulfate, or gluconate salts. Dosing is based on the amount of elemental zinc present in the formulation and should be given fasting. Doses of 50 mg three times a day are typical. It is generally well tolerated with few serious adverse effects and is safe for use during pregnancy. Gastric irritation is common and a dose-limiting adverse effect. Zinc acetate is better tolerated than zinc sulfate. Zinc can rarely lead to sideroblastic anemia. Zinc has not been associated with neurological worsening (Roberts & Schilsky, 2008).

Symptomatic treatment with levodopa–carbidopa, anticholinergics, dopamine blockers, neuroleptics, and targeted botulinum toxin injections has been shown to benefit some patients with WD.

A dreaded complication of WD is fulminant liver failure. This has 100% mortality unless patients can receive a liver transplant. The role of liver transplant in patients with WD-related neurological disability is not yet well defined. Following liver transplant, biliary copper excretion is normalized, and patients no longer require decoppering therapy (Catana & Medici, 2012, Pfeiffer, 2011). Neurological worsening is described following liver transplant, possibly from massive mobilization from the native liver during operation (Litwin, Gromadzka, & Czlonkowska, 2008).

7. TRACKING WD

There are numerous deeply entrenched myths among clinicians about problems encountered in treating WD and expected outcomes (Table 13.5). It is well established that WD is treatable, but so profound is the fear of neurological worsening and drug-related adverse effects that many a clinician are hesitant in administering decoppering drugs in recommended doses and often withdraw treatment at first sign of clinical worsening. Therefore, it cannot be emphasized enough that WD is treatable and WD-related neurological disability is reversible. Following decoppering, liver impairment also stabilize. Inadequate treatment or noncompliance leads to progressive increase in positive copper balance in the body, worsening symptoms, and, if unchecked, death. Decoppering is best titrated with objective clinical monitoring.

Table 13.5 Myths regarding Wilson disease treatment

S. No.	Myth
1	Patients with WD require strict copper-free diets
2	Kayser–Fleisher rings do not regress or resolve with treatment
4	Severe WD-related neurological disability is not reversible
5	Dystonia is not reversible
6	Penicillamine is not a safe drug
7	Neurological worsening with decoppering is permanent or not reversible

Decoppering is a slow process and it takes 1–3 years of aggressive decoppering to remove excessive copper in symptomatic patients. Once adequate decoppering has been achieved, signaled by disappearance of KF rings, neurological recovery, and liver stabilization, patients can be switched to a maintenance phase of decoppering. Maintenance decoppering is aimed at preventing ongoing copper accumulation and reoccurrence of symptoms. Doses of decoppering drugs during the maintenance phase of treatment are much less than those required during the initial decoppering phase (see penicillamine in the preceding text).

Decoppering treatment needs careful titration. Rapid decoppering can lead to neurological worsening from sudden excessive mobilization of copper from liver into blood and thereon into the brain. Too slow, copper chelation increases risk of further clinical deterioration and death.

WD progress and treatment response are best measured by careful clinical assessment. In our WD Clinic, we track patients using the multisystemic, validated WD-specific scale, the Global Assessment Scale for WD (GAS for WD). GAS for WD is a two-tier scale that can be administered by the patient's bedside. Tier 1 measures WD-related disability across four domains: liver (L), cognition and behavior (C), motor (M), and osseomuscular (O). Each domain is scored on an ascending six-point scale (0–5). Tier 2 assesses WD-related neurological dysfunction across 14 items. Each item is graded on an ascending five-point scale (0–4) and summed to obtain the total tier 2 score (0–56) (Aggarwal, 2009).

After initiating treatment, patients are assessed at weekly intervals to track neurological worsening, drug-related adverse effects, and compliance. Decoppering is started at low doses and slowly titrated up (see penicillamine in the preceding text). Generally, no clinical benefit is seen in the first few months of initiation of therapy, though some patients may show improvement in GAS for WD scores within first 2 weeks. Clinical deterioration while on treatment may represent ongoing disease progression, decoppering-induced neurological worsening, noncompliance, or intercurrent illness. Unlike, ongoing disease progression, drug-induced neurological worsening is reflected by an abrupt change in GAS for WD scores. Once compliance is confirmed, neurological worsening is managed by simply downtitrating the decoppering drug. Once the GAS for WD scores return to baseline, decoppering can again be gradually escalated, albeit at a slower pace. Neurological worsening should be differentiated from the phenomenon of emergent psychosis (see earlier text) (Aggarwal et al., 2009).

Improvement in WD facies is the earliest clinical change, seen within 3–5 months of commencing treatment. Reduction in KF rings is observed only after 8–12 months. Complete neurological recovery may take 12–36 months (Aggarwal, 2009). Once it is assured that the patient is tolerating decoppering drugs well and GAS for WD scores are steadily improving, patients' visits can be spaced out to once in 4–12 weeks. Any clinical worsening should prompt careful assessment for neurological worsening or noncompliance.

After clinical recovery, patients may be switched to a maintenance phase with 6–12 monthly follow-up. Worsening of GAS for WD scores on follow-up or reappearance of KF rings, neurological signs, or biochemical liver derangements should signal noncompliance. Patients should be warned about possibility of rapid clinical deterioration if they are non compliant. (Walshe & Dixon, 1986).

REFERENCES

Aggarwal, A., & Bhatt, M. (2008). Emergent psychosis in Wilson's disease. *Movement Disorders, 23*, S181.

Aggarwal, A., & Bhatt, M. (2012). Recovery from severe neurological Wilson's disease with copper chelation. *Movement Disorders, 27*, S71.

Aggarwal, A., Jankharia, G., & Bhatt, M. (2008). Clinical spectrum and pathological correlates of osseomuscular disability in Wilson's disease. *Movement Disorders, 23*, S173–S174.

Aggarwal, A., Nagral, A., Jankharia, G., Aggarwal, N., & Bhatt, M. (2009). A novel global assessment scale for Wilson's disease (GAS for WD). *Movement Disorders, 24*, 509–518.

Aggarwal, A., Chandhok, G., Todorov, T., Parekh, S., Tilve, S., Zibert, A., et al. (2013). Wilson disease mutation pattern with genotype-phenotype correlations from western India: Confirmation of p.C271* as a common Indian mutation and identification of 14 novel mutations. *Annals of Human Genetics*, 77, 8. http://dx.doi.org/10.1111/ahg.12024.

Bearn, A. G. (1953). Genetic and biochemical aspects of Wilson's disease. *The American Journal of Medicine, 15*, 442–449.

Boudin, G., & Pépin, B. (1959). *Dégénérescence hépatolenticulaire: maladie de Wilson, pseudo-sclérose de Westphal-Strumpell.* Paris.

Bramwell, B. (1916). Familial cirrhosis of the liver: Four cases of acute fatal cirrhosis in the same family, the patients being respectively nine, ten, fourteen and fourteen years of age: Suggested relationship to Wilson's progressive degeneration of the lenticular nucleus. *Edinburgh Medical Journal*, *17*, 90–99.

Brewer, G. J., Dick, R. D., Johnson, V. D., Brunberg, J. A., Kluin, K. J., & Fink, J. K. (1998). Treatment of Wilson's disease with zinc: XV long-term follow-up studies. *The Journal of Laboratory and Clinical Medicine*, *132*, 264–278.

Brewer, G. J., Dick, R. D., Johnson, V. D., Fink, J. K., Kluin, K. J., & Daniels, S. (2001). Treatment of Wilson's disease with zinc XVI: Treatment during the pediatric years. *The Journal of Laboratory and Clinical Medicine*, *137*, 191–198.

Bull, P. C., Thomas, G. R., Rommens, J. M., Forbes, J. R., & Cox, D. W. (1993). The Wilson disease gene is a putative copper transporting P-type ATPase similar to the Menkes gene. *Nature Genetics*, *5*(4), 327–337. http://dx.doi.org/10.1038/ng1293-327.

Bull, P. C., & Cox, D. W. (1994). Wilson disease and Menkes disease: New handles on heavy-metal transport. *Trends in Genetics*, *10*, 246–252.

Caca, K., Ferenci, P., Kuhn, H. J., Polli, C., Willgerodt, H., Kunath, B., et al. (2001). High prevalence of the H1069Q mutation in East German patients with Wilson disease: rapid detection of mutations by limited sequencing and phenotype-genotype analysis. *Journal of Hepatology*, *35*(5), 575–581.

Castilla-Higuero, L., Romero-Gomez, M., Suarez, E., & Castro, M. (2000). Acute hepatitis after starting zinc therapy in a patient with presymptomatic Wilson's disease. *Hepatology*, *32*, 877.

Catana, A. M., & Medici, V. (2012). Liver transplantation for Wilson disease. *World Journal of Hepatology*, *4*, 5–10.

Chuang, L. M., Wu, H. P., Jang, M. H., Wang, T. R., Sue, W. C., Lin, B. J., et al. (1996). High frequency of two mutations in codon 778 in exon 8 of the ATP7B gene in Taiwanese families with Wilson disease. *Journal of Medical Genetics*, *33*(6), 521–523.

Coffey, A. J., Durkie, M., Hague, S., McLay, K., Emmerson, J., Lo, C., et al. (2013). A genetic study of Wilson's disease in the United Kingdom. *Brain*, *136*(Pt 5), 1476–1487. http://dx.doi.org/10.1093/brain/awt035.

Coronado, V. A., Damaraju, D., Kohijoki, R., & Cox, D. W. (2003). New haplotypes in the Bedlington terrier indicate complexity in copper toxicosis. *Mammalian Genome*, *14*(7), 483–491. http://dx.doi.org/10.1007/s00335-002-2255-3.

Culotta, V. C., Yang, M., & O'Halloran, T. V. (2006). Activation of superoxide dismutases: Putting the metal to the pedal. *Biochimica et Biophysica Acta*, *1763*, 747–758.

Cumings, J. N. (1948). The copper and iron content of brain and liver in the normal and in hepato-lenticular degeneration. *Brain*, *71*(Pt. 4), 410–415.

Cumings, J. N. (1951). The effects of BAL in hepatolenticular degeneration. *Brain*, *74*, 10–22.

Curtis, D., Durkie, M., Balac, P., Sheard, D., Goodeve, A., Peake, I., et al. (1999). A study of Wilson disease mutations in Britain. *Human Mutation*, *14*(4), 304–311. http://dx.doi.org/10.1002/(SICI)1098-1004(199910)14:4<304::AID-HUMU5>3.0.CO;2-W.

Członkowska, A., Tarnacka, B., Litwin, T., Gajda, J., & Rodo, M. (2005). Wilson's disease-cause of mortality in 164 patients during 1992–2003 observation period. *Journal of Neurology*, *252*, 698–703.

Czlonkowska, A., Rodo, M., & Gromadzka, G. (2008). Late onset Wilson's disease: therapeutic implications. *Movement Disorders*, *23*(6), 896–898. http://dx.doi.org/10.1002/mds.21985.

Czlonkowska, A., Gromadzka, G., & Chabik, G. (2009). Monozygotic female twins discordant for phenotype of Wilson's disease. *Movement Disorders*, *24*(7), 1066–1069. http://dx.doi.org/10.1002/mds.22474.

Dastur, D. K., Manghani, D. K., & Wadia, N. H. (1968). Wilson's disease in India. I. Geographic, genetic, and clinical aspects in 16 families. *Neurology*, *18*, 21–31.

Deguti, M. M., Genschel, J., Cancado, E. L., Barbosa, E. R., Bochow, B., Mucenic, M., et al. (2004). Wilson disease: novel mutations in the ATP7B gene and clinical correlation in Brazilian patients. *Human Mutation*, *23*(4), 398. http://dx.doi.org/10.1002/humu.9227.

Denny Brown, D., & Porter, H. (1951). The effects of BAL (2,3-dimercaptopropanol) on hepatolenticular degeneration (Wilson's disease). *The New England Journal of Medicine*, *245*, 917–925.

Ferenci, P. (2006). Regional distribution of mutations of the ATP7B gene in patients with Wilson disease: Impact on genetic testing. *Human Genetics*, *120*, 151–159.

Ferenci, P., Caca, K., Loudianos, G., Mieli-Vergani, G., Tanner, S., Sternlieb, I., et al. (2003). Diagnosis and phenotypic classification of Wilson disease. *Liver International*, *23*, 139–142.

Firneisz, G., Lakatos, P. L., Szalay, F., Polli, C., Glant, T. T., & Ferenci, P. (2002). Common mutations of ATP7B in Wilson disease patients from Hungary. *American Journal of Medical Genetics*, *108*(1), 23–28.

Gaffney, D., Fell, G. S., & O'Reilly, D. S. (2000). Wilson's disease: Acute and presymptomatic laboratory diagnosis and monitoring. *Journal of Clinical Pathology*, *53*, 807–812.

Gill, H. H., Shankaran, K., & Desai, H. G. (1994). Wilson's disease: Varied hepatic presentations. *Indian Journal of Gastroenterology*, *13*, 95–98.

Glazebrook, A. J. (1945). Wilson's disease. *Edinburgh Medical Journal*, *52*, 83–87.

Golding, D. N., & Walshe, J. M. (1977). Arthropathy of Wilson's disease. Study of clinical and radiological features in 32 patients. *Annals of the Rheumatic Diseases*, *36*, 99–111.

Gromadzka, G., Schmidt, H. H., Genschel, J., Bochow, B., Rodo, M., Tarnacka, B., et al. (2005). Frameshift and nonsense mutations in the gene for ATPase7B are associated with severe impairment of copper metabolism and with an early clinical manifestation of Wilson's disease. *Clinical Genetics*, *68*(6), 524–532. http://dx.doi.org/10.1111/j.1399-0004.2005.00528.x.

Gu, Y. H., Kodama, H., Du, S. L., Gu, Q. J., Sun, H. J., & Ushijima, H. (2003). Mutation spectrum and polymorphisms in ATP7B identified on direct sequencing of all exons in Chinese Han and Hui ethnic patients with Wilson's disease. *Clinical Genetics*, *64*(6), 479–484.

Gupta, A., Aikath, D., Neogi, R., Datta, S., Basu, K., Maity, B., et al. (2005). Molecular pathogenesis of Wilson disease: haplotype analysis, detection of prevalent mutations and genotype-phenotype correlation in Indian patients. *Human Mutation*, *118*(1), 49–57. http://dx.doi.org/10.1007/s00439-005-0007-y.

Hall, H. C. (1921). *La degenerescence hepatolenticulare. Maladie de Wilson-pseudosclerose*. Paris: Mason.

Harris, Z. L., Takahashi, Y., Miyajima, H., Serizawa, M., MacGillivray, R. T. A., & Gitlin, J. D. (1995). Aceruloplasminemia: Molecular characterization of this disorder of iron metabolism. *Proceedings of the National Academy of Sciences*, *92*, 2539–2543.

Haurowitz, F. (1930). Ubereine anamoliedes kupferstoffwechsels. *Hoppe-Seyler's Zeitschrift*, *90*, 72–74.

Kaler, S. G. (1998). Diagnosis and therapy of Menkes syndrome, a genetic form of copper deficiency. *American Journal of Clinical Nutrition*, *67*, 1029S–1034S.

Kenney, S. M., & Cox, D. W. (2007). Sequence variation database for the Wilson disease copper transporter, ATP7B. *Human Mutation*, *28*, 1171–1177.

Kodama, H., & Murata, Y. (1999). Molecular genetics and pathophysiology of Menkes disease. *Pediatrics International*, *41*, 430–435.

Korman, J. D., Volenberg, I., Balko, J., Webster, J., Schiodt, F. V., Squires, R. H., Jr., et al. (2008). Pediatric and Adult Acute Liver Failure Study Groups. Screening for Wilson disease in acute liver failure: A comparison of currently available diagnostic tests. *Hepatology*, *48*, 1167–1174.

Kumar, S., Thapa, B. R., Kaur, G., & Prasad, R. (2005). Identification and molecular characterization of 18 novel mutations in the ATP7B gene from Indian Wilson disease patients: genotype. *Clinical Genetics*, *67*(5), 443–445. http://dx.doi.org/10.1111/j.1399-0004.2005.00440.x.

Lang, C. J., Rabas-Kolominsky, P., Engelhardt, A., Kobras, G., & Konig, H. J. (1993). Fatal deterioration of Wilson's disease after institution of oral zinc therapy. *Archives of Neurology*, *50*, 1007–1008.

Lee, V. D., Northup, P. G., & Berg, C. L. (2006). Resolution of decompensated cirrhosis from Wilson's disease with zinc monotherapy: A potential therapeutic option? *Clinical Gastroenterology and Hepatology*, *4*, 1069–1071.

Lee, B. H., Kim, J. H., Lee, S. Y., Jin, H. Y., Kim, K. J., Lee, J. J., et al. (2011). Distinct clinical courses according to presenting phenotypes and their correlations to ATP7B mutations in a large Wilson's disease cohort. *Liver International*, *31*(6), 831–839. http://dx.doi.org/10.1111/j.1478-3231.2011.02503.x.

Litwin, T., Gromadzka, G., & Czlonkowska, A. (2008). Neurological presentation of Wilson's disease in a patient after liver transplantation. *Movement Disorders*, *23*(5), 743–746. http://dx.doi.org/10.1002/mds.21913.

Lutsenko, S., Barnes, N. L., Bartee, M. Y., & Dmitriev, O. Y. (2007). Function and regulation of human copper-transporting ATPases. *Physiological Reviews*, *87*, 1011–1046.

Mak, C. M., & Lam, C. W. (2008). Diagnosis of Wilson's disease: A comprehensive review. *Critical Reviews in Clinical Laboratory Sciences*, *45*, 263–290.

Marcellini, M., Di Ciommo, V., Callea, F., Devito, R., Comparcola, D., Sartorelli, M. R., et al. (2005). Treatment of Wilson's disease with zinc from the time of diagnosis in pediatric patients: A single-hospital, 10-year follow-up study. *The Journal of Laboratory and Clinical Medicine*, *145*, 139–143.

Martins da Costa, C., Baldwin, D., Portmann, B., Lolin, Y., Mowat, A. P., & Mieli-Vergani, G. (1992). Value of urinary copper excretion after penicillamine challenge in the diagnosis of Wilson's disease. *Hepatology*, *15*, 609–615.

Mishra, D., Kalra, V., & Seth, R. (2008). Failure of prophylactic zinc in Wilson disease. *Indian Pediatrics*, *45*, 151–153.

Miyajima, H., Nishimura, Y., Mizoguchi, K., Sakamoto, M., Shimizu, T., & Honda, N. (1987). Familial apoceruloplasmin deficiency associated with blepharospasm and retinal degeneration. *Neurology*, *37*, 761–767.

Nicastro, E., Ranucci, G., Vajro, P., Vegnente, A., & Iorio, R. (2010). Re-evaluation of the diagnostic criteria for Wilson disease in children with mild liver disease. *Hepatology*, *52*, 1948–1956.

Okada, T., Shiono, Y., Hayashi, H., Satoh, H., Sawada, T., Suzuki, A., et al. (2000). Mutational analysis of ATP7B and genotype-phenotype correlation in Japanese with Wilson's disease. *Human Mutation*, *15*(5), 454–462. http://dx.doi.org/10.1002/(SICI)1098-1004(200005)15:5<454::AID-HUMU7>3.0.CO;2-J.

Pfeiffer, R. F. (2011). Wilson's disease. *Handbook of Clinical Neurology*, *100*, 681–709.

Prashanth, L. K., Taly, A. B., Sinha, S., Arunodaya, G. R., & Swamy, H. S. (2004). Wilson's disease: Diagnostic errors and clinical implications. *Journal of Neurology, Neurosurgery, and Psychiatry*, *75*, 907–909.

Prohaska, J. R., & Gybina, A. A. (2004). Intracellular copper transport in mammals. *Journal of Nutrition*, *134*, 1003–1006.

Roberts, E. A., & Schilsky, M. L. (2008). Diagnosis and treatment of Wilson disease: An update. *Hepatology*, *47*, 2089–2111.

Rosencrantz, R., & Schilsky, M. (2011). Wilson disease: Pathogenesis and clinical considerations in diagnosis and treatment. *Seminars in Liver Disease*, *31*, 245–259.

Rumpel, A. (1913). Uber das Wesen und die Bedeutung der Leberveranderungen und der pigmentierunen bie den damit verbunden Fallen von Pseudosclerose, zugleich ein

Beitrag zur Lehre von der seudosclerose (Westphal-Strumpell). *Deutsche Zeitschrift Nervenheilkunde*, *49*, 54–73.

Santhosh, S., Shaji, R. V., Eapen, C. E., Jayanthi, V., Malathi, S., Chandy, M., et al. (2006). ATP7B mutations in families in a predominantly Southern Indian cohort of Wilson's disease patients. *Indian Journal of Gastroenterology*, *25*(6), 277–282.

Scheinberg, I. H., & Sternlieb, I. (1984). *Wilson's disease. Major problems in internal medicine*. Philadelphia: Elsevier Publishing.

Scheinberg, I. H., & Sternlieb, I. (1995). Treatment of the neurologic manifestations of Wilson's disease. *Archives of Neurology*, *52*, 339–340.

Schilsky, M. L., Scheinberg, I. H., & Sternlieb, I. (1991). Prognosis of Wilsonian chronic active hepatitis. *Gastroenterology*, *100*, 762–767.

Schmidt, H. H. (2009). Role of genotyping in Wilson's disease. *Journal of Hepatology*, *50*, 449–452.

Siemerling, E., & Oloff, H. (1922). Pseudosclerose (Westphal-Strumpell) mit cornealring (Kayser-Fleischer) und doppelseitger scheinkatarakt, die nur bei seitlicherbeleuchtung sichtbar ist und die, der nach verletzung durch kupfersplitterentstehenden katarakt ahnlich ist. *Klinische Wochenschrift*, *1*, 1087–1089.

Sinha, S., & Taly, A. B. (2008). Withdrawal of penicillamine from zinc sulphate-penicillamine maintenance therapy in Wilson's disease: Promising, safe and cheap. *Journal of Neurological Sciences*, *15*, 129–132.

Stapelbroek, J. M., Bollen, C. W., van Amstel, J. K., van Erpecum, K. J., van Hattum, J., van den Berg, L. H., et al. (2004). The H1069Q mutation in ATP7B is associated with late and neurologic presentation in Wilson disease: results of a meta-analysis. *Journal of hepatology*, *41*(5), 758–763. http://dx.doi.org/10.1016/j.jhep.2004.07.017.

Stamelou, M., Lai, S. C., Aggarwal, A., Schneider, S. A., Houlden, H., Yeh, T. H., et al. (2013). Dystonic opisthotonus: A "red flag" for neurodegeneration with brain iron accumulation syndromes? *Movement Disorders*, *28*, 1325–1329. In print.

Sternlieb, I., & Scheinberg, I. H. (1968). Prevention of Wilson's disease in asymptomatic patients. *The New England Journal of Medicine*, *15*, 52–59.

Sternlieb, I., & Scheinberg, I. H. (1972). Chronic hepatitis as a first manifestation of Wilson's disease. *Annals of Internal Medicine*, *76*, 59–64.

Subramanian, I., Vanek, Z. F., & Bronstein, J. M. (2002). Diagnosis and treatment of Wilson's disease. *Current Neurology and Neuroscience Reports*, *2*, 317–323.

Taly, A. B., Meenakshi-Sundaram, S., Sinha, S., Swamy, H. S., & Arunodaya, G. R. (2007). Wilson disease: Description of 282 patients evaluated over 3 decades. *Medicine (Baltimore)*, *86*, 112–121.

Taly, A. B., Prashanth, L. K., & Sinha, S. (2009). Wilson's disease: An Indian perspective. *Neurology India*, *57*(5), 528–540. http://dx.doi.org/10.4103/0028-3886.57789.

Tanzi, R. E., Petrukhin, K., Chernov, I., Pellequer, J. L., Wasco, W., Ross, B., et al. (1993). The Wilson disease gene is a copper transporting ATPase with homology to the Menkes disease gene. *Nature Genetics*, *5*(4), 344–350. http://dx.doi.org/10.1038/ng1293-344.

Thomas, G. R., Forbes, J. R., Roberts, E. A., Walshe, J. M., & Cox, D. W. (1995). The Wilson disease gene: spectrum of mutations and their consequences. *Nature Genetics*, *9*(2), 210–217. http://dx.doi.org/10.1038/ng0295-210.

Todorov, T., Savov, A., Jelev, H., Panteleeva, E., Konstantinova, D., Krustev, Z., et al. (2005). Spectrum of mutations in the Wilson disease gene (ATP7B) in the Bulgarian population. *Clinical Genetics*, *68*(5), 474–476. http://dx.doi.org/10.1111/j.1399-0004.2005.00516.x.

Tümer, Z., & Møller, L. B. (2010). Menkes disease. *European Journal of Human Genetics*, *18*, 511–518.

Vogt, A. (1929). Kupfer und silber aufgespeichert in auge, leber, milz und nieren als symptom der pseudosklerose. *Klinische Monatsblatter Augenheilkunde*, *83*, 417–419.

Walshe, J. M. (1956). Wilson's disease. New oral therapy. *Lancet, 267*, 25–26.
Walshe, J. M. (1962). Wilson's disease. The presenting symptoms. *Archives of Disease in Childhood, 37*, 253–256.
Walshe, J. M. (1973). Copper chelation in patients with Wilson's disease. A comparison of penicillamine and triethylene tetramine dihydrochloride. *Quarterly Journal of Medicine, 42*, 441–452.
Walshe, J. M. (1982). Treatment of Wilson's disease with trientine (triethylene tetramine) dihydrochloride. *Lancet, 1*, 643–647.
Walshe, J. M. (1989). Wilson's disease patients can be decoppered. *Lancet, 22*(2), 228.
Walshe, J. M. (2006). History of Wilson's disease: 1912 to 2000. *Movement Disorders, 21*(2), 142–147. http://dx.doi.org/10.1002/mds.20694.
Walshe, J. M. (2007). Cause of death in Wilson disease. *Movement Disorders, 22*, 2216–2220.
Walshe, J. M. (2009). The conquest of Wilson's disease. *Brain, 132*, 2289–2295.
Walshe, J. M., & Dixon, A. K. (1986). Dangers of non-compliance in Wilson's disease. *Lancet, 1*, 845–847.
Walshe, J. M., & Yealland, M. (1992). Wilson's disease: the problem of delayed diagnosis. *Journal of Neurology, Neurosurgery, and Psychiatry, 55*(8), 692–696.
Walshe, J. M., & Munro, N. A. (1995). Zinc-induced deterioration in Wilson's disease aborted by treatment with penicillamine, dimercaprol, and a novel zero copper diet. *Archives of Neurology, 52*, 10–11.
Weiner, W. J., & Lang, A. E. (1989). Wilson's disease. In W. J. Weiner & A. E. Lang (Eds.), *Movement disorders—A comprehensive survey* (pp. 257–291). New York: Futura.
Wilson, S. A. K. (1912). Progressive lenticular degeneration: A familial nervous disease with cirrhosis of the liver. *Brain, 34*, 295–507.
Wu, Z., Wang, N., Murong, S., & Lin, M. (2000). Identification and analysis of mutations of the Wilson disease gene in Chinese population. *Chinese Medical Journal, 113*(1), 40–43.
Yarze, J. C., Martin, P., Muñoz, S. J., & Friedman, L. S. (1992). Wilson's disease: Current status. *The American Journal of Medicine, 92*, 643–654.

CHAPTER FOURTEEN

An Update on Primary Familial Brain Calcification

Roberta R. Lemos[*,†], **J.B.M.M. Ferreira**[‡], **Matthew P. Keasey**[‡], **Joao R.M. Oliveira**[‡,§,1]

[*]Universidad de Santiago de Compostela, Santiago de Compostela, Spain
[†]Fundación Pública Galega de Medicina Xenómica-SERGAS-IDIS (Hospital Clínico Universitario), Santiago de Compostela, Spain
[‡]Keizo Asami Laboratory, Federal University of Pernambuco, Recife, Pernambuco, Brazil
[§]Neuropsychiatry Department, Federal University of Pernambuco, Recife, Pernambuco, Brazil
[1]Corresponding author: e-mail address: joao.ricardo@ufpe.br

Contents

Abstract

Patients with primary familial brain calcifications (PFBC) present bilateral calcifications, often affecting basal ganglia, thalamus, and cerebellum, inherited in an autosomal dominant pattern of segregation. Affected individuals display a wide variety of motor and cognitive impairments such as parkinsonism, dystonia, migraine, dementia, psychosis, and mood symptoms. Worldwide growth in the availability of neuroimaging procedures, combined with careful screening of patients and their relatives, has increased detection of PFBC. Recently, mutations in the *SLC20A2* gene coding for the inorganic phosphate transporter PiT2 were linked to PFBC, thereby implicating impaired phosphate transport as an underlying disease mechanism. To date, around 20 families of various ethnicities carry different mutations in *SLC20A2* correlate with ~40% of PFBC cases. More recently, two French families were recently reported with mutations in *PDGFRB*: c.1973 T>C, p.L658P and c.2959C>T, p.R987W, a class III tyrosine kinase receptor. Six other families were found with mutations in PDGFB, and, in general, mutations at the *PDGF pathway* add a new dimension to the physiopathology of PFBC so far explained by a disturbance in phosphate homeostasis with *SLC20A2*. The identification of *SLC20A2*, *PDGFRB*, and *PDGFB* provides a new avenue for potential treatments based on compounds such as bisphosphonates and those modulating the *PDGFB* pathway.

International Review of Neurobiology, Volume 110
ISSN 0074-7742
http://dx.doi.org/10.1016/B978-0-12-410502-7.00015-6

1. INTRODUCTION

Ectopic calcifications are often damaging in soft tissues such as vessels, heart valves, lungs, kidney, and brain. This is often taken for granted as an age-dependent process, but recent data suggest that a myriad of molecules might modulate this process in an active fashion (Nitschke & Rutsch, 2012).

Brain calcification (BC) is a nonspecific neuropathology often associated with various chronic and acute brain disorders including Down's syndrome, Lewy body disease, Alzheimer's disease, Parkinson's disease, vascular dementia, brain tumors, and various endocrinologic conditions (Oliveira, 2011).

The pathology of BC is based upon calcium and other mineral deposits in blood vessel walls and tissue parenchyma causing neuronal death and gliosis. Calcification (mineralization) may be triggered by metabolic and inflammatory disorders, but it is also controlled by genetic predisposition. It has been suggested that deposits are not the source of the symptoms but rather a sign of progressive underlying tissue damage (Sobrido, Copolla, Oliveira, Hopfer, & Geschwind, 2013).

Vascular, as well as brain parenchymal, calcification is found in the basal ganglia and other brain areas including the cerebellum, thalamus, and brain stem. Patients presenting such calcifications display a wide variety of motor and cognitive impairments, such as parkinsonism, dystonia, dementia, psychosis, and mood symptoms. Some individuals may be symptom-free, despite extensive signs of calcification; however, studies measuring the total volume of calcification suggest that there are far more deposits in symptomatic relative to asymptomatic subjects. Indeed, this often leads to misdiagnosis (Manyam, 2005).

Despite the different etiologies, BCs often present a similar composition, that is, an amalgam of hydroxyapatite with a major content of calcium phosphate. Electron microscopy suggests that an initial *nidus* is calcified, with a progressive increase in volume and area affected. In some cases, the calcification is self-limited in terms of growth and expansion, and some authors suggest that this process starts with neuronal mitochondria (Maetzler et al., 2009). Calcium deposits found in both human brain and animal models of neurodegeneration have a crystalline structure composed of biological hydroxyapatites similar to those in peripheral tissues. On the other hand, nonatherosclerotic calcium deposits present in vascular dementia, Alzheimer's disease, Parkinson's disease, Lewy body disease, epilepsy, and astrocytoma are very similar, despite the pathology (Honda et al., 1994;

Khan & Hackett, 1986; Kim, 1995; Mahy, Prats, Riveros, Andre's, & Bernal, 1999; Rodríguez, Bernal, Andrés, Malpesa, & Mahy, 2000).

Several different imaging procedures can be used to detect BC, and even traditional head radiography can show anomalies associated with calcification through the skull in advanced cases. Computerized tomography (CT) is the most appropriate neuroimaging technique to detect BC with studies in different populations (controls, affected, young, and senior) showing that its prevalence rate ranges from 1% to 20%, in an age-dependant pattern (Forstl, Krumm, Eden, & Kohlmeyer, 1992; Yamada et al., 2012).

Sporadic cases are reported more often than ever, with the growing availability of neuroimaging exams, suggesting that this phenotype is more common than previously assumed (Oliveira & Steinberg, 2010). Baba, Broderick, Uitti, Hutton, and Wszolek (2005) revised the most frequent familial forms of BC, including conditions with calcium, phosphorus, and parathormone disturbances according to familial forms of isolated hypoparathyroidism, autoimmune polyglandular syndrome, and pseudohypoparathyroidism. However, there are also a number of conditions where there are no obvious metabolic or hormonal changes, that is, Aicardi–Goutieres syndrome, Cockayne syndrome, polycystic lipomembranous osteodysplasia with sclerosing leukoencephalopathy, autosomal dominant chromosome instability syndrome, Krabbe disease, mitochondrial myopathy, encephalopathy, and lactic acidosis.

For diagnosis, a thorough analysis should be done for parathyroid hormone levels in serum as well as for calcium, and phosphorous (including an Ellsworth–Howard test to measure serum and urinary phosphorus after intravenous administration of parathyroid extract, testing for pseudohypoparathyroidism). Additional analysis of serum electrolytes, thyroid and growth hormone, sedimentation rate, renal and liver function tests, cortisol, ceruloplasmin levels, and viral markers should be performed (Oliveira, 2011).

Other examinations including karyotype, muscle biopsy, and mutation screens will be required in some rare conditions such as mitochondrial encephalomyopathy, stroke-like symptoms, lactic acidosis, or other genetic disorders such as Down's syndrome (Sobrido et al., 2013). Several of these conditions manifest themselves with calcifications in various sites, with additional metabolic disturbances. In these instances, the brain is affected not as a primarily organ but as a secondary target. Some authors have tried to find patterns of calcification by neuroimaging phenotypes and comparing different intracranial calcification disorders. For example, Livingston, Stivaros, van der Knaap, and Crow (2013) suggested a classification based on the following categories considering the most affected structures: probable destructive mechanism; calcification associated with polymicrogyria; familial calcification with involvement of the globus pallidus, posterior limb of the internal capsule, genu of the corpus callosum, and deep white matter; and dominant familial calcification "Fahr's disease." However, the authors admit the limitations of such an approach as there are considerable variations in standard procedures and heterogeneous data analysis approaches for neuroimaging protocols. In addition, the authors recognize that this study is biased due to a set of scans sent to the coauthors for a diagnostic opinion that was nonblinded. The authors are currently seeking to improve confidence through the inclusion of additional data sets (more images of patients with and without known diagnosis). However, it is unlikely that a specific pattern could be found as we often see variations within families, even within the same generation. On the other hand, our group reported a small family with identical twins affected by BC and with a noteworthy, almost identical pattern of calcification, anatomical distribution, and predicted volume (Oliveira, Lima, & Zatz, 2009). Curiously, the brain appears to be prone to calcification, even in conditions where there is no associated cause (often referred as Fahr's disease "idiopathic"). This is why it is important to understand this condition in a model such as primary familial brain calcification (PFBC), where the first associated genes were recently identified (Oliveira, Oliveira, Lemos, & Oliveira, 2009).

2. PRIMARY FAMILIAL BRAIN CALCIFICATION

The term *primary familial brain calcification* was recently suggested by Sobrido et al. (2013), after considering the previous terms (Fahr's disease or familial idiopathic basal ganglia calcification) obsolete. Firstly, there are now clear genetic culprits, and secondly, calcifications can also appear in other brain regions. However, the basal ganglia still seems to be a more frequently affected brain region. Diagnosis of PFBC is based on clinical history, neuroimaging analysis, and complementary exams to test for biochemical abnormalities, infections, toxicity, and other causes (Ellie, Julien, & Ferrer, 1989; Manyam, 2005; Moskowitz, Winickoff, & Heinz, 1971).

Defining affected individuals is challenging due to heterogeneity in clinical presentation, age-dependent penetrance, and the fact that many asymptomatic individuals have positive CTs. Thus, we defined affected individuals as those with positive CTs as previously described (Oliveira et al., 2004; Sobrido et al., 2013; Sobrido & Geschwind, 2001). Biochemical investigation should be performed in at least one affected individual per family to rule out abnormalities of calcium regulation and metabolic disorders such as pseudohypoparathyroidism that could underlie brain calcifications. Those with negative CTs who are over the age of 50 are defined as unaffected, whereas those at earlier ages are classified as unknown because of the age-dependent penetrance for calcium deposits (Geschwind, Loginov, & Stern, 1999).

In this chapter, we will focus on the recent genetic finds associated with PFBC.

3. INHERITANCE PATTERNS AND GENES LINKED TO PFBC

Most families reported with PFBC present an autosomal dominant pattern of segregation with only a few kindred reported with recessive inheritance. Several other families have been described more recently, and the list grows as neuroimaging procedures become more widely available, allowing screening of patient's relatives (Baba et al., 2005; Brodaty et al., 2002; Dai et al., 2010; Geschwind et al., 1999; Lhatoo, Perunovic, Love, Houlden, & Campbell, 2003; Oliveira et al., 2003, 2004; Volpato et al., 2009; Wszolek et al., 2006).

Recently, mutations in the *SLC20A2* gene coding for the inorganic phosphate transporter PiT2 were linked to PFBC, thereby implicating impaired phosphate transport as an underlying disease mechanism (Wang et al., 2012). *SLC20A2*, located on chromosome 8, was initially identified as a retrovirus receptor (Kavanaugh & Kabat, 1996). Expression of SLC20A2 occurs in several tissues including bone, vascular smooth muscle cells, parathyroid glands, kidney, intestine, and brain, performing a housekeeping function in inorganic phosphate (P_i) homeostasis (Bøttger & Pedersen, 2011; Villa-Bellosta et al., 2009; Villa-Bellosta & Sorribas, 2010).

Bøttger and Pedersen (2002) reported that p.E575K (a region containing a highly conserved glutamate residue) is critical for P_i transport activity of PiT2. Intriguingly, an independent study later demonstrated this same region contained a number of mutations in PFBC patients (Wang et al., 2012). These same mutations were inserted into *Xenopus* oocytes, resulting in impaired inorganic phosphate uptake. This suggests that these mutations may be associated with haploinsufficiency rather than by encoding dominant-negative activities (Wang et al., 2012). To date, around 20 families of various ethnicities have been reported to possess different mutations in

SLC20A2, and this gene is associated with ~40% of PFBC cases (Hsu et al., 2013; Lemos, Oliveira, & Oliveira, 2013; Nicolas et al., 2013; Schottlaender et al., 2012; Zhang, Guo, & Wu, 2013).

Mutations in *SLC20A2* have been reported as missense, frameshift, deletion, nonsense, synonymous, and in splice site. These rare mutations have not yet been listed in databases of genetic variations such as Exome Variation Analyzer (EVA), dbSNP, 1000 Genomes, Exome Sequencing Project (ESP), or HapMap. *In silico* analysis using PolyPhen-2, SIFT, I-mutation, and Human Splicing Finder predicts most of these mutations to be deleterious (Table 15.1).

Hsu et al. (2013) examined a large cohort of PFBC families in 218 patients from 29 PFBC families with mutations accounting for 41% (or 12 out of 29 of the cases). Interestingly, nine out of the 14 mutations were predicted to introduce a stop codon, pointing also to haploinsufficiency as a mechanism (Hsu et al., 2013). Other mutations in *SLC20A2* were also seen in individuals from nuclear families or unrelated individuals (Lemos et al., 2013; Nicolas et al., 2013).

Importantly, Zhang et al. (2013) assessed the relative expression of *SLC20A2* transcripts by using qRT-PCR in PFBC patients. The deletion of c.510delA (within *SLC20A2*) produced a truncated PiT2 protein with a premature termination codon. This resulted in a reduction in *SLC20A2* mRNA expression (~30%) in tissues from affected individuals relative to biological controls (Zhang et al., 2013). Despite these interesting findings, a deeper understanding of the relationship between clinical expression, findings of calcifications, and genetic mutations in PFBC needs to be established. So far, no obvious genotype–phenotype correlation in study families with *SLC20A2* mutations has been observed. Phosphate imbalance has been associated with vascular calcification *in vivo* (Block, Hulbert-Shearon, Levin, & Port, 1998). In addition, culturing cells in the presence of increasing concentrations of P_i leads to calcium deposition in a dose-dependent fashion (Giachelli, 2003).

The mechanism that underlies P_i-mediated plaque formation is relatively poorly understood. However, it has been reported that high P_i levels activate the transcription factor cbfa1 leading to induction of osteogenic markers (Giachelli et al., 2001).

Curiously, mutations have also been screened in the *SLC20A1* gene, but, as yet, none have been confirmed. *SLC20A1* deletion in a transgenic mouse model is lethal during embryonic development (Beck et al., 2010),

Table 15.1 Descriptions of the mutations in *SLC20A2* and *PDGFRB* genes

No.	cDNA	Amino acid NP_006740.1	Mutation type	Authors	GenBank reference
1	c.1492G>A	p.G498R	Missense	Wang et al. (2012)	NM_006749 (*SLC20A2*)
2	c.1802C>G	p.S601Q	Missense		
3	c.1802C>T	p.S601L	Missense		
4	c.1409delC	p.P470LfsX37	Frameshift		
5	c.1723G>A	p.E575K	Missense		
6	c.1784C>T	p.T595M	Missense		
7	c.124_126delGTG	p.V42del	Deletion		
1	c.1723G>A	p.E575K	Missense	Schottlaender et al. (2012)	NM_006749.4 (*SLC20A2*)
1	c.508delT	p.L170X	Nonsense	Hsu et al. (2013)	NM_006749.3 (*SLC20A2*)
2	c.514A>T	p.K172X	Nonsense		
3	c.583_584delGT	p.V195LfsX61	Frameshift		
4	c.760C>T	p.R254X	Nonsense		
5	c.1101C>G	p.P367P	Synonymous		
6	c.1145G>A	p.R382Q	Missense		
7	c.1506C>A	p.H502Q	Missense		
8	c.1523+1G>A	p.G312VfsX8	Splice site		

Continued

Table 15.1 Descriptions of the mutations in *SLC20A2* and *PDGFRB* genes—cont'd

No.	cDNA	Amino acid NP_006740.1	Mutation type	Authors	GenBank reference
9	c.1652G > A	p.W551X	Nonsense		
10	c.1703C > T	p.P568L	Missense		
11	c.1794 + 1G > A	p.S570RfsX30	Splice site		
12	c.1794 + 1G > C	p.S570RfsX30	Splice site		
13	c.1802C > T	p.S601L	Missense		
14	c.1828_1831delTCCC	p.S610AfsX17	Frameshift		
1	c.1483G > A[a]	p.A495T	Missense	Lemos et al. (2013)	NM_006749.4 (*SLC20A2*)
1	c.551C > T	p.P184L		Nicolas et al. (2013)	NM_006749.3 (*SLC20A2*)
2	c.431-1G > T		Splice site		
1	c.510delA	p.R172fsX19	Frameshift	Zhang et al. (2013)	n.a.[b] (*SLC20A2*)
1	c.1973 T > C	p.L658P	Missense	Nicolas et al. (2013)	NM_002609.3
2	c.2959C > T	p.R987W	Missense		(*PDGFRB*)

[a]Annotation corrected from the position reported in Lemos et al. (2013).
[b]n.a.: Not available.

suggesting that mutations in *SLC20A1* might be too deleterious and incompatible with embryonic survival.

Recently, Nicolas et al. (2013) reported two French families with mutations in *PDGFRB*: c.1973 T > C, p.L658P and c.2959C > T, p.R987W (Nicolas et al., 2013). *PDGFRB* is a class III TK receptor, localized on chromosome 5. Mutations in *PDGFRB* add a new dimension to the pathophysiology of PFBC, so far explained by a disturbance in phosphate homeostasis. Both *PDGFRB* mutations cause amino acid substitutions in the intracellular domain of *PDGFRB*, one of them affecting a conserved residue in the kinase domain. Six other families were recently found with mutations in *PDGFB*, but the functional consequences of these mutations, however, are not currently understood (Keller et al., 2013). The *PDGFB* pathway is involved in many mechanisms including angiogenesis, pericyte survival, and blood–brain barrier (BBB) maintenance. Previous studies have demonstrated that angiogenesis might be crucial to the formation of calcifications and that both processes might act in an adjuvant pathway (Table 15.1).

As a critical regulator of the BBB, mutation of *PDGFRB* is key to homeostasis in the central nervous system (CNS) (Daneman, Zhou, Kebede, & Barres, 2010). The inability to regulate and control the cellular environment in the CNS might represent an important step in the understanding of PFBC. Further, leakage of fluid derived from plasma through the vascular wall and into the CNS may be a factor leading to tissue damage and resulting in progressive calcinosis. This mechanism has been considered for a long time, based on neuroimaging and metabolic studies in idiopathic and nonidiopathic causes of BCs (Baba et al., 2005). Indeed, BBB integrity has been implicated in several neurodegenerative conditions including amyotrophic lateral sclerosis (Garbuzova-Davis et al., 2007), Alzheimer's disease (Zipser et al., 2007), and Parkinson's disease (Kortekaas et al., 2005). However, whether BBB dysfunction is a causative factor in these cases is a matter of debate. In the case of PFBC, why loss of BBB integrity would lead to preferential calcification of specific brain regions, that is, basal ganglia and not others remains to be determined. However, certain regions of the brain are more susceptible to noxious stimuli, for example, CA1 pyramidal neurons of the hippocampus are less resilient to stroke (Smith, Auer, & Siesjö, 1984).

Intriguingly, Giachelli et al. (2001) reported that PDGF-BB, the preferential PDGFRB ligand, can modulate the maximum velocity of P_i uptake by the phosphate transporter PiT1 (*SLC20A1* protein product). This finding provides a possible link between the two genes currently implicated in the pathophysiology of PFBC.

Other rare disorders with vascular calcification such as generalized arterial calcification of infancy, pseudoxanthoma elasticum, and calcification of joints and arteries are linked to mutations in *ENPP1*, *ABCC6*, and *NT5E*, respectively (Nitschke & Rutsch, 2012). These might also represent additional candidate genes for PFBC, especially if we consider that, so far, the two genes linked to PFBC correspond to ~50% of all families studied so far. The unveiling of PFBC neurogenetics might also help to understand other neuropsychiatric conditions with similar clinical outcome.

4. PENETRANCE AND BCs

The study of asymptomatic individuals with extensive brain calcification suggests that the concept of penetrance can be understood in relation to brain lesions (where a threshold for the manifestation of the mutation with formation of calcifications can be observed).

Despite extensive calcifications in some oligosymptomatic and mild cases, the apparent functionality of affected brain structures is remarkably efficient, suggesting an intrinsic resilience mechanism and also a threshold for triggering the symptoms. A previous study suggested that it is necessary to consider two levels of penetrance in PFBC: the penetrance for the calcification formations and the penetrance for the clinical manifestation (Oliveira, Lima, & Zatz, 2009). Curiously, a resilience mechanism modulated by the cerebellum has been recently considered in dystonia. Neurogenetics is a specialty full of such examples, and the advent of imaging studies in association with genetic testing in familial dystonia allowed visualization of significant changes in the brains of asymptomatic carriers of mutations in the *DYT1* gene. The issue of differential levels of penetrance has already been addressed in other movement disorders such as primary torsion disease (PTD) and Huntington disease (HD) (Argyelan et al., 2009; Eidelberg, 2009; Kempton et al., 2009).

In our previous linkage study, we analyzed the ratio between symptomatic and asymptomatic individuals among patients with familial calcifications in the basal ganglia, showing that only 28 patients were clinically affected out of 47 patients with calcifications (Oliveira et al., 2004). This finding reinforces the surprising resilience to this type of brain injury. While both HD and PTD are characterized by autosomal dominant inheritance, they also present with variable penetrance. Compared with the HD carrier state, only a minority of mutated dystonia gene carriers ever develops clinical symptoms. Abnormal metabolic findings in the brains of asymptomatic carriers show that our definition of genetic penetrance may be flawed when dealing with hereditable movement disorders with probable neuroimaging endophenotypes (Argyelan et al., 2009; Eidelberg, 2009).

The fact that such extensive calcified regions may remain asymptomatic over several decades reveals a singular mechanism of behavioral compensation against progressive calcinosis, suggesting continuous resilience and degeneracy of multiple neural systems (Noppeney, Friston, & Price, 2004).

The study of new families will be crucial to confirm current reports, but the discovery of actual genes responsible for PFBC will be the ultimate finding to establish the molecular basis of such intriguing phenotypes.

5. PENDING CANDIDATE LOCI AND OTHER RISK FACTORS

Previously, several candidate genetic regions have been suggested as being associated with the PFBC phenotype. However, at least one of them, such as the IBGC1 locus on chromosome 14, was recently revised and linked to mutations in *SLC20A2* (Hsu et al., 2013; Wang et al., 2012).

The first locus linked to this condition (IBGC1) was localized on the long arm of chromosome 14 (Geschwind et al., 1999). The first screening for possible mutations identified a heterozygous variation (rs36060072) at the *MGEA6/CTAGE5* gene, potentially pathogenic, present in all affected individuals of a large American pedigree and absent in controls (Oliveira et al., 2007). This missense substitution at exon 20 leads to proline to alanine substitution at position 521 (P521A). Exon 20 of the *MGEA6* gene is commonly spliced, generating the isoform MGEA 11, also expressed in the brain (Usener et al., 2003). Considering that the population frequency of a given variation is an indirect indicator of potential pathogenicity, we screened a control population and found two samples heterozygous for P521A among 174, bringing the minor allelic frequency (MAF) of this single nucleotide polymorphism (SNP) to 0.0058.

We recently screened 200 chromosomes in a random control set of Brazilian samples and in two nuclear families, comparing with our previous analysis in a US population. Our genetic screen found no P521A carriers, and pooling this data together with previous studies in the United States, we now have a MAF of 0.0036 (Lemos, Oliveira, Zatz, & Oliveira, 2011). The exon 20 SNP should be considered a rare variation based on the study of Freudenberg-Hua et al. (2003) who analyzed 65 candidate genes for CNS disorders and concluded that rare SNPs have MAF <0.05.

Computer-based modeling is often used to simulate pathogenicity of genetic mutations. However, the lack of protein data, including crystallography and structural modeling in the case of *MGEA6*, limits the strength of such simulations

Proline residues play an important role in the structure and function of various proteins. The insertion of an alanine (a nonpolar side-chain amino acid) is likely to have significant implications to the tertiary and quaternary structure of this protein (Macias et al., 1996; Macias, Wiesner, & Sudol, 2002; Sansom & Weinstein, 2000). In addition, we performed analyses through bioinformatics programs to predict the pathogenicity of such variations, which provided conflicting findings (Lemos et al., 2011).

Several aspects make *MGEA6* a great candidate for PFBC. *MGEA6* is a coil-coiled protein with a proline-rich region expressed in several tissues including brain. This type of proline-rich "signature" is related to regulatory pathways, through interaction with domains of other proteins that have affinity for sequences rich in prolines, namely, the Src homology 3 (SH3) and tryptophan-rich domains (WW). These are involved in a wide variety of cellular activities such as growth, cell cycle, transcription, synaptic signaling, and cell motility (Comtesse et al., 2002; Sudol et al., 2001; Zarrinpar et al., 2003). *MGEA6* is also highly expressed in meningioma, the most common benign brain tumor that often presents calcifications visible in neuroimaging studies, especially CTs (Comtesse et al., 2002; Comtesse, Reus, & Meese, 2001; Usener et al., 2003). Another interesting molecular overlap between meningiomas and bone-related proteins is seen in a study by Hirota et al. (1995). Hirota et al. (1995) suggested a possible involvement of osteopontin in the development of psammoma bodies, which are microscopic, calcified, extracellular bodies commonly found in meningioma.

Previous preliminary linkage studies performed by our group have highlighted other candidate regions. A 10cM genome-wide scan was performed in four families with a total of 47 subjects using an ABI MD 10 marker panel. An additional family with 14 more subjects is still under analysis.

The largest single-point LOD score was observed on chromosome 7 with a LOD score of 3.023 at marker D7S519 in the FB2 family (10 affected). The minimal interval defined by haplotype analysis is between the markers D7S484 and D7S506, spanning about 20cM. Three smaller families, with a total of 16 affected, had a maximum multipoint LOD score of 3.487 on chromosome 9 at a position roughly midway between D9S157 and D9S171. Haplotype analysis shows that these families share a common region around 16cM between the markers D9S171 and D9S1817 (Oliveira et al., 2003). These results reinforce preliminary studies demonstrating genetic heterogeneity in PFBC and report two potential new candidate regions, one of them shared by three families. Additional fine-mapping analysis, SNP genotyping, and exome analysis will confirm further mutations.

Comparing the regions from three chromosomes linked by our group to PFBC (7, 9, and 14), we realized that some genes could be grouped based on a similar biochemical role. Linkage studies by Wszolek et al. (2006), in a large kindred with autosomal dominant dystonia and brain calcification, listed different regions of chromosomes 8 and 9 as PFBC-associated loci.

Additionally, Volpato et al. (2009) reported loci on chromosome 2, identifying several genes related to calcium metabolism located in 2q37, such as *INPP5* and *EFHD1* in a South Tyrolean family. Curiously, two families with Fried syndrome (X-linked mental retardation with hydrocephalus and calcifications in basal ganglia) have been reported in France and Scotland and bearing *AP1S2* mutations (Saillour et al., 2007).

The continuing research suggests more PFBC gene candidates might be located on chromosomes 2, 7, 8, and 9 demonstrating the genetic heterogeneity of this disease (Oliveira & Lemos, 2009; Oliveira, Lima, & Zatz, 2009; Oliveira et al., 2007; Volpato et al., 2009).

With the growing report of incidental findings of patients with PFBC (Oliveira & Steinberg, 2010), there is also a high chance that more families will be reported in the near future and with new mutations at *SLC20A2*, *PDGFRB*, *PDGFB*, or in other genes.

Recently, Da Silva, Pereira, and Oliveira (2013) described the gene expression pattern across the whole brain for SLC20A2 using the Allen Institute Human Brain Atlas database. Microarray analysis provided evidence that the neuroanatomical pattern of expression for this gene is highest in regions most commonly affected in PFBC. The following genes demonstrated high correlation with *SLC20A2* expression: *PCDH12*, *RHOBTB2*, *FLJ43860*, *VAMP1*, *EPB41L3*, *AGPAT9*, *NEFH*, *PVALB*, *CDR2L*, *EMC10*, and *SCN1B*. On the other hand, the following genes (anti-correlated) presented an "inverted" pattern of expression compared with *SLC20A2*: *FAM171B*, *GABRB1*, *CDH4*, *GABRA2*, *PNCK*, *MYH7*, *RBP4*, *SCN3B*, *C1orf173*, *ACTN1*, *MMD*, *LPPR3*, *C1orf194*, *FRRS1L*, *PKIA*, and *LMO3*. These genes likely point to a molecular network with potential to explain the relatively localized neuroanatomical distribution of calcifications in PFBC.

6. MODELING BCs

The most common animal models for BCs include knockouts, chemically induced calcification, or aged animals without any genetic or chemically interventions. The major challenge is to find a model that can actually present the calcifications spontaneously without an external triggering agent.

Previous reports of basal ganglia calcification in aged animals show cerebral mineralization in horses, cattle, monkeys, and aged dogs, with cognitive deficits together with histological alterations resembling those of Alzheimer's disease (Gavier-Widen, Wells, Simmons, Wilesmith, & Ryan, 2001; Petegnief et al., 1999; Ramonet et al., 2002; Towfighi, Yager, Housman, & Vannucci, 1996; Yanai et al., 1996, 1994; Yasui et al., 1997; Yasui, Yano, Tsuda, & Ota, 1991).

Calcium precipitation is part of the ongoing degenerative process in rodent brains and can be induced by microinjection of glutamate agonists such as ibotenic acid (IBO) or quisqualate (QUIS). Calcinosis is mainly composed of calcium and phosphorous amalgamate in a hydroxyapatite like composition (Bernal, Saura, Ojuel, & Mahy, 2000; Rodríguez et al., 2000; Rodríguez, Pugliese, & Mahy, 1999, 2009; Rodríguez, Robledo, Andrade, & Mahy, 2005; Saura, Boatell, Bendahan, & Mahy, 1995). As a result of calcification, cytoplasmic calcium ions precipitate as solid, spherical concretions that grow progressively in size and number (Mahy et al., 1999). Similarly, hydroxyaluminosilicates have also been observed in Alzheimer's disease and Parkinson's disease (Birchall, 1992; Dowe, Heitzman, & Larkin, 1992; Dusart, Marty, & Peschanski, 1991; Ellie et al., 1989; Hohmann, Wenk, Lowenstein, Brown, & Coyle, 1987; Mahy et al., 1999). There is a marked tendency of large deposits to develop into a spherical shape, and the presence of templates over which supercritical nuclei can be formed is also an important contributor to this mechanism. However, pathological analysis using electron microscopy shows that such aggregates undergo structural modifications over time, suggesting also that the aggregation *per se* might not be the main cause but rather a result of the neuronal damage process.

Conversely, other studies support the theory that such aggregates might have a protective effect, avoiding additional free Ca^{2+} (Maetzler et al., 2009). After microinjection of glutamate agonists, extensive analysis demonstrated a limited neuronal loss and a normal morphology of oligodendrocytes in rats, with good preservation of cytoplasm and processes. Reactive astrocytes and neurons frequently presented cytoplasmic, needle-shaped calcium deposits on a matrix close to cellular organelles, such as mitochondria, microtubules, or vesicles. Large extracellular deposits were surrounded by reactive microglia in a phagocytic-like process. Their formation does not depend on the glutamate receptor subtype initially stimulated. However, their size, number and distribution vary not only by excitotoxin but also by the location within the CNS. For example, the development of calcium deposits over time in rats with lesions induced in basal ganglia was more pronounced in lesions induced by IBO than with QUIS (Mahy et al., 1995, 1999).

A study with transgenic mice expressing the calcium binding protein parvalbumin in neurons showed that excitotoxicity with IBO accelerated the calcification process. It is also demonstrated that mitochondria may be the initial structure involved in triggering calcium deposits, leading to the hypothesis that these processes may be causative in the normally aging brain and in neurodegeneration (Maetzler et al., 2009).

The intracellular mucopolysaccharides shown in human and rodent calcification play an important role as modulators for mineralization (Adachi & Volk, 1968; Beall, Patten, Mallette, & Jankovic, 1989). It has been postulated that hydroxyapatite formation with further reductions in free Ca^{2+} would take place as an alternative homeostatic step to reduce excitotoxicity (Ramonet et al., 2002).

Maetzler et al. (2010) studied knockout osteopontin mice and demonstrated an exacerbated co-occurrence of progressive secondary neurodegeneration and microcalcification induced by ibotenate, another glutamate analog. The authors suggested that osteopontin could have a protective role in this process.

Other alternative models were investigated by Lammie et al. (2005) who characterized basal ganglia calcification in BB/E rats with diabetes. Alternatively, thalamic calcification in vitamin D receptor knockout mice was detected and represents another potential target; however, patients with PFBC, by definition, show no vitamin D disturbances (Kalueff et al., 2006).

More recently, Jensen et al. (2013) showed that homozygous mice for the knockout cassette, L1L2-PGK-P, inserted after exon 2 and flanking exon 3 at the Slc20a2 gene, present calcifications in the thalamus, basal ganglia, and other brain regions, compatible with PFBC, demonstrating that reduced PiT2 expression causes BCs. This is the first Slc20a2 knockout reported as a model for loss-of-function mutations, associated with BCs.

More recently, Keller et al. (2013) reported a hypomorphic knockout mice for the PDGFB gene, presenting BCs in a similar pattern compared to humans, reinforcing the involvement of the BBB in such phenotype. Developing additional knockout mice for other mutations in *SLC20A2, PDGFRB, and PDGFB* is a major priority for understanding PFBC physiopathology and for testing new drugs.

7. NEW THERAPIES BASED ON NEW GENES

The report of three new genes associated with PFBC might help to enlighten the current guidelines for medical treatment. So far, there is no specific treatment for familial BC. The huge clinical heterogeneity shows that the symptoms are likely too variable to be treated by a single agent. Currently, treatment might be managed based on the main symptoms, and there are case reports of patients treated with various medications such anticonvulsants, mood stabilizers, antipsychotics, antiparkinsonism, anti-incontinence, analgesics, antidepressants, and benzodiazepines. The prognosis is also variable with some reports stating cases where the symptoms were transient after treatment. Yoshikawa et al. (2003) reported a pediatric case in which a 12-year-old girl with substantial calcinosis presented transient motor and mood symptoms and recovered completely after symptomatic treatment. A CT screening of other siblings showed asymptomatic subjects with calcinosis. However, most patients show a progressive deterioration, despite continuous treatment.

Reports of mutations in the inorganic phosphate transporter (*SLC20A2*) associated with PFBC support the concept of designing treatments according to a primary cause and not only the major symptoms. However, the most intuitive treatments, based on the possible blockage of calcium or phosphate transport, have limited descriptions, and so far, no consistent clinical trial was specifically oriented to treat this rare condition. Calcium channel blockers such as nimodipine have failed as a treatment according to anecdotal reports (Manyam, 2005).

In very few cases, bisphosphonates (BP) such as etidronate disodium had positive effects in reducing symptoms but failed to reduce deposits according to CT imagery. Following treatment of patients presenting parkinsonism, ataxia, spasticity, and dystonia, there were improvements in speech and gait, without improvements in spasticity, dystonia, ataxia, and calcifications (Loeb, 1998). In addition, two cases including: (i) An 8-year-old boy with headaches and seizures treated with etidronate disodium had alleviated symptoms but with no apparent reduction in calcifications, and (ii) 45 patients were treated with the same medication to control seizures, and psychotic symptoms lead to an improvement of symptoms but no apparent loss of calcified deposits as imaged by CT (Loeb, Sohrab, Huq, & Fuerst, 2006). The authors proposed that neurological symptoms associated with BCs can be reduced with etidronate disodium, which is known to inhibit the microcrystal growth of hydroxyapatite. BP are more commonly used for the prevention and treatment of postmenopausal osteoporosis, bone metastases, Paget's disease, and prevention and treatment of glucocorticoid-induced osteoporosis. Other osteoporosis-related conditions such as immobility are also currently treated with this drug (Räkel, Boucher, & Ste-Marie, 2011). Most BP show affinity for mineralized bone, binding to hydroxyapatite, with preferential localization to sites of intensive bone remodeling helping to reduce resorption mediated by osteoclasts. Independent studies confirm that the synthesis of farnesyl pyrophosphate, an enzyme involved in lipid biosynthesis, is the main target for various BP. BP rapidly decrease the rate of bone turnover through reduction of bone resorption and osteogenesis. There was no progressive reduction of bone turnover markers with repeated annual administration, in the case of Aclasta®, and age does not seem to influence this effect (Räkel et al., 2011).

Prospective and longitudinal studies that directly relate neurological symptoms to the formation of macroscopic calcifications as well as large-scale clinical trials with BP to both prevent and treat cerebral calcifications could be an effective approach to treat patients with brain calcinosis.

The identification of *SLC20A2*, *PDGFRB*, and *PDGFB* provides a new avenue for potential treatments based on the use of BP and drugs modulating the *PDGFRB* pathway such as imatinib. Imatinib is a tyrosine kinase inhibitor currently approved for the treatment of various cancers in the United States, and PDGFRB represents an interesting target for therapy as it is a kinase receptor activating many downstream pathways.

There are still several questions regarding this unusual neuropsychiatric condition; however, the reports of new genes as well as increased availability of detection methods bring a promising perspective for patients, clinicians, and basic scientists. Hopefully, this will improve our basic knowledge concerning ectopic calcification mechanisms and eventually benefit other conditions that share molecular and clinical overlapping with PFBC.

REFERENCES

Adachi, M., & Volk, B. W. (1968). Protracted form of spongy degeneration of the central nervous system (van Bogaert and Bertrand type). *Neurology, 18*, 1084–1092.

Argyelan, M., Carbon, M., Niethammer, M., Ulug, A. M., Voss, H. U., Bressman, S. B., et al. (2009). Cerebellothalamocortical connectivity regulates penetrance in dystonia. *The Journal of Neuroscience, 29*, 9740–9747.

Baba, Y., Broderick, D. F., Uitti, R. J., Hutton, M. L., & Wszolek, A. K. (2005). Heredofamilial brain calcinosis syndrome. *Mayo Clinic Proceedings, 80*, 641–651.

Beall, S. S., Patten, B. M., Mallette, L., & Jankovic, J. (1989). Abnormal systemic metabolism of iron, porphyrin, and calcium in Fahr's syndrome. *Annals of Neurology, 26*, 569–575.

Beck, L., Leroy, C., Beck-Cormier, S., Forand, A., Salauen, C., Paris, N., et al. (2010). The phosphate transporter PiT1 (Slc20a1) revealed as a new essential gene for mouse liver development. *PloS One, 5*(2), e9148.

Bernal, F., Saura, J., Ojuel, J., & Mahy, N. (2000). Differential vulnerability of hippocampus, basal ganglia, and prefrontal cortex to long-term NMDA excitotoxicity. *Experimental Neurology, 161*, 686–695.

Birchall, J. D. (1992). The interrelationship between silicon and aluminium in the biological effects of aluminium. *Ciba Foundation Symposium, 169*, 50–61.

Block, G. A., Hulbert-Shearon, T. E., Levin, N. W., & Port, F. K. (1998). Association of serum phosphorus and calcium × phosphate product with mortality risk in chronic hemodialysis patients: A national study. *American Journal of Kidney Diseases, 31*(4), 607–617.

Bøttger, P., & Pedersen, L. (2002). Two highly conserved glutamate residues critical for type III sodium-dependent phosphate transport revealed by uncoupling transport function from retroviral receptor function. *The Journal of Biological Chemistry, 277*(45), 42741–427417.

Bøttger, P., & Pedersen, L. (2011). Mapping of the minimal inorganic phosphate transporting unit of human PiT2 suggests a structure universal to PiT-related proteins from all kingdoms of life. *BMC Biochemistry, 12*, 21.

Brodaty, H., Mitchell, P., Luscombe, G., Kwok, J. J., Badenhop, R. F., McKenzie, R., et al. (2002). Familial idiopathic basal ganglia calcification (Fahr's disease) without neurological, cognitive and psychiatric symptoms is not linked to the IBGC1 locus on chromosome 14q. *Human Genetics, 110*, 8–14.

Comtesse, N., Niedermayer, I., Glass, B., Heckel, D., Maldener, E., Nastainczyk, W., et al. (2002). MGEA6 is tumor-specific overexpressed and frequently recognized by patient-serum antibodies. *Oncogene, 21*, 239–247.

Comtesse, N., Reus, K., & Meese, E. (2001). The MGEA6 multigene family has an active locus on 14q and at least nine pseudogenes on different chromosomes. *Genomics, 75*, 43–48.

Da Silva, R. J., Pereira, I. C., & Oliveira, J. R. (2013). Analysis of gene expression pattern and neuroanatomical correlates for SLC20A2 (PiT-2) shows a molecular network with potential impact in idiopathic basal ganglia calcification ("Fahr's disease"). *Journal of Molecular Neuroscience, 50*, 280–283.

Dai, X., Gao, Y., Xu, Z., Cui, X., Liu, J., Li, Y., et al. (2010). Identification of a novel genetic locus on chromosome 8p21.1-q11.23 for idiopathic basal ganglia calcification. *American Journal of Medical Genetics, 153B*(7), 1305–1310.

Daneman, R., Zhou, L., Kebede, A. A., & Barres, B. A. (2010). Pericytes are required for blood–brain barrier integrity during embryogenesis. *Nature, 468*(7323), 562–566.

Dowe, D. A., Heitzman, E. R., & Larkin, J. J. (1992). Human immunodeficiency virus infection in children. *Clinical Imaging, 16*, 145–151.

Dusart, I., Marty, S., & Peschanski, M. (1991). Glial changes following an excitotoxic lesion in the CNS—II. Astrocytes. *Neuroscience*, *45*, 541–549.

Eidelberg, D. (2009). Metabolic brain networks in neurodegenerative disorders: A functional imaging approach. *Trends in Neurosciences*, *32*, 548–557.

Ellie, E., Julien, J., & Ferrer, X. (1989). Familial idiopathic striopallidodentate calcifications. *Neurology*, *39*, 381–385.

Forstl, H., Krumm, B., Eden, S., & Kohlmeyer, K. (1992). Neurological disorders in 166 patients with basal ganglia calcification: A statistical evaluation. *Journal of Neurology*, *239*(1), 36–38.

Freudenberg-Hua, Y., Freudenberg, J., Kluck, N., Cichon, S., Propping, P., & Nöthen, M. M. (2003). Single nucleotide variation analysis in 65 candidate genes for CNS disorders in a representative sample of the European population. *Genome Research*, *13*, 2271–2276.

Garbuzova-Davis, S., Saporta, S., Haller, E., Kolomey, I., Bennett, S. P., Potter, H., et al. (2007). Evidence of compromised blood-spinal cord barrier in early and late symptomatic SOD1 mice modeling ALS. *PloS One*, *2*(11), e1205.

Gavier-Widen, D., Wells, G. A., Simmons, M. M., Wilesmith, J. W., & Ryan, J. (2001). Histological observations on the brains of symptomless 7-year-old cattle. *Journal of Comparative Pathology*, *124*, 52–59.

Geschwind, D. H., Loginov, M., & Stern, J. M. (1999). Identification of a locus on chromosome 14q for idiopathic basal ganglia calcification (Fahr disease). *American Journal of Human Genetics*, *65*(3), 764–772.

Giachelli, C. M. (2003). Vascular calcification: In vitro evidence for the role of inorganic phosphate. *Journal of the American Society of Nephrology*, *14*(9 Suppl 4), S300–S304.

Giachelli, C. M., Jono, S., Shioi, A., Nishizawa, Y., Mori, K., & Morii, H. (2001). Vascular calcification and inorganic phosphate. *American Journal of Kidney Diseases*, *38*(4), S34–S37.

Hirota, S., Nakajima, Y., Yoshimine, T., Kohri, K., Nomura, S., Taneda, M., et al. (1995). Expression of bone-related protein messenger RNA in human meningiomas: Possible involvement of osteopontin in development of psammoma bodies. *Journal of Neuropathology and Experimental Neurology*, *54*, 698–703.

Hohmann, G. F., Wenk, G. L., Lowenstein, P., Brown, M. E., & Coyle, J. T. (1987). Age-related recurrence of basal forebrain lesion-induced cholinergic deficits. *Neuroscience Letters*, *82*, 253–259.

Honda, T., Fujiwara, N., Abe, T., Kumanishi, T., Yoshimura, M., & Shimoji, K. (1994). Prior mechanical injury inhibits rise in intracellular Ca^{2+} concentration by oxygen-glucose deprivation in mouse hippocampal slices. *Brain Research*, *666*, 263–269.

Hsu, S. C., Sears, R. L., Lemos, R. R., Quintáns, B., Huang, A., Spiteri, E., et al. (2013). Mutations in SLC20A2 are a major cause of familial idiopathic basal ganglia calcification. *Neurogenetics*, *14*(1), 11–22.

Jensen, N., Schrøder, H., Hejbøl, E. K., Füchtbauer, E. M., Oliveira, J. R. M., & Pedersen, L. (2013). Loss of function of Slc20a2 associated with familial idiopathic basal ganglia calcification in humans causes brain calcifications in mice. *Journal of Molecular Neuroscience*, in press.

Kalueff, A., Loseva, E., Haapasalo, H., Rantala, I., Keranen, J., Lou, Y. R., et al. (2006). Thalamic calcification in vitamin D receptor knockout mice. *Neuroreport*, *17*, 717–721.

Kavanaugh, M. P., & Kabat, D. (1996). Identification and characterization of a widely expressed phosphate transporter/retrovirus receptor family. *Kidney International*, *49*(4), 959–963.

Keller, A., Westenberger, A., Sobrido, M. J., García-Murias, M., Domingo, A., Sears, R. L., et al. (2013). Mutations in the gene encoding PDGF-B cause brain calcifications in humans and mice. *Nature Genetics*, *45*, 1077–1082.

Kempton, M. J., Haldane, M., Jogia, J., Grasby, P. M., Collier, D., & Frangou, S. (2009). Dissociable brain structural changes associated with predisposition, resilience, and disease expression in bipolar disorder. *The Journal of Neuroscience*, *29*, 10863–10868.

Khan, S. R., & Hackett, R. L. (1986). Histochemistry of colloidal iron stained crystal associated material in urinary stones and experimentally induced intrarenal deposits in rats. *Scanning Electron Microscopy*, *Pt2*, 761–765.

Kim, K. M. (1995). Apoptosis and calcification. *Scanning Microscopy*, *9*, 1137–1178.

Kortekaas, R., Leenders, K. L., van Oostrom, J. C. H., Vaalburg, W., Bart, J., Willemsen, A. T. M., et al. (2005). Blood–brain barrier dysfunction in parkinsonian midbrain in vivo. *Annals of Neurology*, *57*(2), 176–179.

Lammie, G. A., Kelly, P. A. T., Baird, J. D., Smith, W., Chatterjee, S., Frier, B. M., et al. (2005). Basal ganglia calcification in BB/E rats with diabetes. *Journal of Clinical Neuroscience*, *12*, 49–53.

Lemos, R. R., Oliveira, M. F., & Oliveira, J. R. M. (2013). Reporting a new mutation at the SLC20A2 gene in a Brazilian family with idiopathic basal ganglia calcification. *European Journal of Neurology*, *20*, e43–e44.

Lemos, R. R., Oliveira, D. F., Zatz, M., & Oliveira, J. R. M. (2011). Population and computational analysis of the MGEA6 P521A variation as a risk factor for familial idiopathic basal ganglia calcification (Fahr's disease). *Journal of Molecular Neuroscience*, *43*(3), 333–336.

Lhatoo, S. D., Perunovic, B., Love, S., Houlden, H., & Campbell, M. J. (2003). Familial idiopathic brain calcification—A new and familial alpha-synucleinopathy? *European Neurology*, *49*, 223–226.

Livingston, J. H., Stivaros, S., van der Knaap, M. S., & Crow, Y. J. (2013). Recognizable phenotypes associated with intracranial calcification. *Developmental Medicine and Child Neurology*, *1*, 46–57.

Loeb, J. A. (1998). Functional improvement in a patient with cerebral calcinosis using a bisphosphonate. *Movement Disorders*, *13*, 345–349.

Loeb, J. A., Sohrab, S. A., Huq, M., & Fuerst, D. R. (2006). Brain calcifications induce neurological dysfunction that can be reversed by a bone drug. *Journal of Neurological Sciences*, *243*, 77–81.

Macias, M. J., Hyvönen, M., Baraldi, E., Schultz, J., Sudol, M., Saraste, M., et al. (1996). Structure of the WW domain of a kinase-associated protein complexed with a proline-rich peptide. *Nature*, *382*, 646–649.

Macias, M. J., Wiesner, S., & Sudol, M. (2002). WW and SH3 domains, two different scaffolds to recognize proline-rich ligands. *FEBS Letters*, *513*, 30–37.

Maetzler, W., Berg, D., Funke, C., Sandmann, F., Stünitz, H., Maetzler, C., et al. (2010). Progressive secondary neurodegeneration and microcalcification co-occur in osteopontin-deficient mice. *The American Journal of Pathology*, *177*, 829–839.

Maetzler, W., Stünitz, H., Bendfeldt, K., Vollenweider, F., Schwaller, B., & Nitsch, C. (2009). Microcalcification after excitotoxicity is enhanced in transgenic mice expressing parvalbumin in all neurones, may commence in neuronal mitochondria and undergoes structural modifications over time. *Neuropathology and Applied Neurobiology*, *35*, 165–177.

Mahy, N., Bendahan, G., Boatell, M. L., Bjelke, B., Tinner, B., Olson, L., et al. (1995). Differential brain area vulnerability to long-term subcortical excitotoxic lesions. *Neuroscience*, *65*, 15–25.

Mahy, N., Prats, A., Riveros, A., Andre's, N., & Bernal, F. (1999). Basal ganglia calcification induced by excitotoxicity: An experimental model characterized by electron microscopy and X-ray microanalysis. *Acta Neuropathologica*, *98*, 217–225.

Manyam, B. V. (2005). What is and what is not 'Fahr's disease'. *Parkinsonism & Related Disorders*, *11*(2), 73–80.

Moskowitz, M. A., Winickoff, R. N., & Heinz, E. R. (1971). Familial calcification of the basal ganglions: A metabolic and genetic study. *The New England Journal of Medicine*, *285*, 72–77.

Nicolas, G., Pottier, C., Maltête, D., Coutant, S., Rovelet-Lecrux, A., Legallic, S., et al. (2013). Mutation of the PDGFRB gene as a cause of idiopathic basal ganglia calcification. *Neurology*, *80*(2), 181–187.

Nitschke, Y., & Rutsch, F. (2012). Genetics in arterial calcification: Lessons learned from rare diseases. *Trends in Cardiovascular Medicine*, *22*, 145–149.

Noppeney, U., Friston, K. J., & Price, C. J. (2004). Degenerate neuronal systems sustaining cognitive functions. *Journal of Anatomy*, *205*, 433–442.

Oliveira, J. R. M. (2011). *Managing Idiopathic Basal Ganglia calcification (Fahr's disease)*. New York: Nova Publishing.

Oliveira, J. R. M., Hopfer, S., Papp, J., Spiteri, E., Klepper, J., Gilbert, J., et al. (2003). A genome wide scan for familial idiopathic basal ganglia calcification (Fahr's disease) identifies new candidate regions and confirms genetic heterogeneity. In: *ASHG Annual Meeting*, Program Nr: 2354.

Oliveira, J. R. M., & Lemos, R. R. (2009). Linkage studies in familial idiopathic basal ganglia calcification: Separating the wheat from the chaff. *American Journal of Medical Genetics*, *150B*(4), 594–595.

Oliveira, J. R., Lima, F. J., & Zatz, M. (2009). Identical twins with idiopathic basal ganglia calcification ("Fahr's disease") presenting with a remarkably similar pattern of neuroimaging findings. *Parkinsonism & Related Disorders*, *15*(5), 396.

Oliveira, J. R., Oliveira, M. F., Lemos, R. R., & Oliveira, M. L. (2009). The current status of "Fahr's disease" nosology. *Practical Neurology*, *9*, 302.

Oliveira, J., Sobrido, M. J., Spiteri, E., Hopfer, S., Meroni, G., Petek, E., et al. (2007). Analysis of candidate genes at the IBGC1 locus associated with idiopathic basal ganglia calcification ("Fahr's disease"). *Journal of Molecular Neuroscience*, *33*(2), 151–154.

Oliveira, J. R. M., Spiteri, E., Sobrido, M. J., Hopfer, S., Klepper, J., Voit, T., et al. (2004). Genetic heterogeneity in familial idiopathic basal ganglia calcification (Fahr disease). *Neurology*, *63*(11), 2165–2167.

Oliveira, J. R. M., & Steinberg, S. S. (2010). Far from rare: Revisiting the relevance of idiopathic basal ganglia calcifications. *Neurological Sciences*, *31*, 679.

Petegnief, V., Saura, J., Dewar, D., Cummins, D. J., Dragunow, M., & Mahy, N. (1999). Long-term effects of α-amino-3-hydroxy-5-methyl-4-isoxazole propionate and 6-nitro-7 sulphamoylbenzo(f)quinoxaline-2,3-dione in the rat basal ganglia: Calcification, changes in glutamate receptors and glial reactions. *Neuroscience*, *94*, 105–115.

Räkel, A., Boucher, A., & Ste-Marie, L. G. (2011). Role of zoledronic acid in the prevention and treatment of osteoporosis. *Clinical Interventions in Aging*, *6*, 89–99.

Ramonet, D., Pugliese, M., Rodrýguez, M. J., De Yebra, L., Andrade, C., Adroer, R., et al. (2002). Calcium precipitation in acute and chronic brain diseases. *Journal of Physiology*, *96*, 307–312.

Rodríguez, M. J., Bernal, F., Andrés, N., Malpesa, Y., & Mahy, N. (2000). Excitatory amino acids and neurodegeneration: A hypothetical role of calcium precipitation. *International Journal of Developmental Neuroscience*, *18*, 299–307.

Rodríguez, M. J., Pugliese, M., & Mahy, N. (1999). Drug abuse, brain calcification and glutamate-induced neurodegeneration. *Current Drug Abuse Reviews*, *2*, 99–112.

Rodríguez, M. J., Pugliese, M., & Mahy, N. (2009). Drug abuse, brain calcification and glutamate-induced neurodegeneration. *Current Drug Abuse Reviews*, *2*, 99–112.

Rodríguez, M. J., Robledo, P., Andrade, C., & Mahy, N. (2005). In vivo coordinated interactions between inhibitory systems to control glutamate-mediated hippocampal excitability. *Journal of Neurochemistry*, *95*, 651–661.

Saillour, Y., Zanni, G., Des Portes, V., Heron, D., Guibaud, L., Iba-Zizen, M. T., et al. (2007). Mutations in the AP1S2 gene encoding the sigma 2 subunit of the adaptor protein 1 complex are associated with syndromic X-linked mental retardation with hydrocephalus and calcifications in basal ganglia. *Journal of Medical Genetics*, *44*, 739–744.

Sansom, M. S., & Weinstein, H. (2000). Hinges, swivels and switches: The role of prolines in signalling via transmembrane alpha-helices. *Trends in Pharmacological Sciences*, *21*, 445–451.

Saura, J., Boatell, M. L., Bendahan, G., & Mahy, N. (1995). Calcium deposit formation and glial reaction in rat brain after ibotenic acid-induced basal forebrain lesion. *European Journal of Neuroscience*, 7, 1569–1578.

Schottlaender, L. V., Mencacci, N., Koepp, M., Hanna, M., Hardy, J., Lees, A. J., et al. (2012). Interesting clinical features associated with mutations in the SLC20A2 gene. In: *EFNS*, Abstract Book: 44–44.

Smith, M. L., Auer, R. N., & Siesjö, B. K. (1984). The density and distribution of ischemic brain injury in the rat following 2–10 min forebrain ischemia. *Acta Neuropathologica*, *64*, 319–332.

Sobrido, M. J., Copolla, G., Oliveira, J. R. M., Hopfer, S., & Geschwind, D. H. (2013). Familial primary brain calcification. In R. A. Pagon, M. P. Adam, T. D. Bird, C. R. Dolan, C. T. Fong, & K. Stephens (Eds.), *Genereviews™*. Seattle: University of Washington.

Sobrido, M. J., & Geschwind, D. H. (2001). Genetics of familial idiopathic basal ganglia calcification (FIBGC). In S.-M. Pulst (Ed.), *Genetics of movement disorders* (pp. 443–448). San Diego: Academic Press.

Sudol, M., Sliwa, K., & Russo, T. (2001). Functions of WW domains in the nucleus. *FEBS Letters*, *490*, 190–195.

Towfighi, J., Yager, J. Y., Housman, C., & Vannucci, R. C. (1996). Neuropathology of remote hypoxic-ischemic damage in the immature rat. *Acta Neuropathologica*, *81*, 578–587.

Usener, D., Gerhardt, A., Schadendorf, D., & Eichmüller, S. (2003). Seroreactivity against MAGE-A and LAGE-1 proteins in melanoma patients. *The British Journal of Dermatology*, *149*, 282–288.

Villa-Bellosta, R., Ravera, S., Sorribas, V., Stange, G., Levi, M., Murer, H., et al. (2009). The Na^{+}–Pi cotransporter PiT-2 (SLC20A2) is expressed in the apical membrane of rat renal proximal tubules and regulated by dietary Pi. *American Journal of Physiology Renal Physiology*, *296*(4), F691–F699.

Villa-Bellosta, R., & Sorribas, V. (2010). Compensatory regulation of the sodium/phosphate cotransporters NaPi-IIc (SCL34A3) and Pit-2 (SLC20A2) during Pi deprivation and acidosis. *European Journal of Physiology*, *459*, 499–508.

Volpato, C. B., De Grandi, A., Buffone, E., Facheris, M., Gebert, U., Schifferle, G., et al. (2009). 2q37 as a susceptibility locus for idiopathic basal ganglia calcification (IBGC) in a large South Tyrolean family. *Journal of Molecular Neuroscience*, *39*, 346–353.

Wang, C., Li, Y., Shi, L., Ren, J., Patti, M., Wang, T., et al. (2012). Mutations in SLC20A2 link familial idiopathic basal ganglia calcification with phosphate homeostasis. *Nature Genetics*, *44*, 254–256.

Wszolek, Z. K., Baba, Y., Mackenzie, I. R., Uitti, R. J., Strongosky, A. J., Broderick, D. F., et al. (2006). Autosomal dominant dystonia-plus with cerebral calcifications. *Neurology*, *67*, 620–625.

Yamada, M., Asano, T., Okamoto, K., Hayashi, Y., Kanematsu, M., Hoshi, H., et al. (2012). High frequency of calcification in basal ganglia on brain computed tomography images in Japanese older adults. *Geriatrics and Gerontology International*, *13*(3), 706–710.

Yanai, T., Masegi, T., Ishikawa, K., Sakai, H., Iwasaki, T., Moritomo, Y., et al. (1996). Spontaneous vascular mineralization in the brain of horses. *The Journal of Veterinary Medical Science*, *58*, 35–40.

Yanai, T., Masegi, T., Ueda, K., Manabe, J., Teranishi, M., Takaoka, M., et al. (1994). Vascular mineralization in the monkey brain. *Veterinary Pathology*, *31*, 546–552.

Yasui, N., Sato, M., Ochi, T., Kimura, T., Kawahata, H., Kitamura, Y., et al. (1997). Three modes of ossification during distraction osteogenesis in the rat. *Journal of Bone and Joint Surgery (British Volume)*, *79*, 824–830.

Yasui, M., Yano, I., Tsuda, T., & Ota, K. (1991). Effects of 4-(o-benzylphenoxy)-N-methylbutylamine hydrochloride (bifemelane) on deposition of calcium and magnesium in the central nervous system tissues of rats maintained on low-calcium diets. *Nō to Shinkei*, *43*, 1133–1137.

Yoshikawa, H., & Abe, T. (2003). Transient parkinsonism in bilateral striopallidodentate calcinosis. *Pediatric Neurology*, *29*, 75–77.

Zarrinpar, A., Bhattacharyya, R. P., & Lim, W. A. (2003). The structure and function of proline recognition domains. *Science's STKE*, *2003*(179), RE8.

Zhang, Y., Guo, X., & Wu, A. (2013). Association between a novel mutation in SLC20A2 and familial idiopathic basal ganglia calcification. *PloS One*, *8*(2), e57060.

Zipser, B. D., Johanson, C. E., Gonzalez, L., Berzin, T. M., Tavares, R., Hulette, C. M., et al. (2007). Microvascular injury and blood–brain barrier leakage in Alzheimer's disease. *Neurobiology of Aging*, *28*, 977–986.

INDEX

Note: Page numbers followed by "*f*" indicate figures and "*t*" indicate tables.

CONTENTS OF RECENT VOLUMES

Volume 39

Volume 40

Volume 41

Volume 42

Volume 43

Volume 44

Volume 45

Volume 46

Volume 47

Volume 48

Volume 49

Volume 50

Volume 53

Volume 54

Volume 60

Volume 61

Volume 62

Volume 63

Volume 64

Volume 65

Volume 66

Volume 67

Volume 68

Volume 69

Volume 70

Volume 71

Volume 72

Volume 73

Volume 74

Volume 75

Volume 76

Volume 77

Volume 78

Volume 79

Volume 80

Volume 81

Volume 82

Volume 83

Volume 84

Volume 85

Volume 86

Volume 87

Volume 88

Volume 89

Volume 90

Volume 91

Volume 92

Volume 93

Volume 94

Volume 95

Volume 96

Volume 97

Volume 98

Volume 99

Volume 100

Volume 101

Volume 102

Volume 103

Volume 104

Volume 105

Volume 106

Volume 107

Volume 108

Volume 109

FSC
www.fsc.org
MIX
Papier | Fördert gute Waldnutzung
FSC® C083411

Kolibri 360 GmbH
Ferdinand-Jühlke-Straße 7
99095 Erfurt, Deutschland
produktsicherheit@kolibri360.de